Oxford Series in Ecology and Evolution

Edited by Paul H. Harvey, Robert M. May, H. Charles J. Godfray, and Jennifer A. Dunne

Parasites and the Behavior of Animals
Janice Moore
Evolutionary Ecology of Birds
Peter Bennett and Ian Owens
The Role of Chromosomal Change in Plant Evolution
Donald A. Levin
Living in Groups
Jens Krause and Graeme Ruxton
Stochastic Population Dynamics in Ecology and Conservation
Russell Lande, Steiner Engen and Bernt-Erik Sæther
The Structure and Dynamics of Geographic Ranges
Kevin J. Gaston
Animal Signals
John Maynard Smith and David Harper
Evolutionary Ecology: The Trinidadian Guppy
Anne E. Magurran
Infectious Diseases and Primates Socioecology
Charles L. Nunn and Sonia M. Altizer
Computational Molecular Evolution
Ziheng Yang
The Evolution and Emergence of RNA Viruses
Edward C. Holmes
Aboveground–Belowground Linkages: Biotic Interactions, Ecosystem Processes, and Global Change
Richard D. Bardgett and David A. Wardle

Aboveground–Belowground Linkages

Biotic Interactions, Ecosystem Processes, and Global Change

RICHARD D. BARDGETT

*Soil and Ecosystem Ecology Laboratory, Lancaster
Environment Centre, Lancaster University, Lancaster, UK*

DAVID A. WARDLE

*Department of Forest Ecology and Management, Swedish
University of Agricultural Sciences, Umeå, Sweden*

OXFORD
UNIVERSITY PRESS

Great Clarendon Street, Oxford OX2 6DP
United Kingdom

Oxford University Press is a department of the University of Oxford.
It furthers the University's objective of excellence in research, scholarship,
and education by publishing worldwide. Oxford is a registered trade mark of
Oxford University Press in the UK and in certain other countries

British Library Cataloguing in Publication Data
Data available

Library of Congress Cataloging in Publication Data
Data available

ISBN 978-0-19-954688-6

Printed and bound by CPI Group (UK) Ltd, Croydon, CR0 4YY

Contents

Preface

The principal aim of this book is to synthesize recent advances in our understanding of interactions between aboveground and belowground communities, and how these interactions regulate the functioning of terrestrial ecosystems and their responses to global change. The last two decades, and especially the last 5 years, have witnessed vastly increased activity on this topic, and as this has led to several important conceptual advances in this field; hence, we felt that a comprehensive synthesis of the topic would be timely. Even since we first began to plan this book, during a field excursion in the mountains of New Zealand's South Island in 2007, we have been astonished by the growing number of studies emerging on this topic, several of which have led to major leaps in conceptual understanding. As a consequence, during the writing of this book we have had to constantly update chapters as new and important studies have emerged. We have attempted to capture this recent and ongoing interest, right up to the last day of writing in December 2009. If the recent and rapid development of the topic does not leap out of these pages, then we have failed to do it justice.

A key message of this book is that an improved understanding of how terrestrial ecosystems function and respond to global change can be gained only by considering the aboveground and belowground subsystems in tandem. We illustrate this by considering four key aspects of aboveground–belowground linkages: biotic interactions in the soil; plant community effects; the role of aboveground consumers; and the influence of species gains and losses. We then draw together this information and identify a number of cross-cutting themes, which include how aboveground–belowground feedbacks manifest themselves at vastly different spatial and temporal scales, the consequences of these feedbacks for community and ecosystem processes, and how aboveground–belowground interactions link to human-induced global change. In doing so, we have attempted to identify areas of significant recent development as well as productive avenues for future research. In writing this book, we have drawn on literature from a wide range of disciplines and from studies that cross many types of ecosystems around the world. However, it is not our intention to provide an exhaustive analysis of the vast literature on aboveground–belowground linkages. Rather, we have selected representative studies and examples to illustrate key concepts and significant advances in this field, and to suggest interesting detours that the reader may wish to explore themselves.

There are many people we would like to thank who helped us in various ways in writing this book. First, we thank Robert May of Oxford University for originally suggesting that we write the book, and Ian Sherman from Oxford University Press for giving us the initial encouragement needed to get it off the ground. We are also most grateful to Helen Eaton from Oxford University Press for her considerable assistance throughout this endeavour, and for ensuring that the developmental process of this book was smooth and problem-free. We are also very grateful to our many colleagues who willingly provided us with in-depth comments on individual chapters and portions of text, directed us towards additional relevant literature that we had missed, and made constructive suggestions on improving the content as well as the clarity of text. Colleagues who provided valuable feedback on various chapters include Hans Cornelissen, Franciska de Vries, Doug Frank, Micael Jonsson, Paul Kardol, Marie-Charlotte Nilsson, Kate Orwin, Heikki Setälä, Carley Stevens, Wim Van der Putten, and Rene Van der Wal. We are also very grateful to the many colleagues with whom we have each interacted over the past 20 or more years, too numerous to list here, who have all contributed in multiple ways to the development of our own ideas, research programs, and understanding of how terrestrial communities and ecosystems function. Without a range of interesting colleagues with whom to interact, collaborate, and discuss, a book of this type could not have been written. Finally, we are indebted to our families, and especially Jill and Anna, who supported, tolerated, and encouraged us throughout the development and writing of this book.

Richard D. Bardgett
and David A. Wardle

December 2009

1

Introduction

Our understanding of how organisms, and their interactions with each other and their abiotic environment, govern the functioning of terrestrial ecosystems has advanced rapidly over the last two decades. This has been driven by a number of issues, the most significant being the need to better understand the consequences for ecosystems and the Earth-system of the dramatic changes in biological communities that have resulted from human activity (Pimm et al. 1995; Vitousek et al. 1997c; Sala et al. 2000; Millennium Ecosystem Assessment 2005). An important development in this area has been a growing recognition that terrestrial ecosystems consist of both aboveground and belowground subsystems, and that feedbacks between these subsystems play a crucial role in regulating community structure and ecosystem functioning (e.g. Hooper et al. 2000; Van der Putten et al. 2001; Wardle 2002; Wardle et al. 2004a; Bardgett 2005). As a consequence, it is now widely understood that biotic interactions between aboveground and belowground communities play a fundamental role in regulating the response of terrestrial ecosystems and the Earth-system to human-induced global change (Wolters et al. 2000; Wardle et al. 2004a; Bardgett et al. 2008; Van der Putten et al. 2009).

The principal aim of this book is to synthesize recent advances in understanding of the roles that biotic interactions between aboveground and belowground communities play in regulating the structure and function of terrestrial ecosystems, and the response of ecosystems to global change. We also evaluate what can be learned from the recent proliferation of studies on the ecological and biogeochemical significance of biotic interactions between aboveground and belowground communities. In doing so, we draw together a large body of literature that tackles this topic from a range of perspectives and at vastly different spatial and temporal scales. These studies range from those that explore the significance of root–microbial interactions for nutrient acquisition and plant growth at the individual plant level, to those which examine how biome-level shifts in vegetation composition resulting from global change influence carbon-cycle feedbacks at a global scale. We do not aim to present new facts, but we do attempt to bring some order and synthesis to the topic of aboveground–belowground interactions as terrestrial ecosystem drivers.

There is no shortage of recent reviews on the topic of aboveground–belowground interactions, including our own books (Wardle 2002; Bardgett 2005). However, these books do not capture the many new insights into this topic that have emerged in the past half-decade, and other recent overviews about aboveground–belowground interactions have generally focused on specific components of the topic, for example in the context of multitrophic interactions (Scheu 2001; Van der Putten et al. 2001), herbivory (Bardgett et al. 1998b; Bardgett and Wardle 2003), soil biodiversity (Hooper et al. 2000; De Deyn and Van der Putten 2005; Wardle 2006; Fierer et al. 2009), plant community dynamics (Van der Putten 2003, 2009; Van der Heijden et al. 2008; Van der Heijden and Horton 2009), ecosystem nutrient cycling and carbon dynamics (Schimel and Bennett 2004; Bardgett et al. 2005; De Deyn et al. 2008; Frank and Groffman 2009), and climate-feedback processes (Wardle et al. 1998c; Wolters et al. 2000; Bardgett et al. 2008; Wookey et al. 2009). Here, we bring such information together to provide a comprehensive synthesis about the role of aboveground–belowground interactions as terrestrial ecosystem drivers in a changing world.

The complexity of biotic interactions between aboveground and belowground organisms, the interactions of these organisms with their abiotic environment, and their variations in spatial and temporal scales at which they operate, is enormous (Wardle 2002). Hence, we have divided the book into what we believe to be a logical and palatable format. In the next three chapters we examine what we consider to be the three main groups of biotic drivers of terrestrial ecosystem functioning: Chapter 2 focuses on the belowground consumers; Chapter 3 focuses on the plant community; and Chapter 4 focuses on the aboveground consumers. Chapter 5 then draws upon the concepts developed in these preceding chapters to examine how biotic interchange, or human-induced species loss and gain, affects ecosystems through affecting aboveground–belowground linkages. Finally, in Chapter 6 we bring together the main threads from the preceding chapters to generate overall conclusions, emphasize emerging developments, and highlight what we believe to be productive avenues for future work. One important feature of the book is that the issue of global change is considered in depth throughout all the chapters. This issue might have been considered as an individual chapter (as was done in Wardle 2002 and Bardgett 2005), but given that global change is of fundamental importance to each of the three groups of biotic drivers considered in this book, and that there has been a substantial body of recent literature produced in this area, we argue that it is most effective to consider each group separately in the context of global change.

The aim of this introductory chapter is to place our examination of aboveground–belowground linkages in the broader context of ecosystem ecology, and explore the historical development of this topic and its relationship to other fields of ecology. We begin by providing a historical perspective on ecosystem ecology, emphasizing the development of understanding of the role of abiotic and biotic drivers of ecosystem processes. We then explore this theme further by considering developments in our understanding of how species and species interactions act as important drivers of ecosystem processes. Following this, we discuss more recent advances in the topic of how interactions and feedbacks between aboveground and belowground biota may serve as drivers of ecosystem processes. Finally, we place this discussion in a global change context, by emphasizing how aboveground and belowground interactions play a fundamental role in regulating the response of terrestrial ecosystems and the Earth-system to human-induced global change.

1.1 Controls on terrestrial ecosystem processes: an historical perspective

Ecosystem science, which addresses the interactions between organisms and their physical environment as an integrated system, is a relatively young discipline. Indeed, the term *ecosystem* was not used in the literature until 1935, when the British plant ecologist Arthur Tansley coined it to emphasize the importance of transfers of materials between organisms and their abiotic environment (Tansley 1935). This systems approach to studying ecosystems was further developed by Raymond Lindeman, whose classic work on trophic dynamics and ecosystem energy flow (Lindeman 1942) paved the way for many basic concepts in ecosystem ecology, and by Eugene and Howard Odum who pioneered the use radioactive tracers to document patterns of energy flow through ecosystems. Following this, Eugene Odum published his classic strategy-based theory for understanding ecosystem development (Odum 1969). Since this time, and especially during the last twenty years, the study of ecosystem science has grown rapidly. Moreover, the growth of ecosystem-level experiments, including whole-ecosystem manipulations (e.g. Likens et al. 1977; Carpenter et al. 1985) and the use of well-defined environmental gradients (e.g. Vitousek 2004), has served to greatly increase understanding of the factors that regulate the structure and function of terrestrial ecosystems, and provide the basis for critical decision-making based on ecosystem management. The demands of global change and Earth-system research, which concerns human influences on interactions among terrestrial ecosystems, the atmosphere, and the oceans, are likely to increase the importance of ecosystem science in the future.

Another key step in ecosystem ecology was the development of the state factor concept of Jenny (1941), which recognizes that the structure and function of terrestrial ecosystems is governed by a set of state factors, namely climate, parent material, topography, time, and biota. This concept has provided a logical framework for ecologists to study the relative influence of each state factor on ecosystem processes, thereby greatly advancing the field. For example, the chronosequence approach and associated space-for-time substitution has been successfully used as a tool for studying temporal dynamics of plant and microbial communities, soil development, and ecosystem processes, as will be considered in Chapter 3 (e.g. Crocker and Major 1955; Whittaker 1956; Chapin et al. 1994; Crews et al. 1995; Wardle et al. 2004b; Bardgett et al. 2007a). Likewise, different types of environmental gradients, for example of topographic position, climate, and geologic substrate, have been used to determine the relative role of landscape versus biotic controls, such as herbivory, on ecosystem processes, as will be covered in Chapter 4 (e.g. Tracy and Frank 1998; Augustine and McNaughton 2006; Anser et al. 2009). Meanwhile, climatic gradients have been used to inform on the consequences for plant production and carbon dynamics of woody plant invasion in grass-dominated ecosystems (e.g. Jackson et al. 2002; Knapp et al. 2008), an issue that will be considered in Chapter 5. Another related development is the recognition that natural ecosystems contain complex networks of interacting feedbacks, which can strongly regulate the internal dynamics of terrestrial ecosystems (DeAngelis and Post 1991). For instance, and as will be discussed in Chapter 3, there is now growing evidence that negative feedbacks, which occur when two components of an ecosystem have opposite effects on one another, play an important role in regulating the dynamics of plant communities (e.g. Bever 1994; Van der Putten et al. 1993; Klironomos 2002; Bezemer et al. 2006; Kardol et al. 2007). Likewise, positive feedbacks, such as those which operate between herbivores, soil nutrient cycles, and plants, are now recognized as being key drivers of ecosystem dynamics (McNaughton 1983, 1985; Bardgett et al. 1998b; Bardgett and Wardle 2003). Further, carbon-cycle feedbacks between the land surface and atmosphere are increasingly seen as having major implications for the climate and Earth-system (Jenkinson et al. 1991; Cox et al. 2000; Heimann and Reichstein 2008; Chapin et al. 2009).

Although organisms and their interactions are fundamental to the concept of ecosystem ecology, the inclusion of biotic interactions in ecosystem and Earth-system science has been relatively recent. This most likely stems from community and ecosystem ecology having to some extent developed independently of one another. Here, community ecology has been primarily concerned with understanding how communities are affected by environmental (e.g. climate, disturbance, and soil fertility) and biotic factors (e.g. competition, predation, and mutualisms), while ecosystem ecology has been mostly focused on studying rates and flows of energy and nutrients within ecosystems. However, the last two decades have witnessed a growing recognition of the roles that species play in ecosystems, and studies that consider elements of both community ecology and ecosystem ecology are now commonplace in the ecological literature. As will be discussed in the following sections, this shift has been motivated by many factors, including a growing recognition of the roles that aboveground and belowground biota and multitrophic interactions play in driving ecosystem processes (Grime 1979; Coley et al. 1985; Lawton and Jones 1995; van der Putten et al. 2001), and a vast research effort aimed at connecting biodiversity to ecosystem functioning (Hooper et al. 2005). Further, the growing desire to predict how species responses to global change influence ecosystem processes such as decomposition and the cycling of nutrients has generated much interest in the concept of functional classification, and especially the role of plant species traits in ecosystem functioning (e.g. Grime 1998; Lavorel and Garnier 2002; Vile et al. 2006; Diaz et al. 2007).

Another important development in ecosystem ecology has been in the use of mathematical models to predict the consequences of changes in abiotic and biotic conditions for ecosystem processes, including nutrient and carbon fluxes. For instance, a number of modelling approaches have been developed for understanding the controls on carbon cycling, ranging from global circulation models (GCMs) to soil-carbon process models that can be parameterized at the plot, core, or microsite scale. Further, progress has been made in coupling GCMs with dynamic global vegetation models (DGVMs) for studying impacts of broad-scale shifts in vegetation (at the level of coarse functional types) on feedbacks to the climate system. This has in turn enabled simulation of impacts of climate change on vegetation cover and soil-carbon storage (Cox et al. 2000; Sitch et al. 2003; Woodward and Lomas 2004). Also, while most soil-carbon models have historically represented soil carbon as pools of organic matter (Paustian 1994), attempts have been made to include more detailed understanding of the role of soil food webs in regulating soil-carbon fluxes. For example, the pioneering soil food-web model developed by Hunt et al. (1987) for shortgrass steppe in North America used functional groupings of soil organisms to calculate carbon and nitrogen fluxes and rates of nitrogen mineralization. This model explicitly incorporated estimates of various parameters for these organisms including their carbon-to-nitrogen ratio, rate of turnover, assimilation and production efficiency, and population size. This modelling has also been used to examine functional responses of shifts in food-web energy channels (Moore and Hunt 1988) and the ecosystem effects of losses of functional groups and their combinations within soil food webs (Hunt and Wall 2002), as will be discussed in Chapter 2. There is still much to be done to accurately represent biological interactions in ecosystem models (Van der Putten et al. 2009), although such integration is crucial to improve predictions of global change on ecosystem processes and feedbacks to the Earth-system.

1.2 Species and biotic interactions as ecosystem drivers

The issue of how species and differences among species affect ecosystems has been long recognized by ecologists. For example, Müller (1884) proposed that plant species and variants within plant species were powerful determinants of the types of soils that develop under them (i.e. 'mull', 'moder' or 'mor'), the soil invertebrate communities that they support, and nutrient supply from the soil for plant growth. Further, Handley (1954, 1961) performed a comprehensive series of experiments to show that the ericaceous shrub *Calluna vulgaris* could form tannin–protein complexes that reduce the availability of nitrogen to coexisting plant species. However, despite this historical literature on how plant species differences affect ecosystems, widespread recognition by ecologists that species effects are powerful ecosystem drivers, or the question of 'what do species do in ecosystems' (Lawton 1994), is comparatively recent. This topic has become an increasingly dominant theme in ecology, especially following publication of the influential book *Linking Species and Ecosystems* edited by Jones and Lawton (1995).

It has long been recognized that some species in a community have greater effects on community or ecosystem properties than do others, either because they simply produce more mass (i.e. dominant species) or because they have some attribute that causes them to have disproportionate effects relative to their mass (i.e. keystone species) (Paine 1969; Power et al. 1996). With regard to plant communities, Grime (1998) proposed through his Mass Ratio Hypothesis that the effects of each species in a community on ecological processes should scale with its relative contribution to total community biomass. Conversely, there is strong evidence that some plant species, notably those capable of symbiotic nitrogen fixation, may substantially alter community and ecosystem properties even when occupying only a small proportion of total community biomass, as shown through the classic work of Vitousek et al. (1987) and Vitousek and Walker (1989) on the invasion of nitrogen-poor native forests of Hawai'i with the exotic nitrogen-fixer *Myrica faya*. Differences between species on ecosystem properties are also apparent through the rapidly expanding literature on plant strategies and traits. It has long been recognized, including through the classic *r-K* continuum model (Macarthur and Wilson 1967), that a fundamental trade-off exists between those organisms with rapid growth rates which are adapted for resource acquisition, and those which grow more slowly and are adapted for resource conservation. More complex models for characterizing plant strategies and traits have since been developed, the best known being the C-S-R (competitor, stress tolerator, ruderal) model of Grime (1977). Subsequently, analysis of plant trait databases for large numbers of species over regional or global scales (e.g. Grime et al. 1997; Díaz et al. 2004; Wright et al. 2004) has confirmed the existence of a primary axis of evolutionary specialization that discriminates between species having traits suited for resource acquisition and those suited for resource conservation. These traits are also increasingly recognized as being important in determining differences between species in the quality of resources that they produce, and ultimately their effects on ecosystem processes, both aboveground and belowground (Wardle et al. 2004a), as we explore in Chapter 3.

A major component of understanding how plant species differ in terms of their ecosystem effects involves interspecific differences in plant litter quality: different plant species differ greatly in terms of the decomposability of their litter by soil microbes and fauna, and the pattern of nutrient release from this litter. This in turn has a major influence on nutrient cycling and plant nutrition, soil organic-matter formation and dynamics, the quality of soil organic matter, communities of soil organisms, and ecosystem carbon storage. There is a long history of work on this topic, which has gained increasing attention from soil biologists and ecosystem scientists, especially following publication of the influential synthesis *Decomposition in Terrestrial Ecosystems* by Swift et al. (1979). As such, there has been much interest in the chemical and physical controls of decomposition and nutrient mineralization of plant litter (e.g. ratios of carbon to nutrients or nutrients to lignin; concentrations of lignin or polyphenolics), and how this varies among species (e.g. Taylor et al. 1989; Berg and Ekbohm 1991). The issue of plant-litter quality as an ecological driver gained renewed vigour at around the time of publication of the book *Driven by Nature: Plant Litter Quality and Decomposition* edited by Cadisch and Giller (1997). As such, the past 15 years have witnessed a sharp increase in recognition of the importance of plant traits (notably those that distinguish species with a resource-acquisitive strategy in contrast to those with a conservative strategy) as powerful drivers of litter quality and ecosystem processes driven by litter quality (e.g. Cornelissen 1996; Grime et al. 1996; Wardle et al. 1998a; Cornwell et al. 2008; Fortunel et al. 2009).

Another aspect of how plant species and traits affect ecosystem properties is through their interactions with herbivores. There is a long history of trophic dynamic theory regarding how herbivores and their predators both respond to and influence plant productivity (e.g. Hairston et al. 1960; Menge and Sutherland 1976; Oksanen et al. 1981), and some historical awareness that forage quality might influence these interactions (e.g. White 1978; Lawton and McNeill 1979). Following this has been a growing recognition that plant traits can play an important role in mediating the interaction between plants and herbivores (e.g. Grime 1979; Coley et al. 1985; Díaz et al. 2006), and that plants that possess traits associated with an acquisitive strategy and high-quality litter also produce foliage that is more palatable than plants which have traits characteristic of a conservative strategy (Grime et al. 1996). Coupled with this has been a growing awareness that palatable fast-growing plants may enter a positive feedback with their herbivores which involves enhanced soil fertility, as shown by classical studies on grasslands in the Serengeti (McNaughton 1983, 1985). Similarly, there is also recognition that herbivory in less productive ecosystems can lead to a negative feedback whereby palatable plant species are replaced by less palatable species which produce poorer-quality litter, leading to reduced soil fertility and ecosystem productivity, as shown in classical studies involving moose (*Alces alces*) browsing in Isle Royale, Michigan, USA (Pastor et al. 1988). Herbivores and their ecological effects are in turn governed by their predators, and although historically the importance of predator-induced trophic cascades in terrestrial ecosystems has been questioned (Strong 1992; Polis 1994), the last decade has witnessed increasing recognition that trophic cascades in at least some terrestrial environments have a significant role in modulating herbivore effects on plant communities and therefore upon ecosystem processes (e.g. Pace et al. 1999; Terborgh et al. 2001)

Plant communities consist of multiple coexisting plant species, and an issue that has long fascinated ecologists is how combinations of plant species influence ecosystem processes (Odum 1969). For instance, many early studies have explicitly investigated how multiple species combinations and species diversity influence ecosystem stability (McNaughton 1977), productivity (Trenbath 1974; Austin and Austin 1980), and soil biota (Christie et al. 1974, 1978; Chapman et al. 1988). However, from the mid-1990s, notably following publication of the book *Biodiversity and Ecosystem Function* edited by Schulze and Mooney (1993), this field has attracted considerable attention, and many experimental studies have been performed since the mid-1990s to demonstrate the role of species diversity as a driver of ecosystem processes (e.g. Naeem et al. 1994; Tilman et al. 1996; Hector et al. 1999), although the interpretation of these studies has caused much debate (Aarssen 1997; Huston 1997; Kaiser 2000; Hooper et al. 2005). Other studies have addressed this issue from a combined aboveground–belowground perspective, either by considering how plant biodiversity influences belowground organisms and processes (e.g. Hooper and Vitousek 1997, 1998; Wardle et al. 1997b, 1999), or how biodiversity of belowground organisms affects aboveground organisms and processes (e.g. Van der Heijden et al. 1998b; Laakso and Setälä 1999a) Many issues remain unresolved about the importance of species diversity in driving ecosystem processes in real (natural or non-experimental) ecosystems, as will be discussed in Chapter 5, and multiple points of view about this topic persist (e.g. Tilman 1999; Leps 2004; Grace et al. 2007; Duffy 2009; Wardle and Jonsson 2010).

1.3 Aboveground–belowground interactions as drivers of ecosystem processes

Traditionally, the aboveground and belowground components of ecosystems have frequently been considered in isolation from one another. However, as noted above, the last two decades have witnessed a proliferation of studies exploring the influence that these components exert on each other, and the fundamental role that aboveground–belowground interactions play in controlling the structure and functioning of terrestrial ecosystems. As will be demonstrated throughout this book, this has led to an increased recognition of the role that aboveground–belowground interactions play as terrestrial ecosystem drivers, and their increasing inclusion as a fundamental element of ecosystem ecology (Wardle et al. 2004a). This development has been driven in part by issues discussed above, such as the growing understanding of the roles that plant species and litter quality play in regulating decomposition processes (e.g. Swift et al. 1979; Cadisch and Giller 1997). However, other issues have also contributed to this development, some of which we highlight here.

As has been documented elsewhere (e.g. Wardle 2002), the discipline of soil science, including soil biology, has historically often given relatively little attention to what goes on aboveground. However, in the late 1970s and early 1980s, soil ecologists focused increasingly on the significance of trophic interactions in soil for decomposition and nutrient-cycling processes (Coleman et al. 1977; Anderson et al. 1981; Anderson and Ineson 1984; Clarholm 1985; Coleman 1985). These trophic interactions were shown to influence the supply of plant-available nutrients from the soil, and subsequent studies pointed to their role in influencing plant nutrient acquisition and growth. For example, the classic experimental study by Ingham et al. (1985) showed that the addition of microbial-feeding nematodes to soil enhanced nitrogen uptake by and growth of the grass *Bouteloua gracilis*. Similarly, Setälä and Huhta (1991) showed that leaf, stem, and shoot biomass of birch (*Betula pendula*) seedlings was increased when they were grown in the presence of a diverse soil fauna. These studies laid the foundation for many subsequent experiments which have shown multitrophic interactions in soil to serve as drivers of plant nutrient acquisition and growth (e.g. Alphei et al. 1996; Bardgett and Chan 1999; Laakso and Setälä 1999a, 1999b; Cole et al. 2004), plant community structure (Brown and Gange 1990; Bradford et al. 2002), and consumers of plant material (e.g. Gange and Brown 1989; Scheu et al. 1999; Bonkowski et al. 2001). Moreover, as will be discussed in Chapter 2, this research has led to the recognition that non-nutritional links also occur between soil and plant communities through soil animals affecting bacterial production of plant growth-promoting hormones (Jentschke et al. 1995; Alphei et al. 1996; Bonkowski 2004). Consequently, effects of belowground communities on plant growth are now increasingly recognized as being highly complex, and involving a variety of nutritional and non-nutritional responses to microbial grazing (Bonkowski 2004).

It has long been known that plant and soil communities are linked through the actions of root-associated biota, including symbionts, parasites, pathogens, and root herbivores. Mycorrhizal fungi, which associate with roots of some 80% of all terrestrial plant species (Smith and Read 1997), have received a particularly large amount of attention. An important development has been the recognition that the major mycorrhizal types differ both in the types of ecosystem in which they dominate and the types of function that they perform (Harley 1969; Read 1994). For example, trees and shrubs of temperate and boreal forests and shrublands, especially those that are nutrient-limited, are known to have an obligate need for ectomycorrhizae and ericoid mycorrhizae, which perform a crucial role in the acquisition organic nutrients contained in litter and humus in these ecosystems (Leake and Read 1997; Read and Perez-Moreno 2003). Meanwhile, arbuscular mycorrhizal fungi, which dominate in grasslands and many temperate and tropical forests, have been shown to play a significant role in structuring plant communities. For example, in a classic study, Grime et al. (1987) manipulated arbuscular mycorrhizal fungi in grassland microcosms and found that their presence caused a reduction in dominance of competitive grass species in favour of several subordinate herb species, in turn increasing plant species diversity. This study laid the foundation for several other studies which have manipulated mycorrhizal communities and revealed their role as powerful drivers of plant community dynamics (e.g. Gange et al. 1993; Newsham et al. 1995a; Van der Heijden et al. 1998a; Hartnett and Wilson 1999). At around that time, an increasing number of studies began to explore the role of nitrogen-fixing symbionts as drivers of vegetation dynamics and ecosystem functioning in natural ecosystems (e.g. Vitousek and Walker 1989; Olff et al. 1993; Chapin et al. 1994), leading to new discoveries about roles that they play in the nitrogen cycle of particular ecosystems. For example, nitrogen-fixing cyanobacteria which inhabit the incurves of leaves of the feather moss *Pleurozium schreberi* (the most common moss on Earth) have been found to fix significant quantities of nitrogen in boreal forests (De Luca et al. 2002b). Also, leaf cutter ants, which are prevalent in neotropical ecosystems, have been shown to engage in mutualistic associations with nitrogen-fixing bacteria in their belowground fungal gardens, leading to significant ecosystem inputs of nitrogen (Pinto-Tomás et al. 2009).

Another topic that has contributed to the development of interest in aboveground–belowground interactions involves the nature of feedbacks between the two subsystems. One aspect that is attracting increasing attention involves the role of soil antagonists—that is, root pathogens and herbivores—in regulating vegetation dynamics via negative feedback. The mechanism involved, namely the accumulation of specific plant antagonists that inhibit the growth of the host plant, is well established in agricultural science, and forms the basis of crop rotations for the prevention of build up of crop-specific pathogens in soil. However, the role of these antagonists in natural ecosystems only began to be appreciated in the last two decades through studies pointing to their role in driving vegetation dynamics (e.g. Brown and Gange

1989, 1992; Van der Putten et al. 1993; Bever 1994). These studies laid the foundation for many subsequent studies, particularly in the past five years, which have placed the concept of negative feedback firmly within the ecological literature (Kulmatiski et al. 2008; Van der Putten 2009). Meanwhile, other studies have plant–soil feedbacks in which plants provide resources for decomposers as plant litter and rhizosphere materials, and the decomposers in turn release nutrients from these materials for plant growth. These include studies which have shown plant species to differentially influence soil biological communities (e.g. Rovira et al. 1974; Grayston et al. 1998; Bardgett et al. 1999c) and nutrient cycling in their immediate vicinity (e.g. Northup et al. 1995; Berendse 1998), and to preferentially select for decomposer taxa that enhance the decomposition of their own litter (e.g. Hunt et al. 1988; Hansen 1999; Vivanco and Austin 2008; Ayres et al. 2009). Moreover, classic studies on aboveground herbivore effects on ecosystems have shown herbivores to alter soil fertility in a manner that can feed back to plant growth both positively (McNaughton 1983, 1985) and negatively (Pastor et al. 1988, 1993). Other studies have revealed that plant–soil feedbacks involving herbivory can also operate at the individual plant level (Seastedt 1984). As such, there is growing evidence that defoliation causes a short-term pulse of root exudates which stimulates soil biotic activity (Holland et al. 1996; Mawdsley and Bardgett 1997), leading to increased soil nitrogen mineralization and plant nitrogen acquisition, and ultimately benefiting plant growth (Hamilton and Frank 2001).

1.4 Aboveground–belowground interactions and global change

Human activities have a substantial and ever-growing impact on ecosystems, to the extent that much of the Earth's land surface has been transformed by a suite of global-change phenomena (Vitousek et al. 1997c). The most obvious human impact on Earth is through the increasing transformation of land for food production and forestry, which is presently the primary driving force behind the loss of biological diversity on both local and global scales (Sala et al. 2000; Millennium Ecosystem Assessment 2005). However, every terrestrial ecosystem on Earth is also simultaneously affected by a suite of other global change phenomena and their interactions, including global climate change, the invasion of alien species into new territories, losses of existing species, and increasing rates of nitrogen deposition. The last two decades have witnessed a vast number of studies exploring the impacts of these global change phenomena on the structure and functioning of terrestrial ecosystems. Concomitant with this has been a developing awareness that understanding the consequences of global phenomena requires explicit consideration of linkages between aboveground and belowground biota (e.g. Wardle et al. 1998c; Wolters et al. 2000). This is because, with the exception of disturbances that directly impact on belowground biota and the processes that they drive, the influence of global change phenomena on terrestrial ecosystems is largely indirect via changes that occur aboveground. These include shifts in plant community composition, spectra of plant traits, carbon-allocation patterns, and the quantity and quality of plant-derived organic matter entering soil. In turn, such belowground responses to global change, both direct and indirect, create feedbacks that affect aboveground biota, ecosystem nutrient and carbon dynamics, and the flux of carbon dioxide and other greenhouse gases from land to the air.

The driver of global change that has perhaps attracted the most attention in this context is climate change. As such, a number of studies have contributed to the now widespread recognition that an understanding of the impacts of climate change on terrestrial ecosystems and Earth-system feedbacks requires explicit consideration of belowground biota and their linkages aboveground (Bardgett et al. 2008; Tylianakis et al. 2008). For instance, several studies in the 1990s revealed that elevated concentrations of carbon dioxide in the atmosphere can indirectly impact on soil organisms through changing the quantity and quality of plant litter returned to soil, the rate of root turnover, and the exudation of carbon from roots into soil (Billes et al. 1993; Jones et al. 1998; Coûteaux et al. 1999). These effects on soil biota were also found to have important consequences for the dynamics of nutrients in soil and hence their availability to plants, thereby enforcing either positive or negative interactions with plant growth depending on context (Zak et al. 1993; Díaz et al. 1993). Another area that has attracted much attention concerns the effects of climate warming on organic matter decomposition in soil. Of particular relevance is the pioneering work of Jenkinson et al. (1991), who demonstrated that rising temperatures could accelerate rates of soil microbial respiration, thereby increasing the transfer of carbon dioxide to the atmosphere and promoting a positive feedback on climate change. Although this study was primarily concerned with a direct effect of climate warming on the activity of heterotrophic microbes, subsequent work has shown the temperature dependence of decomposition to vary with organic matter quality (e.g. Luo et al. 2001; Melillo et al. 2002; Fierer et al. 2005), and thus the quality of plant material entering the soil.

Human activities have also lead to major increases in global emissions of nitrogen to the atmosphere, leading to dramatic increases in the deposition of atmospheric nitrogen to the terrestrial biosphere (Holland et al. 1999; Bobbink and Lamers 2002; Galloway et al. 2004). As with climate change, it is now well known that such nitrogen enrichment can greatly modify terrestrial ecosystems by affecting aboveground–belowground feedbacks. As will be discussed in Chapter 2, a significant body of literature has developed showing that effects of nitrogen enrichment can directly impact on belowground organisms (e.g. Scheu and Schaefer 1998; Donnison et al. 2000; Egerton-Warburton and Allen 2000) and extracellular enzymes involved in decomposition processes (e.g. Carreiro et al. 2000; Frey et al. 2004). However, as will be discussed in Chapter 3, a growing number of studies have also suggested that impacts of nitrogen enrichment may be indirect, by altering the composition of the plant community (Bobbink 1991; Bowman et al. 1993; Wedin and Tilman 1993), although explicit tests of this are still scarce (e.g. Manning et al. 2006, Suding et al. 2008). Another important development in this area is the recognition that interactive effects on the belowground subsystem of nitrogen enrichment and climate change have the potential to strongly influence ecosystem functioning. For instance, it is now widely recognized that the response of terrestrial ecosystems to elevated carbon dioxide concentrations is limited by nitrogen availability (e.g. Luo et al. 2004) and that nitrogen enrichment can amplify climate-change-driven changes in vegetation composition, thereby increasing the potential for indirect effects of vegetation change on the belowground subsystem (Wookey et al. 2009). Also, as we will discuss in Chapter 2, nitrogen deposition can have substantial effects on decomposition processes, thereby influencing the response of soil respiration to warming and the capacity of soils to sequester carbon under climate change (Davidson and Janssens 2006; Bardgett et al. 2008).

Another element of human-driven global change that has attracted much recent attention is how invasion of alien species into new territories affect ecosystem properties. For instance, a body of literature has been developing for over 20 years, which includes the classic studies of Vitousek et al. (1987) and Vitousek and Walker (1989) described above, revealing that invasion of alien plants into new territories can have strong effects both aboveground and belowground, when the invading species has vastly different physiological traits from the native flora. As described in Chapter 5, there is now much evidence showing that invasive plants may greatly affect the quantity and quality of resource inputs to decomposer organisms and the processes that they drive, ultimately affecting soil nutrient availability (Ehrenfeld 2003; Liao et al. 2008). Further, starting with the work of Klironomos (2002), several recent studies have also focused on aboveground–belowground feedbacks which operate via interactions between invasive species and root-associated biota, for example in terms of escape of the invader from natural soil-borne enemies (e.g. Reinhart et al. 2003; Callaway et al. 2004). Moreover, as will be discussed in Chapter 5, a handful of studies over the past decade have shown major transformations by invasive animals of both the aboveground and belowground components of ecosystems; some of the most spectacular examples involve invasive herbivores such as moths (Lovett et al. 2006), deer (e.g. Wardle et al. 2001; Vázquez 2002), and beavers (*Castor canadensis*) (Anderson et al. 2006), and cascading effects of invasive predators such as rats (*Rattus* spp.) (Fukami et al. 2006), foxes (Croll et al. 2005), and ants (O'Dowd et al. 2003). Finally, as will also be highlighted in Chapter 5, recent evidence suggests that range-expanding species under climate change can have important effects on ecosystem properties and, potentially, carbon-cycle feedbacks by affecting aboveground–belowground feedbacks (e.g. Knapp et al. 2008; Kurz et al. 2008), and that belowground biota might play a role in affecting range expansion of species under climate change (Engelkes et al. 2008).

1.5 Emerging issues and trends

As highlighted throughout this chapter, aboveground–belowground interactions and feedbacks are highly complex and operate at a wide range of temporal and spatial scales. Despite the rapid development of this field, many challenges remain, as we will highlight in this book. Perhaps one of the most significant challenges, however, is the need for a stronger theoretical basis for the subject, in that almost all studies on aboveground–belowground interactions have been carried out from an empirical perspective and modelling approaches are still in their infancy (Van der Putten et al. 2009). As a consequence, our ability to make predictions about the role of aboveground–belowground interactions and feedbacks in regulating terrestrial ecosystem processes and their response to global change remains limited. A related issue that is emerging is the need to demonstrate the significance of aboveground–belowground interactions as drivers of community dynamics and functioning of real ecosystems, and the importance of their role relative to that of abiotic factors. As will be discussed in Chapter 5, this point is especially relevant for understanding the importance of consequences of species loss (and gain) for the functioning of real terrestrial ecosystems, and stems partly from the fact that relatively few studies have considered the impacts of non-random species losses in real ecosystems and how they vary in different environment contexts. As we argue in Chapter 5, the most product-ive way forward here would be the widespread use of experimental or theoretical approaches that directly investigate how loss of species, such as may result from global environmental change, directly impacts upon other organisms and ecosystem processes in real ecosystems.

Another emerging issue that we consider throughout this book concerns soil-carbon dynamics and especially the role that soils play in the global carbon cycle and climate change. This interest has arisen because soils absorb and release green-house gases (notably carbon dioxide and methane), and act as a major global carbon reservoir, storing some 80% of the Earth's terrestrial carbon stock (IPCC 2007). Despite the importance of soils for carbon cycling, remarkably little is known about the factors that regulate the fluxes of carbon to and from soil, or about the role that interactions between plants and soil biota play in regulating soil-carbon cycling (Wardle et al. 2004a; Bardgett et al. 2009; Peltzer et al. 2010). The last few years have witnessed a proliferation of studies into the ecological and biogeochemical dynamics of carbon-cycling processes in soils, including investigations into plant and rhizosphere effects on soil respiration, effects of soil microbial communities on plant productivity and soil-carbon turnover, and interactions among element cycles (e.g. carbon, nitrogen, and phosphorus). However, current models of the global carbon cycle seldom include these processes; rather, they simply treat net carbon emissions from ecosystems as the balance between net primary production (NPP) and heterotrophic respiration. An emerging challenge is therefore to use our advancing understanding of plant and soil microbial processes involved in carbon cycling to improve the representation of aboveground–belowground interactions in carbon-cycle models.

Finally, an issue that has attracted much attention in recent years is the value of ecosystems in providing ecosystem goods and services, or the benefits that people derive from ecosystems (Ehrlich and Mooney 1983; Tscharntke et al. 2005; Hooper et al. 2005). These benefits are obvious and wide-ranging, and include carbon sequestration, climate regulation, the provision of clean water, and the maintenance of soil fertility and primary production (food, fodder, and fibre). This issue has moved the science of ecosystem ecology beyond the scientific community to the global policy community, as politicians increasingly recognize ecosystems as natural capital assets that supply life-support services of great value (Daily and Matson 2008). A major impetus for this change was the publication of the Millennium Ecosystem Assessment (2005), which presented a new conceptual framework for documenting, analysing, and understanding the effects of global change on ecosystems and human well-being, viewing ecosystems in terms of the services that they provide for society. This approach has been adopted widely among the scientific and policy communities and has led to new approaches for research, conservation, and sustainable development (Carpenter et al. 2009; Daily and Matson 2009). Moreover, this approach presents many new challenges for ecosystem ecologists, most notably the need to understand biotic drivers of ecosystems, including aboveground–belowground interactions, in the context of ecosystem services and human well-being. As noted by Carpenter et al. (2009), this requires a new kind of interdisciplinary science that considers the interactions of social and ecological constituents of the Earth-system. While the issue of ecosystem service provision is not a primary focus of this book, our synthesis does provide a basis for understanding the biotic factors that influence their provision and how this provision may be driven by aboveground–belowground linkages and their responses to global change.

2

Biotic interactions in soil as drivers of ecosystem properties

2.1 Introduction

It has long been appreciated that soil organisms are central to the maintenance of soil fertility and the functioning of terrestrial ecosystems. For example, the pioneering work of Winogradsky (1856–1953) established some of the fundamental roles that soil microbes play in the nitrogen and sulphur cycles, and in 1881 Darwin published his last book, *The Formation of Vegetable Mould through the Action of Worms and Observations on their Habits*, which recognized the important roles that earthworms perform in biogeochemical cycles. Despite this historical recognition of the contribution of soil organisms to soil fertility, it is only in recent years that community and ecosystem ecologists have more fully recognized the key role that belowground organisms and biotic interactions play in determining a range of important ecosystem processes, including decomposition, nutrient and carbon cycling, and soil formation. Moreover, it has only relatively recently been recognized that the soil biological community represents a major, but largely unexplored, reservoir of global biodiversity (Whitman et al. 1998; Torsvik et al. 2002; Bardgett 2005). For instance, it has been estimated that 1 g of soil can contain up to 50 000 bacterial species (Curtis et al. 2002; Torsvik et al. 2002), and up to 200 m of fungal hyphae (Read 1992; Bardgett et al. 1993a). Further, as many as 89 nematode species have been found in a single soil core in Cameroon tropical forest (Bloemers et al. 1997) and 159 mite species were identified in a patch of prairie grassland soil in Kansas (St John et al. 2006). Recently, the analysis of soil DNA by Wu et al. (2009) revealed an estimated 1320 and 2010 operational taxonomic units (OTUs) of fauna in Alaskan boreal forest and tundra, respectively. Despite this growing recognition of both the diversity and functional importance of soil organisms for ecosystem processes, our understanding of the impact of soil biotic interactions on community and ecosystem properties is still in its infancy.

As noted in Chapter 1, the last decade has seen a dramatic increase in the number of studies showing that aboveground and belowground communities are intimately linked, and that feedbacks between the aboveground and belowground subsystems play fundamental roles in controlling ecosystem processes and properties (e.g. Hooper et al. 2000; Van der Putten et al. 2001, 2009; Scheu and Setälä 2002; Wardle 2002; Wardle et al. 2004a; Bardgett et al. 2005; Van der Heijden et al. 2008). The complexity of possible interactions between aboveground and belowground communities, and hence their consequences for ecosystem properties, is enormous (Wardle 2002; Wardle et al. 2004a; Van der Heijden et al. 2008). However, as a general framework (Fig. 2.1), plants regulate the quantity and quality of resources available for the functioning of the decomposer community and for obligate belowground biotrophs, such as root herbivores, pathogens, and symbiotic mutualists. In turn, the belowground community regulates plant growth and community composition indirectly through the activity of the decomposer food web (which regulates the supply of available soil nutrients to plants), and directly through the action of root-associated biotrophs (Wardle et al. 2004a). In this way, biotic interactions between aboveground

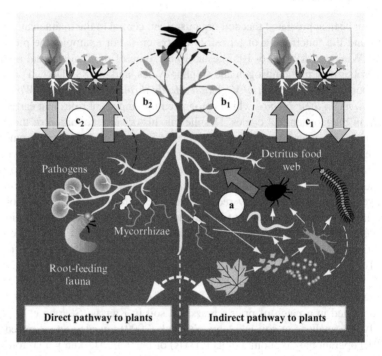

Fig. 2.1 Schematic diagram showing interactions of aboveground communities with soil food-web organisms. On the right-hand side, the feeding activities in the detritus food web (solid arrows) stimulate nutrient turnover (large grey arrow), plant nutrient acquisition (a), plant performance, and thereby indirectly influence aboveground herbivores (broken arrows) (b_1). On the left-hand side, soil biota exert direct effects on plants by feeding upon roots and forming antagonistic or mutualistic relationships with their host plants. Such direct interactions with plants influence not just the performance of the host plants themselves, but also that of the herbivores (b_2) and ultimately their predators. Further, the soil food web can control the successional development of plant communities both directly (c_2) and indirectly (c_1), and these plant community changes can in turn influence soil biota. From Wardle et al. (2004a), with permission from the American Association for the Advancement of Science.

and belowground communities act as powerful drivers of community and ecosystem properties, and play an important role in regulating the response of terrestrial ecosystems to human-induced global change.

In this chapter, we explore one part of this feedback, namely how interactions between the vast diversity of organisms that live in soil regulate important ecosystem processes, and how they indirectly and directly influence the productivity, diversity, and composition of plant communities. The other component of this feedback, namely how plants in turn affect those organisms that inhabit the soil, is considered in Chapter 3. First, we describe the roles that the soil biological community play in regulating ecosystem processes and the productivity and composition of plant communities through their influence on the mineralization of nutrients. Second, we explore how root-associated organisms affect plant productivity and community structure, including their role in driving succession and invasion. Third, we examine how belowground engineers modify soil nutrient dynamics and plant communities through altering the physical environment in the soil. Finally, we discuss the role of soil biotic interactions in a global change context, focussing on their influence on ecosystem carbon dynamics and contribution to climate change via carbon-cycle feedbacks. The ultimate aim of this chapter is to highlight the significance of soil biotic interactions for ecosystem processes, the productivity and diversity of plant communities, and carbon and nutrient dynamics under global change.

2.2 Influence of decomposers on aboveground communities and ecosystem processes

Ecosystem ecologists have long recognized the importance of dead organic matter, or detritus, for community organization and ecosystem energy flow (Odum 1969; Swift et al. 1979; Moore et al. 2004). As such, it is well established that detritus provides both a resource and habitat for the various components of the decomposer food web, including microbial decomposers, detritivores, microbivorous organisms, and predators (Wardle 2002; Moore et al. 2004) (Fig. 2.2). In turn, the decomposer community strongly influences aboveground community organization and ecosystem processes through its role in the breakdown and transformation of organic matter, and liberation of plant growth-limiting nutrients (Fig. 2.2). In this section, we demonstrate the various mechanisms by which biotic interactions within the decomposer food web influence plant communities indirectly through their affect on decomposition processes and nutrient dynamics.

2.2.1 Free-living soil microbes, nutrient availability, and plant growth

Plant growth is primarily limited by the availability of nutrients (Vitousek 2004), and hence any biotic interaction in soil that alters the rate and timing of nutrient supply to plants (whether positive or negative) will influence plant growth and potentially plant community dynamics. Biotic interactions that occur in the rhizosphere are of

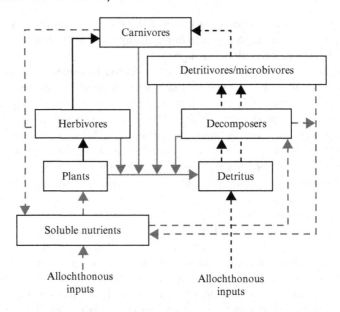

Fig. 2.2 Features of a generalized food web incorporating detrital dynamics, including resource inputs into the web and internal cycling within the web. The solid black arrows represent the flow of matter through the grazer pathway that originates from primary producers, whereas the broken black arrows represent the decomposer pathway that originates from detritus (external imports and internal pools). The solid grey arrows depict the autochthonous flow of matter to the detritus pool that results from death of all living organisms and from unassimilated prey. The broken grey arrows signify the mineralization and immobilization of soluble nutrients. From Moore et al. (2004), with permission from Wiley-Blackwell.

importance for plant nutrient supply, as this is the zone of soil that is most strongly influenced by plant roots and their exudates (Bardgett 2005). Here, microbial processes of nutrient mineralization occur whereby soluble and insoluble organic matter is broken down and converted to inorganic plant available forms, strongly regulating plant nutrient supply (Kaye and Hart 1997; Jones et al. 2005; Schimel and Bennett 2004).

Most soil nitrogen (96–98%) is contained in dead organic matter as complex insoluble polymers such as proteins, nucleic acids, and chitin. These polymers are broken down into dissolved organic nitrogen by extracellular enzymes produced by soil microbes (Schimel and Bennett 2004). This dissolved organic nitrogen can constitute a significant portion of the total soluble nitrogen pool (Jones and Kielland 2002) and can be absorbed by free-living soil microbes, mineralized by the microbial biomass (i.e. converted from an organic to inorganic form, with inorganic nitrogen being liberated into the soil environment), or taken up directly by plant roots in the form of amino acids and bypassing the microbial mineralization step. Direct uptake of amino acids by plants appears to be especially important in strongly nitrogen-limited ecosystems, such as in Arctic and alpine tundra (Raab et al. 1999; Nordin et al. 2004), boreal (Näsholm et al. 1998; Nordin et al. 2001) and temperate forest (Finzi and Berthrong 2005), and low-fertility grasslands (Bardgett et al. 2003; Weigelt et al. 2005; Harrison et al. 2007). Recent recognition of the importance of direct uptake by plants has led to a radical rethink of terrestrial nitrogen cycling and of the processes that control plant N availability (Schimel and Bennett 2004; Jones et al. 2005). In a similar way, microbes also influence the mineralization of organic phosphorus through the production of phosphatases, which cleave ester bonds in organic matter and liberate inorganic phosphorus for plant uptake. Unlike nitrogen, most reactions that govern phosphorus availability to plants are geochemical rather than biological, although, as will be discussed later in this chapter, mycorrhizal fungi play an important role in the uptake of phosphorus by plants.

Microbes in soil also compete with plants for nutrients in soil solution, with potentially negative effects on plant nutrient acquisition and growth. This is most likely to occur in strongly nutrient-limited ecosystems such as Arctic and alpine tundra (Schimel and Chapin 1996; Nordin et al. 2004). Further soil microbes have been shown to compete effectively with plants for nitrogen in nutrient-poor grasslands (Bardgett et al. 2003; Harrison et al. 2007), and this can lead to as much nitrogen being stored in the microbial pool as in the plant pool (Jonasson et al. 1999; Bardgett et al. 2007b). However, in some ecosystems, soil microbes are not a significant sink for nitrogen, especially when the availability of mineral nitrogen and rates of nitrification are high. For instance, ^{15}N-tracer studies in tropical humid forests reveal that those ecosystems which typically lack nitrogen limitation are characterized by very rapid rates of nitrification and negligible microbial assimilation of nitrogen (Silver et al. 2001, 2005; Templer et al. 2008), especially when concentrations of mineral nitrogen in soil are high (Vitousek and Matson 1988). Similarly, ^{15}N-tracer studies in temperate grasslands in England reveal that microbial nitrogen uptake and retention is substantially less in high-fertility (e.g. fertilized) grassland where rates of nitrogen mineralization and nitrification are relatively high, than in low-fertility grassland with low rates of nitrogen turnover (Bardgett et al. 2003). Collectively, these studies point to the microbial community acting as greater nitrogen sink in nitrogen-limited ecosystems than in nitrogen-rich ecosystems. Moreover, they indicate the different roles that soil microbes play in nitrogen retention and loss: microbial nitrogen uptake appears to play a significant role in ecosystems where nitrogen limits plant growth (Jackson et al. 1989; Fisk et al. 2002; Zogg et al. 2000; Bardgett et al. 2003), whereas nitrification plays a key role in ecosystems that are nitrogen replete (Silver et al. 2001, 2005; Templer et al. 2008).

Plants and soil microbes compete for soil nitrogen (Kaye and Hart 1997; Dunn et al. 2006), through close linkages that occur in the rhizosphere (Frank and Groffman 2009). However, they can also avoid competition through partitioning of the ecosystem nitrogen pool over seasons (Jaeger et al. 1999; Bardgett et al. 2005). In particular, studies in alpine regions reveal the existence of a strong link between the temporal dynamics of microbial communities and plant nutrient supply. Here, as in other seasonal ecosystems, it has traditionally been assumed that soil microbes are inactive during the winter. However, it has recently been shown that the biomass of microbes in alpine soils is maximal during late winter when the soil is frozen (Schadt et al. 2003), and that seasonal changes in microbial biomass are associated with shifts in microbial community composition (Lipson and Schmidt 2004). Specifically, fungi that utilize complex plant residues dominate in the winter, whereas bacteria that thrive on root exudates are more active in the summer (Lipson and Schmidt 2004). Moreover, there is an almost complete turnover of the microbial community between winter and summer, with a large number of novel DNA sequences (Schadt et al. 2003) associated with specific functional attributes occurring only in the winter or only in the summer (Lipson and Schmidt 2004). These seasonal dynamics in microbial communities are of importance because they control the temporal partitioning of nitrogen between plants and soil microbes over seasons. For example, in an alpine meadow in Colorado, Jaeger et al. (1999) found that the dominant plant species *Kobresia myosuroides* took up nitrogen maximally after snowmelt, whereas the soil microbial community immobilized nitrogen maximally in the autumn following plant senescence, and then retained this nitrogen throughout the winter. The release of microbial nitrogen for plant uptake in spring appears to be facilitated by a decline in microbial biomass after snowmelt, leading to a pulse of soluble nitrogen in the form of protein into soil (Lipson and Schmidt 2004). This pulse coincides with a peak in soil protease activity, which facilitates the supply of amino acids for plant uptake at a time of high plant nitrogen demand (Raab et al. 1999). Together, these studies indicate that in nitrogen-limited ecosystems the cycling of labile nitrogen pools over seasons relies on intimate, temporal coupling between plants and microbes and their seasonal resource demands (Fig. 2.3) (Bardgett et al. 2005; Frank and Groffman 2009).

Free-living microbes can also influence plant community dynamics through altering the availability of different chemical forms of nitrogen for plant uptake. The proposed mechanism here is that coexisting plant species partition a limited nitrogen pool (thereby avoiding competition for nitrogen), through the uptake of different chemical forms of soil nitrogen which have been produced by microbial enzymes and mineralization processes. Two lines of evidence support this idea. First, it is known that microbial activities produce a wide variety of nitrogen forms in soil solution, including inorganic nitrogen forms and a variety of amino acids of varying complexity (Kielland 1994), which provides a variety of possible resources for plant uptake. Second, several studies show that plant species are highly versatile in their ability to uptake different chemical forms of nitrogen (Weigelt et al. 2005; Harrison et al. 2007, 2008), and that species differ in their ability to uptake nitrogen forms, pointing to them having niches based on nitrogen form (Miller and Bowman 2002, 2003; Weigelt et al. 2005; Harrison et al. 2008). Moreover, coexisting species of nitrogen limited Arctic tundra plants have been shown to differ in the chemical forms of nitrogen that they take up, with the dominant plant species using the most abundant nitrogen form that is present in soil (McKane et al. 2002). This is indicative of species niche partitioning based on the chemical form of nitrogen (McKane et al. 2002). Meanwhile, ^{15}N-labelling studies of low-fertility grasslands in Germany show that different plant functional groups rely on different nitrogen pools to meet their nitrogen demands, also suggesting that nitrogen-uptake patterns across functional groups are

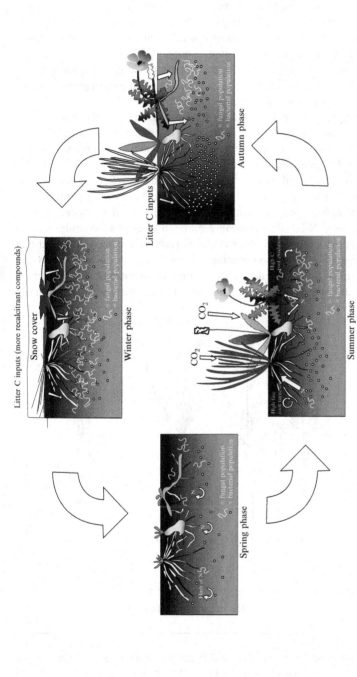

Fig. 2.3 Seasonal dynamics of plant and microbial resource demands in strongly nitrogen-limited alpine ecosystems. In autumn, senescing plants provide a pulse of labile carbon to support microbial growth, and in winter the microbial biomass, and especially fungi which degrade recalcitrant polyphenolic compounds, continues to increase as carbon and nitrogen in plant litter is consumed and mineralized by the microbial community. In spring, rapid changes in microclimate and the exhaustion of labile carbon compounds lead to enhanced turnover of microbial community, with concomitant release of labile N for plant uptake, and in summer plant uptake of nitrogen continues to meet growth demands, followed by a period of C sequestration and loss to soil microbes. From Bardgett et al. (2005), with permission from Elsevier.

driven by different niches based on the chemical form of nitrogen (Kahmen et al. 2006) (Fig. 2.4). A similar idea has been proposed for partitioning of the soil phosphorus pool: a large number of biologically available phosphorus compounds occur in soil and there are likely to be a variety of mechanisms, some involving soil microbes, through which plants can access them (Bardgett 2005; Turner 2008). This is thought to occur widely in terrestrial environments, but especially where productivity is limited by the availability of soil phosphorus, such as in humid temperate regions and ecosystems developed on strongly weathered soils (Turner 2008).

Although it is tempting to conclude that partitioning on the basis of chemical form might provide a mechanism for plants to efficiently partition a limited soil nutrient pool, thereby facilitating species coexistence and the maintenance of plant diversity (McKane et al. 2002; Reynolds et al. 2003; Kahmen et al. 2006), not all studies support this idea. A recent field-based ^{15}N-labelling study in relatively fertile temperate grassland by Harrison et al. (2007) found that while coexisting plant species varied markedly in uptake rates of different chemical forms of nitrogen, they all preferentially took up ^{15}N-labelled inorganic nitrogen over more complex amino acids, and more simple amino acids over complex ones. It has been proposed that variation in the capacity of plant communities to uptake and partition the nitrogen pool based on chemical form is related to the shifting dominance of nitrogen forms along gradients of soil fertility and ecosystem productivity, which is driven primarily

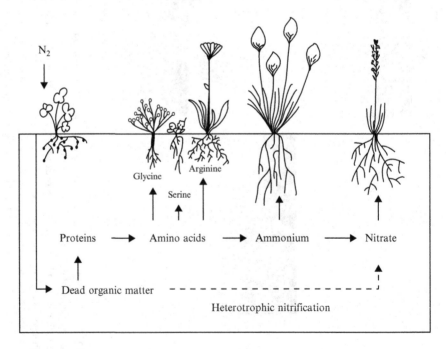

Fig. 2.4 Individual plant species access different chemical forms of nitrogen enabling species to coexist in terrestrial ecosystems. From Bardgett 2005, with permission from Oxford University Press.

by the rate of microbial turnover of components of dissolved organic nitrogen. This idea has been proposed by a number of authors (Schimel and Bennett 2004; Jones et al. 2005; Harrison et al. 2007), but as far as we are aware remains untested.

As discussed above, microbes can influence ecosystem nitrogen availability and hence plant productivity by rapid transformation of nitrogen to more mobile forms, such as nitrate through the bacterial process of nitrification, and through denitrification whereby nitrate is lost to the atmosphere as nitrogen gases under anaerobic conditions. Both these microbial processes cause net loss of nitrogen from soil, especially in ecosystems such as humid tropical forests and fertilized agricultural systems with high mineral nitrogen availability. In these systems, hydrologic throughput and fluctuating redox conditions stimulate nitrate loss through leaching and provide conditions optimal for denitrification (Schellekens et al. 2004; Silver et al. 2005; Houlton et al. 2006). The impact of denitrification on plant productivity and community composition is poorly understood, although worldwide an estimated 109–124 megatonnes of nitrogen are lost from the land surface yearly through denitrification (Galloway et al. 2004; Seitzinger et al. 2006; Schlesinger 2009), which may impair total global terrestrial productivity by around 7% (Schlesinger 1997). Another important recent development in this area is the discovery that archaeal ammonium oxidizers (which perform the first step in the process of nitrification that was previously thought to be carried out exclusively by bacteria) are the most abundant ammonium oxidizers in soil, and are therefore likely to play a substantial role in the nitrogen cycle (Leininger et al. 2006).

2.2.2 Trophic interactions in soil, nutrient availability, and plant growth

Plant nutrient availability is strongly regulated by microbial activities. However, it is important to recognize that the soil also hosts an abundant and highly complex community of invertebrates, including nematodes, collembolans, mites, and enchytraeids, which derive most of their nutrition and carbon directly from roots and exudates, or indirectly from feeding on the microbes that live in the root zone (Pollierer et al. 2007). These faunal consumers of microbes, and their predators, strongly regulate nutrient and plant community dynamics by altering the balance between microbial immobilization and mineralization, and hence the availability of nutrients for uptake by plants. An understanding of how belowground communities influence aboveground community properties therefore requires explicit consideration of the role of predation as a control point in the rhizosphere (Moore et al. 2003).

As mentioned above, microbes can immobilize significant quantities of nitrogen and phosphorus, and thereby render these nutrients temporally unavailable for plant use. For example, in strongly nutrient-limited ecosystems such as Arctic and alpine tundra, the microbial biomass can contain up to 10% of the total ecosystem nitrogen pool (Jonasson et al. 1999; Bardgett et al. 2002, 2007b) and 30% of the total soil organic phosphorus pool (Jonasson et al. 1999), although as discussed above these quantities show considerable seasonal variation. Nutrient liberation from the microbial biomass can result from environmental factors, such as dry/wet or freeze/thaw cycles, which subject microbes to physiological stress (Groffman et al. 2001; Schimel et al. 2007; Gordon et al. 2008). However, a primary mechanism by which nutrients are liberated from microbes and made locally available for uptake by plants is through predation of microbes by protozoa, nematodes, and microarthropods, which excrete nutrients that are in excess of their own requirements into the soil environment and in forms that are biologically available. This remobilization of nutrients is termed the 'microbial loop' (Clarholm 1985) and is considered to be a primary control point of nutrient availability in the rhizosphere according to the nutrient-enrichment model of Moore et al. (2003). In this model, which is based on a simple relationship between trophic structures and detritus input (Fig. 2.5) (Moore and de Ruiter 2000), root exudates stimulate microbial populations in the rhizosphere, in turn prompting an increase in consumers and their predators. This increase in predation in turn leads to increased N availability in the root zone, and enhanced plant growth due to excretion of excess N by consumers. These changes in turn stimulate microbial growth and turnover, which further promotes more consumers and attracts more predators (Moore et al. 2003). Hence, nutrient availability to plants can be controlled several trophic steps away from root through top-down control.

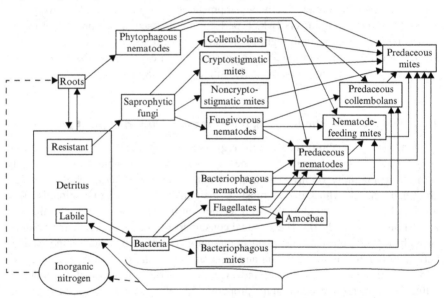

Fig. 2.5 Schematic of the belowground food web from the shortgrass steppe of Colorado (Hunt et al. 1987). Species are aggregated into functional groups based on food choice and life-history parameters. Material flows (carbon and nitrogen) are represented by solid arrows, and net nitrogen flows are represented by dashed arrows. Plant roots, through the rhizodeposition of labile carbon exudates and the turnover of root cells and hairs rich in resistant cell walls, initiate the dominant materials flows in the rhizosphere. Material flows to the detritus pools and the inorganic nitrogen pool, for example through the death rates and the excretion of waste products, are represented as a single flow at the base of the diagram. From Moore et al. (2003), with permission from the Ecological Society of America.

Several experimental studies provide support for this concept, through reporting positive effects of microbial predation by soil fauna on nutrient availability and uptake by plants (Anderson et al. 1983; Clarholm 1985; Ingham et al. 1985; Bardgett and Chan 1999), and on plant growth (Ingham et al. 1985; Setälä and Huhta 1991; Setälä 1995; but see Bardgett and Chan 1999). Also, it has been shown that grazing activities of soil animals, which can be highly selective as in the case of fungal-feeding microarthopods (Moore et al. 1985; Newell 1984a; Bardgett et al. 1993b), can lead to significant changes in the composition, biomass, and activity of their microbial communities. This can in turn have both positive and negative consequences for decomposition processes and nutrient availability in soil (Mikola et al. 2002; Cole et al. 2006), although a recent meta-analysis of litter-bag studies suggests that negative effects may be more common, at least for microarthropods (Kampichler and Bruckner 2009). However, non-nutritional effects of predation on plant growth can also occur, for example through animals affecting bacterial production of plant growth-promoting hormones (Jentschke et al. 1995; Alphei et al. 1996; Bonkowski 2004). Therefore, effects of predation on plant growth are likely to be highly complex, and involve a combination of nutritional and non-nutritional responses to microbial grazing (Bonkowski 2004).

Most studies on effects of soil fauna on soil processes and plants have focused on specific interactions between particular animal taxa and microbes, mostly in very simple microcosm systems. In nature, however, soil food webs consist of multiple assemblages of species of several trophic groups. For this reason, multitrophic interactions in soil have the potential to serve as a control of nutrient cycling and plant growth. Relatively few studies have examined effects of varying the complexity of soil communities on plant growth, although Bradford et al. (2002) showed that the manipulation of soil fauna on the basis of body size in a model grassland ecosystem brought about changes in the composition of the plant community. When soil macrofauna were present, significant changes were detected in microbial biomass and organic-matter decomposition, resulting in altered nutrient dynamics and an increase in the abundance of grasses relative to herbs (Bradford et al. 2002). In another study, Alphei et al. (1996) found that protists and nematodes enhanced plant production, whereas earthworms reduced production. Furthermore, the positive effect of protists on plant productivity only occurred when earthworms were absent, which is consistent with the finding of Bradford et al. (2002) that the presence of larger-bodied soil organisms reduced the aboveground effects of smaller-sized soil organisms.

Other studies have considered effects of the complexity of soil communities on nutrient mineralization and plant growth in the context of diversity effects. Of note is the classic study of Laakso and Setälä (1999a) who manipulated soil faunal communities on the basis of trophic group structure and species richness within trophic groups to examine effects on nutrient mineralization and plant growth. They found that variation in species composition within trophic groups was far less important for nutrient mineralization and plant growth than was variation across trophic groups. Further, they found that certain species, namely the enchytraeid worm *Congettia sphagnetorum*, had a functionally disproportionate role in these processes. These findings, together with those of more recent microcosm studies of the effects of animal diversity on plant nutrient uptake and growth (e.g. Cole et al. 2004), support the notion that the effects of animal communities on plant growth are context-dependent and are driven by the presence or absence of certain functionally irreplaceable organisms. Modelling studies have also explored linkages between complex soil food webs and pathways of nutrient cycling, such as the pioneering soil food-web model of Hunt et al. (1987) of shortgrass steppe in North America. This model is based on functional groupings of soil organisms (Fig. 2.5) and enables calculation of carbon and nitrogen fluxes and rates of nitrogen mineralization, based on estimates of various parameters such as the carbon-to-nitrogen ratio, rate of turnover, assimilation and production efficiency, and population size (Hunt et al. 1987; Moore et al. 1988). This modelling approach to the functional classification of the soil food web has been used to examine functional responses of shifts in food-web energy channels (de Ruiter et al. 1995) and the loss of functional groups and their combinations within soil food webs. For example, Hunt and Wall (2002) used a simulation model based on a short-grass prairie to show that the loss of individual functional groups of soil organisms had little effect on nitrogen mineralization and plant production, due largely to compensatory responses in the abundances of remaining functional groups.

2.2.3 Functional consequences of trophic cascades in the soil food web

Most of the above examples of predation focus on consumers of microbes. However, decomposer food webs also include higher-level predators that consume fauna of lower trophic levels, such as those that consume fungi, bacteria, and plant litter. These predators can induce trophic cascades through changing the density of prey fauna, and thereby indirectly influence the abundance of bacteria and fungi (Fig. 2.6). Trophic cascades have long been recognized by soil ecologists to indirectly influence soil processes driven by soil microorganisms. For instance, Santos et al. (1981) showed that experimental reduction of predatory tydeid mites enhanced densities of bacterial feeding nematodes (their main prey source), which in turn reduced rates of microbial decomposition of plant litter. Several subsequent studies have shown that manipulation

	MANIPULATION		TROPHIC EFFECTS			PROCESS EFFECTS
Santos *et al.* (1981)	Reduce tydeid mites	→	Increase bacterial feeding nematodes	→	Reduce bacteria	→ Reduce decomposition rate
Kajak *et al.* (1993)	Reduce epigeic predators	→	Increase litter feeding fauna	→	Increase microflora	→ Increase decomposition and N mineralisation
Mikola and Setälä (1998)	Add top predatory nematodes	→	Reduce microbe feeding nematodes	→	No effect on microbial biomass	→ Reduce C and N mineralisation rate
Hedlund and Öhrn (2000)	Add predatory mites	→	Likely reduction of springtails	→	Enhance hyphal length	→ Enhance C mineralisation

Fig. 2.6 Contrasting studies investigating trophic cascades in the decomposer food web, and the consequences for belowground ecosystem processes.

of top predators in the soil food web can affect decomposition and mineralization of carbon and nitrogen positively (e.g. Allen-Morley and Coleman 1989; Hedlund and Öhrn 2000), neutrally (e.g. Martikainen and Huhta 1990) or negatively (e.g. Wyman 1998). Moreover, indirect cascading effects of top predators on soil microorganisms and processes have been shown to occur for a range of types of predators operating at different spatial scales, including predatory nematodes (Allen-Morley and Coleman 1989), predatory mites (Hedlund and Öhrn 2000; Lenoir et al. 2007), spiders (Kajak et al. 1993; Lensing and Wise 2006) and salamanders (Wyman 1998). These effects can occur even when lower trophic levels of the soil food web are unresponsive to predators, so long as the predators influence the turnover and production of organisms in the lower trophic levels (Mikola and Setälä 1998a).

The indirect effects of predator-induced trophic cascades on decomposition and nutrient mineralization in turn have the potential to regulate the availability of nutrients for plants, and therefore conceivably affect plant nutrient acquisition and growth. As discussed earlier in this chapter, it is well established that saprotrophic and microbe-feeding soil fauna have marked effects on plant nutrition and productivity. However, the indirect effect of their predators on plant growth has only been investigated in two studies by Laakso and Setälä (1999a, 1999b). In both studies, experimental microcosms containing planted seedlings of *Betula pendula* were set up with different soil faunal communities, and these included microcosms with or without top predatory mites that feed upon microbial-feeding nematodes and saprophagous mesofauna. In both studies, there was no effect of these predators on plant production, although one of these studies (Laakso and Setälä 1999b) also found no effect of predators on the basal trophic level of the soil food web (i.e. soil microbes), indicative of a weak or non-existent trophic cascade in the microcosms. It is possible that in those situations in which trophic cascades in soil food webs are much stronger and have more pronounced effects on soil nutrient availability, plant growth might indeed be responsive to top soil predators.

The variable response of belowground processes to predator-induced trophic cascades points to these responses being determined by environmental context. Consistent with this, Lensing and Wise (2006) found that cascading effects of spiders on plant litter decomposition depended on moisture regime, and were important at a high-rainfall site with reduced moisture addition, but not at a high-rainfall site with ambient moisture addition or at a low-rainfall site. Further, Lenoir et al. (2007) showed that trophic cascades involving top predators, notably gamasid mites, were important in indirectly influencing soil microflora in soil from a low-fertility site, but not in soil from a high fertility site. Environmental context can also influence the occurrence and importance of trophic cascades in soil food webs through influencing the types of soil organisms present. For example, there is evidence that trophic cascades are more important in soil food webs dominated by bacteria and bacterial-feeding organisms than in those dominated by fungi and fungal-feeding organisms (Wardle and Yeates 1993). Further, with the exception of obligate predators such as spiders and centipedes, soil animals with large body sizes are unlikely to induce trophic cascades because they interact with lower trophic organisms in a mutualistic rather than an antagonistic (i.e. predator–prey) manner. Given that there is much variation across habitats in the body-size distribution of soil fauna (Wardle 2002), there is also likely to be much variation in the importance of predator-induced cascades. In any case, much remains unknown about the types of environmental conditions that are most likely to favour a dominant role of predator-induced trophic cascades in driving soil processes, or in which types of ecosystems they should be the most important.

2.2.4 Bacterial-based and fungal-based energy channels and nutrient cycling

Another approach to the study of soil food webs and their significance for nutrient cycling and plant production is focused on food web energy channels; these channels differ in how they influence the availability of nutrients. Moore et al. (1988) proposed the existence of root, bacterial and fungal energy channels: the root channel consists of root-associated fauna and symbiotic microbes, while the bacterial and fungal energy channels consist of organisms that derive their energy from detritus (Fig. 2.7) (Moore et al. 1998). Moreover, it has been proposed that bacterial and fungal energy channels have distinct functions, representing 'fast' and 'slow' cycles of nutrient availability, respectively (Coleman et al. 1983; Moore and Hunt 1988). As a consequence, ecosystems dominated by bacterial channels are characterized by having high nutrient availability and low amounts of nutrient-rich organic matter (e.g. narrow carbon-to-nitrogen ratio), resulting in part from elevated biological activity, whereas those dominated by fungal channels often have acid soils of high organic matter content and low resource availability and quality (e.g. high carbon-to-nitrogen ratio). This was confirmed by a recent meta-analysis of global data on soil microbial communities by Fierer et al. (2009), who showed that coniferous forest soils had the greatest fungal to bacterial ratios, whereas soils from deserts and grasslands had the lowest. Moreover, across the range of soils examined, the soil fungal to bacterial ratio was positively correlated with the soil carbon-to-nitrogen ratio (Fig. 2.8), although this might reflect the integrated effects of a suite of other soil characteristics, including soil pH, soil organic matter content, and the quality of plant carbon inputs entering the soil (Bardgett and McAlister 1999; Van der Heijden et al. 2008; Fierer et al. 2009). Importantly, these types of energy channels are interchangeable: the fungal channel often becomes increasingly important as primary succession proceeds (Ohtonen et al. 1999; Neutel et al. 2002; Bardgett et al. 2007a) and following land abandonment (Zeller et al. 2001; Van der Wal et al. 2006), whereas the bacterial channel becomes increasingly important following intense disturbance, nutrient enrichment, heavy grazing by aboveground herbivores, tillage, and intensive farming (Hendrix et al. 1986; Bardgett and McAlister 1999; Bardgett et al. 2001b; Van der Wal et al. 2006; R. Smith et al. 2008).

The functional significance of shifts between the fungal- and bacterial-based energy channel for plant community dynamics is poorly understood. However, it has been proposed that the bacterial-based energy channel enhances rates of nutrient mineralization and the availability of nutrients to plants, whereas the fungal-based

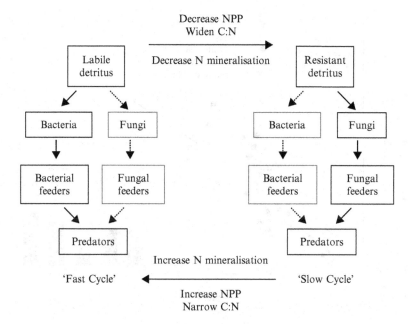

Fig. 2.7 Simplified bacterial and fungal energy channels of the belowground food web. The bacterial energy channel represents the 'fast cycle' due to the higher turnover rates of bacteria and their consumers relative to the fungi and their consumers, whereas the fungal energy channel represents the 'slow cycle'. Changes in the C:N ratio of the detritus, net primary production (NPP), disturbance, or rates of nitrogen mineralization have been associated with shifts in the relative dominance of one channel to the other. From Moore et al. (2003), with permission from the Ecological Society of America.

energy channel promotes 'slow' and highly conservative cycling of nutrients (Coleman et al. 1983; Wardle 2002). This is supported by evidence that decomposer fungi are highly conservative in their use of nutrients and immobilize substantial quantities of nutrients in their hyphal networks (Boddy 1999), whereas bacteria have a more exploitative nutrient use strategy through rapid use of newly produced labile substrates, for example from plant roots (Bardgett 2005). Further, because bacterial tissues have low carbon-to-nutrient ratios and concentrations of defence chemicals relative to fungal tissues, bacterial tissues are consumed by their predators to a much greater extent, leading to greater nutrient release and turnover (Wardle and Yeates 1993; Wardle 2002).

A growing number of studies show that increases in the abundance of fungi relative to bacteria within the microbial community are associated with reduced rates of nutrient mineralization, and *vice versa*. For instance, Bardgett et al. (2006) showed that the presence of the hemiparasite *Rhinanthus minor* in temperate grassland increased plant diversity and caused a shift in the composition of the microbial community towards increasing dominance of bacteria relative to fungi, which was paralleled by a substantial increase in rates of nitrogen mineralization in soil (Fig. 2.9).

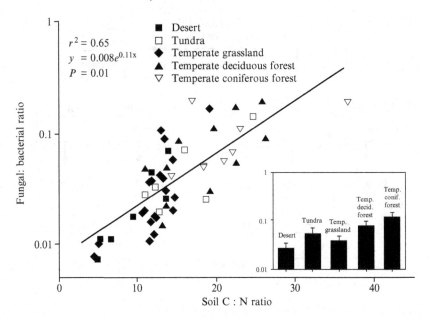

Fig. 2.8 Relationships between fungal to bacterial ratios estimated using quantitative poly-merase chain reaction and soil C:N ratios across five biomes. A fungal:bacterial ratio of 1 means that fungal and bacterial rRNA gene copies are in equal abundance. The bar chart in the insert shows averages and standard errors of the fungal:bacterial ratios determined for each of the five biomes. From Fierer et al. (2009), with permission from Wiley-Blackwell.

Consistent with this, Högberg et al. (2007) studied a range of boreal forests of contrasting nitrogen availability and nitrogen-loading, and found a strong negative correlation between gross nitrogen mineralization and the fungal-to-bacterial biomass ratio. Further, Wardle et al. (2004b) studied a series of long-term chronosequences around the globe (spanning from 6000 to over 4 million years) for which declines in standing plant biomass had occurred over time, and found that this decline was associated with increasing substrate phosphorus limitation for microbes and an increase in dominance of fungi relative to bacteria. In that study, these changes were associated with reduced rates of litter decomposition and nutrient mineralization, indicative of a negative feed-back involving further intensified nutrient limitation and ultimately reduced ecosystem productivity.

Studies are also emerging to support the view that the fungal-based energy channel is more effective than the bacterial energy channel in promoting nutrient retention, and this has implications for the nutrient balance of soil. For example, an *in situ* ^{15}N-labelling study by Bardgett *et al.* (2003) showed that the uptake and retention of added ^{15}N-glycine and ammonium by the microbial community was significantly greater in unfertilized low-productivity grassland with fungal-dominated microbial communities than in adjacent fertilized grassland that had a greater relative abundance

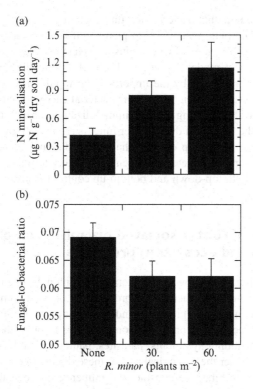

(a)

(b)

Fig. 2.9 The impact of the hemiparasite *Rhinanthus minor* on (a) rates of soil nitrogen mineralization and (b) the fungal-to-bacterial biomass ratio, calculated using phospholipid fatty acid analysis (PLFA). Values are means and standard errors. From Bardgett et al. (2006), with permission from Macmillan Publishers Ltd.

of bacteria. Using the same soils, Gordon et al. (2008) found that the fungal-rich soil was better able to retain nutrients under drying/wetting disturbance than was the soil with a greater bacterial contribution. Moreover, the increased nutrient loss from improved soils was associated with a marked decline (45% relative to control) in microbial biomass nitrogen, suggesting that bacterial-dominated communities are more sensitive to disturbance than are their fungal-rich counterparts. Other evidence that shifts in microbial composition relate to changes in nutrient cycling at the field scale comes from the study by de Vries et al. (2006) of a range of grassland sites in The Netherlands. These authors found that a higher fungal biomass in soil was associated with reduced nitrogen leaching and improved nitrogen balance in soil, suggesting a positive effect of fungi on nutrient retention. While other soil physico-chemical factors might contribute to the patterns explained above, these studies collectively point to grassland soils with fungal-dominated microbial communities being more efficient in retaining nutrients than are their bacterial-dominated counterparts.

It is important to note that bacterial and fungal energy channels operate simultaneously, and hence nutrient availability to plants will be controlled by their relative dominance, rather than by the influence of just one channel. This is recognized within the nutrient-enrichment model (Moore et al. 2003) described above, which asserts that plant growth is regulated by the importance of top-down control by predation affecting nutrient mineralization through the bacterial energy channel relative to that of bottom-up control influencing nutrient immobilization through the fungal energy channel. Because the growth of decomposer fungi is regulated more by the availability of resources (i.e. bottom-up control) than by predation (i.e. top-down control), while the reverse is true for bacteria, nutrient availability to plants will be regulated simultaneously by both top-down and bottom-up control (Wardle 2002).

2.3 Influence of root-associated organisms on plant communities and ecosystem processes

The above discussion shows how biotic interactions within the decomposer food web can indirectly influence the structure and productivity of plant communities through their effects on decomposition processes and nutrient dynamics. Soil organisms also have substantial direct effects on vegetation dynamics through the actions of root-associated biota, including symbionts, parasites, pathogens, and root herbivores. In this section, we consider the ecological role of these direct effects as drivers of plant community dynamics. First, we consider the influence on vegetation dynamics of interactions between plant roots and mutualistic symbionts, namely mycorrhizal fungi and nitrogen-fixers. Second, we examine the role of parasites, pathogens, and root herbivores as drivers of vegetation dynamics.

2.3.1 Microbial symbionts and plant community dynamics

Perhaps the most well-known belowground symbiotic relationship is that which occurs between plant roots and nitrogen-fixing bacteria; these bacteria form nodules in plant roots and convert atmospheric nitrogen into ammonium, thereby making the nitrogen available to the plants. The most well-known nitrogen-fixing plants are the legumes which associate with bacteria of the genus *Rhizobium*, but many non-leguminous nitrogen-fixing plants also exist, such as those of the genera *Casuarina*, *Myrica*, *Hippophae*, and *Alnus*, which associate with actinomycetes of the genus *Frankia*. The beneficial effects of nitrogen-fixing bacteria for nitrogen cycling and plant productivity and nitrogen capture are well documented in both agricultural and natural settings (Tilman et al. 1997; Cleveland et al. 1999; Spehn et al. 2002; Rochon et al. 2004; Hopkins and Wilkins 2006; Van der Heijden et al. 2006a). However, their contribution to plant productivity is believed to be greatest in tropical savannah, and some grasslands and tropical forests that are dominated by legumes; in these situations, nitrogen-fixing bacterial symbionts of the legumes can contribute up to 20%

of all plant nitrogen that is annually acquired by vegetation (Cleveland et al. 1999; Van der Heijden et al. 2008).

It was recently proposed by Houlton et al. (2008) that the high abundance of symbiotic nitrogen-fixing plants in the tropics compared to mature nitrogen-limited forests of temperate and boreal zones was due to the ability of nitrogen-fixers to invest nitrogen into phosphorus acquisition, which is a clear advantage in phosphorus-limited tropical savannas and lowland tropical forests. In contrast, it was suggested that modern-day temperatures should constrain nitrogen-fixation rates and exclude nitrogen-fixing species from mature forests at high latitudes (Houlton et al. 2008). However, as described below, relatively high rates of nitrogen fixation performed by cyanobacteria associated with feathermosses may be widespread in old-growth, high-latitude boreal forests (Zackrisson et al. 2004). It was also recently suggested by Menge et al. (2008) that the absence of symbiotic nitrogen-fixers in nitrogen limited temperate and boreal ecosystems could be explained by evolutionary trade-offs, whereby low nutrient use efficiency, high mortality (or turnover) rate and low losses of plant-unavailable nitrogen all increase the likelihood that nitrogen fixation will be selected against (Menge et al. 2008). However, as stressed by these authors, better characterization of these parameters in multiple ecosystems is necessary to determine whether these mechanisms explain the lack of symbiotic nitrogen-fixers and thus the maintenance of nitrogen limitation in old-growth forests.

Symbiotic associations between nitrogen-fixing bacteria and plants also play a key role in vegetation succession, by raising soil nitrogen contents during early succession to levels needed to support later successional species (Chapin et al. 1994; Kohls et al. 2003; Walker and del Moral 2003) (Fig. 2.10). For instance, during primary succession on glacial moraine in Alaska, nitrogen-fixing *Alnus* shrubs are known to increase soil nitrogen and organic matter content, thereby creating a more favourable environment for later successional tree species, especially *Picea* (Crocker and Major 1955; Chapin et al. 1994). Further, in New Zealand sand dunes, *Lupinus* herbs have been shown to facilitate the growth of *Pinus* trees through a similar mechanism of nutrient enhancement (Gadgill 1971), while the shrub *Hippophäe* has been found to promote plant biomass of later successional species in coastal Dutch sand dunes (Olff et al. 1993). Further, Bellingham et al. (2001) used field and glasshouse studies to show that litter from the pioneer nitrogen-fixing legume *Carmichaelia odorata*, as well as soil modified by the legume, both significantly promoted growth and nitrogen uptake by later successional tree species in a New Zealand montane river valley. As will be discussed further in Chapter 5, the invasion of nitrogen-fixing plant species into nitrogen-limited ecosystems can also dramatically enhance soil nitrogen availability and plant productivity, with far-reaching consequences for community and ecosystem properties (Vitousek and Walker 1989; Sprent and Parsons 2000). The influence of nitrogen-fixers on vegetation dynamics is also potentially affected by other organisms, both aboveground and belowground. For instance, the transfer of nitrogen between legumes and grasses has been shown to occur through hyphal networks of arbuscular mycorrhizal fungi (Haystead et al. 1988), and it has been suggested that this route of nitrogen transfer is of most importance in nitrogen-limited

(a)   (b)   (c)

Fig. 2.10 Dominant N-fixing plants of different primary seres: (a) *Alnus sinuata*, which forms thickets on deglaciated terrain at Glacier Bay, south-east Alaska; (b) *Coriaria arborea* colonizing recently deglaciated terrain at the Fox Glacier, South Island, New Zealand; and (c) *Coriaria arborea* colonizing floodplains near Kaikoura, South Island, New Zealand. Images by Richard D. Bardgett.

pioneer plant communities where legumes are often strongly mycorrhizal (Haystead et al. 1988). The transfer of nitrogen from nitrogen-fixing legumes to neighbouring grasses has also been shown to be enhanced by infestation of legume roots by host-specific parasitic nematodes, resulting in increased growth of the neighbouring grass (Bardgett et al. 1999a; Denton et al. 1999; Dromph et al. 2006; Ayres et al. 2007) with potential consequences for plant competitive interactions.

Free-living and associative nitrogen-fixing bacteria, which are ubiquitous in terrestrial ecosystems, can also contribute greatly to ecosystem nitrogen budgets. Although this input is usually relatively small (<3 kg nitrogen ha^{-1} year^{-1}) (Cleveland et al. 1999), it may nevertheless represent the main ecosystem nitrogen input in ecosystems that experience low nitrogen deposition and lack symbiotic nitrogen-fixers. For example, nitrogen-fixing cyanobacteria can greatly contribute to the nitrogen economy of terrestrial ecosystems, such as recently exposed glacial terrain (Schmidt et al. 2008) and boreal forests, where nitrogen fixation has traditionally been thought to be extremely limited. For example, DeLuca et al. (2002b) showed that nitrogen-fixing cyanobacteria which inhabit the incurves of leaves of the feather moss *Pleurozium schreberi* fix significant quantities of nitrogen (1.5–2.0 kg nitrogen ha^{-1} year^{-1}) and acts as a major contributor to nitrogen accumulation and cycling in boreal forests, especially in late successional systems and old-growth forests (Zackrisson et al. 2004; Lagerström et al. 2007; Gundale et al. 2010) (Fig. 2.11). This finding is of high significance because *P. schreberi* is the most common moss on Earth, and accounts

(a)　　　　　　　　　(b)

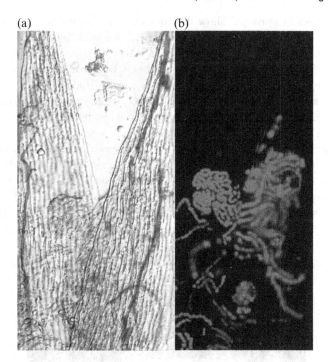

Fig. 2.11 Nitrogen-fixing cyanobacteria which inhabit the incurves of leaves of the feather moss *Pleurozium schreberi* fix significant quantities of nitrogen and acts as a major contributor to nitrogen accumulation and cycling in boreal forests, especially in late successional systems and old-growth forests. (a) Under light microscope; and (b) ultraviolet-fluorescence micrograph with a green filter. Coiled chains of *Nostoc* are hidden in the leaf under light microscopy, but are readily observed as the red cells under ultraviolet-fluorescence microscopy with a green filter. Images by P. Lundgren and U. Rasmussen, with permission from Macmillan Publishers Ltd.

for as much as 80% of the ground cover in boreal forests (DeLuca et al. 2002b). Biological soil crusts, which are assemblages formed by lichens, mosses, liverworts, cyanobacteria, and other microorganisms, also play key roles in the nitrogen economy of arid and semi-arid ecosystems worldwide (Belnap 2003; Belnap and Lange 2003; Bowker et al. 2010). Not only do biological soil crusts fix nitrogen in these ecosystems, but they also influence processes of nitrification and other transformations of nitrogen (Belnap 2003; Maestre et al. 2005; Bowker et al. 2010).

Another important group of plant symbionts are the mycorrhizal fungi, which are widespread and associate with roots of some 80% of all terrestrial plant species (Smith and Read 1997). There are three main groups of mycorrhizal fungi, namely the arbuscular mycorrhizal fungi, the ectomycorrhizal fungi, and the ericoid mycorrhizal fungi. Numerous studies demonstrate their capacity to enhance nutrient supply to plants, as well as to provide resistance to pathogens, insectivorous herbivores, and drought (Gange and West 1994; Newsham et al. 1995b; Smith and Read 1997). For instance, studies in grassland show that arbuscular mycorrhizal fungi can enhance plant productivity by up to two-fold (Van der Heijden et al. 1998b; Vogelsang et al. 2006), a response that is mostly attributed to enhanced plant uptake of phosphorus plants (Van der Heijden et al. 1998, 2006b), although acquisition of nitrogen through arbuscular mycorrhizal fungi might also play a role (Hodge et al. 2001; but see Reynolds et al. 2005; Van der Heijden et al. 2006b). Likewise, ectomycorrhizal fungi play a crucial role in plant nutrient acquisition in many boreal and temperate forest ecosystems (Read and Perez-Moreno 2003). Here, organic nutrients contained in litter and humus are mobilized by extensive ectomycorrhizal hyphal networks that forage for nutrients and excrete a range of extracellular enzymes that degrade complex organic compounds (Fig. 2.12) (Leake and Read 1997). While it is difficult to experimentally determine the contribution of ectomycorrhizal fungi to nutrient acquisition *in situ*, several studies suggest that some 80% of all plant nitrogen in

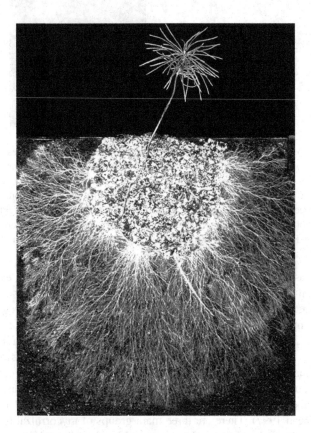

Fig. 2.12 Extensive ectomycorrhizal hyphal networks forage for nutrients and excrete a range of extracellular enzymes that degrade complex organic compounds. Image by Damian P. Donnelly and Jonathan R. Leake, University of Sheffield Department of Animal & Plant Sciences, UK.

boreal forests could be taken up through ectomycorrhizal fungi (Simard et al. 2002; Hobbie and Hobbie 2006).

In general, the role of different types of mycorrhizal fungi changes with vegetation succession. It has been proposed that in early succession most colonizing plants are non-mycorrhizal, and that in mid-succession the dominant herbaceous plants often have a facultative requirement for arbuscular mycorrhizal fungi, whereas in late succession, trees and shrubs (which typically dominate the vegetation) often have an obligate need for ectomycorrhizae (Read 1994). However, this pattern is far from universal, and in particular many late successional forests are also dominated largely or exclusively by arbuscular mycorrhizal tree species. Importantly, interactions between mycorrhizal fungi and other symbionts might also act as important regulators of plant production. For instance, legumes commonly form associations with both arbuscular mycorrhizal fungi and rhizobia (Scheublin et al. 2004), potentially benefitting from the characteristics of both (Pacovsky et al. 1986). The importance of such mycorrhizal–rhizobia interactions for plant community dynamics has seldom been explored.

Mycorrhizal fungi also play a significant role in structuring plant communities, and in many cases (though not all, e.g. Stampe and Daehler 2003; Van der Heijden et al. 2006b) enhance plant species diversity through reducing plant competition intensity and promoting more equitable distribution of resources within the plant community. For instance, Grime et al. (1987) manipulated arbuscular mycorrhizal fungi in grassland microcosms and found that their presence caused a reduction in dominance of competitive grass species in favour of several subordinate herb species which benefited from arbuscular mycorrhizal infection, thereby increasing plant species diversity. Likewise, Gange et al. (1993) showed that selective fungicides which suppressed arbuscular mycorrhizal fungi in early successional grassland increased the abundance of competitively dominant grasses at the expense of subordinate herb species, thereby reducing plant species richness. Variations in mycorrhizal fungal diversity have also been shown to affect plant productivity and diversity, especially in nutrient poor situations. For instance, Van der Heijden et al. (1998b) found that plant diversity and productivity were positively related to arbuscular mycorrhizal fungal diversity in experimental grassland, a response that was attributed to more efficient exploitation and partitioning of soil phosphorus reserves in high arbuscular mycorrhizal diversity treatments. In support, Maherali and Klironomos (2007) discovered that there is much functional complementarity between coexisting arbuscular mycorrhizal fungal taxa, which acts to enhance plant production. In contrast, Vogelsang et al. (2006) manipulated arbuscular mycorrhizal fungal diversity in microcosms and found that plant biomass in treatments inoculated with six arbuscular mycorrhizal fungi did not differ from that of the most effective single mycorrhizal fungus, indicative of little functional complementarity among arbuscular mycorrhizal fungal species. Further, Jonsson et al. (2001) performed an experiment in which ectomycorrhizal fungal diversity was varied from one to eight species, and found that the effects of fungal diversity on tree seedling growth were context-dependent and influenced by both host plant species and soil fertility. Consistent with other diversity experiments of soil organisms, studies to date indicate that positive responses of plants to mycorrhizal fungal infection are likely to driven largely by fungal species identity rather than by species richness *per se.*

It should be noted that not all studies show arbuscular mycorrhizal fungi to promote plant productivity and diversity. For instance, in tallgrass prairie, Hartnett and Wilson (1999) found that the suppression of arbuscular mycorrhizal fungal infection through the application of a fungicide enhanced plant species diversity. This response was attributed to a reduction in the abundance of competitively dominant and obligately mycotrophic C_4 tall grasses, and consequent competitive release of subordinate facultative mycotrophs. Likewise, it was suggested by Connell and Lowman (1989) that in tropical rainforests, ectomycorrhizal fungal associations can encourage dominance of particular tree species at the expense of other (arbuscular mycorrhizal) tree species which are less able to acquire nutrients and tolerate pathogen attack, thereby reducing species coexistence. These studies point to an alternative mechanism through which mycorrhizal fungi can influence plant communities. Here, differences in host plant responses to fungal colonization by mycorrhizal fungi can result in changes in plant diversity if the dominant competitors are either more strongly or more weakly mycotrophic than are their neighbours. Also, it is important to note that the costs and benefits of maintaining symbiosis with mycorrhizal fungi vary greatly among plant species: individual plant responses to infection can range from positive (mutualism) to neutral (commensalism), and even negative (parasitism). As a consequence, it is has been proposed that symbiosis might be best defined as a continuum from parasitism to mutualism (Johnson et al. 1997; Klironomos 2003), and that inter-species differences in the direction and intensity of response to mycorrhizal infection along this continuum might act as important (though little explored) structuring forces in plant communities (Klironomos 2003).

2.3.2 Belowground pathogens, herbivores, and plant community dynamics

A growing body of literature points to the power of soil pathogens to act as drivers of spatial and temporal patterns in vegetation (Van der Putten 2003, 2009). The mechanism involved is negative feedback, which operates through the accumulation of specific plant pathogens that inhibit the growth of the host plant more than that of neighbouring, non-infected plant species, thereby causing vegetation change (Bever et al. 1997; Packer and Clay 2000; Van der Putten 2003). It has also been suggested that soil pathogens can contribute to the maintenance of plant diversity through negative feedback when it suppresses the growth of dominant plants (Bever et al. 1997). Conversely, a study by Klironomos (2002) showed that rare plant species were affected more than dominant plants by negative plant–soil feedback, thereby suggesting that soil pathogens contribute to plant rarity and reduce plant diversity. Several experimental studies have demonstrated the powerful role of pathogens as drivers of vegetation dynamics (e.g. Van der Putten et al. 1993; Packer and Clay 2000), and these are discussed in Chapter 3 in the context of how plant communities affect pathogens and vice versa. With regards to vegetation succession, it has been proposed that effects of soil pathogens via negative feedback are most prominent during late succession when plant host densities are greatest, thereby enabling the build up of pathogens in soil (Reynolds et al. 2003; but see Kardol et al. 2006), as discussed in depth in Chapter 3. Plant pathogens have also been shown to play an important role in plant invasion (Wolfe and Klironomos 2005; Reinhart and Callaway 2006; Van der Putten et al. 2007), as will be described further in Chapter 5.

The soil also contains a wide variety of root-feeding animals, including parasitic nematodes and insects, which selectively feed on roots mostly on the basis of their palatability. As with other root-associated organisms, such selective feeding differentially influences the performance of plant species in mixed communities, thereby altering vegetation dynamics. A classic example of the capacity of belowground insect herbivores to influence vegetation dynamics is the work of Brown and Gange (1989; 1992), which used selective insecticides to exclude both aboveground and belowground insects from grasslands differing in successional stage in southern England. Their results showed that while aboveground insect herbivory slowed down secondary succession by preventing perennial grass development, root herbivory enhanced it by reducing early succession annual forbs. In a similar field experiment in Germany, Schädler et al. (2004) found that while aboveground insects had no effect on plant species replacement, belowground herbivory reduced the competitive ability of resident species, thereby facilitating colonization by late successional species and secondary succession. In a grassland mesocosm experiment in The Netherlands, De Deyn et al. (2003) found that belowground nematodes, wireworms, and microarthropods enhanced the replacement of early by later secondary succession plant species and promoted plant diversity, which was attributed to the suppression of dominant plant species by root-feeding fauna which allowed subordinate plant species to proliferate.

The influence of root-feeding animals on vegetation dynamics also has the potential to be manifested through their influence on aboveground herbivores and *vice versa*. The topic of aboveground multitrophic interactions as drivers of vegetation dynamics and ecosystem processes is discussed in Chapter 4, but it should be noted here that empirical studies have shown root herbivory to strongly influence the performance of foliage-feeding insects (Gange and Brown 1989; Masters and Brown 1992; Masters et al. 2001; Bezemer et al. 2004; but see Salt et al. 1996). Similarly, the performance of root herbivores has been shown to be affected by foliar herbivores (Masters and Brown 1992; Salt et al. 1996; but see Masters et al. 1993). Moreover, there is potential for both aboveground and belowground herbivores to influence vegetation dynamics indirectly through their influence on free-living soil organisms that regulate nutrient availability to plants. For instance, microcosm studies show that root-feeding nematodes and foliar herbivory can enhance root exudation, which in turn stimulates rhizosphere microbial growth and activity, and alters microbial community composition (Denton et al. 1999; Guitian and Bardgett 2000; Ayres et al. 2007), with likely consequences for nutrient availability. However, De Deyn et al. (2007) found no effects of aboveground and belowground herbivory by wireworms on decomposer nematodes, although the wireworms did facilitate root-feeding nematodes, which was thought to be driven by shifts in plant species composition. Most studies to date on this topic have been done under specific and often artificial conditions, and so the consequences of these complex aboveground–belowground interactions for vegetation dynamics in real ecosystems are little understood.

2.4 Soil ecosystem engineers and plant community dynamics

Soil fauna also affect nutrient cycling and the plant community through physical alteration of decomposing material and of the soil habitat (Bardgett 2005). Two groups of soil organisms play a key role. First, the litter transformers consume plant detritus and egest this material into soil as fecal pellets with a higher surface-to-volume ratio; this provides a more favourable environment for microbial growth than does the original leaf litter, and therefore stimulates rates of decomposition and nutrient release (Webb 1977; Hassall et al. 1987; Zimmer and Topp 2002). Second, the ecosystem engineers build physical structures in soil that provide habitats for microbes and other organisms, and also alter the movement of materials through soils and across ecosystems (Lavelle et al. 1995, 1997; Bardgett et al. 2001a). The role of litter transformers will be discussed later in this chapter in the context of carbon dynamics. Here, we focus on how the activities of soil ecosystem engineers, which include mostly macrofauna, influence vegetation dynamics in terrestrial ecosystems (Fig. 2.13). We also briefly discuss the role of ecosystem engineers in modifying soil physical properties, and in affecting the transfer of water and nutrients within soil and to ground-waters.

The most widely known ecosystem engineers in soil are the earthworms and there is a huge literature documenting their importance for soil fertility and plant growth in both natural and agricultural systems (e.g. Lee 1985; Edwards and Bohlen 1996; Scheu 2003; Edwards 2004). These effects arise largely from their role in consuming vast quantities of plant litter and through their production of surface and subsurface casts, which provide a more favourable environment for microbial activity and hence nutrient availability. Benefits also arise through their burrowing activities which integrate plant litter into the soil and improve its physical structure. The activity of earthworms in soil can also influence the abundance and distribution of other soil fauna, such as microbe-feeders and litter transformers (Yeates 1981; Alphei et al. 1996; Binet et al. 1998; Ilieva-Makulec and Makulec 2002; Tao et al. 2009), with indirect consequences for soil microbial communities, nutrient availability, and plant growth.

While numerous studies have explored the benefits of such activities for the growth of single plant species (see Scheu 2003), only a handful have considered the consequences for vegetation dynamics; most of these have been done in grassland systems and show that, in general, the presence of earthworms favours the growth of fast-growing grasses over forbs and legumes. For instance, Hopp and Slater (1948) found

(a)   (b)

(c)   (d)

Fig. 2.13 Soil ecosystem engineers and the structures that they produce: (a) mounds of the soil-feeding termite *Cubitermes severus* in moist savanna, Nigeria (image by Reine Leuthold); (b) ant nest in boreal forest, Sweden (image by Richard Bardgett); (c) dung beetles (image by Rosa Menendez); and (d) casts produced by the earthworm *Aporrectodea longa* (image by Kevin Butt).

that earthworms increased the dominance of grasses relative to legumes in experimental grassland systems. Similarly, Hoogerkamp et al. (1983) showed that the introduction of earthworms into polder soils in which they were previously absent increased the performance of grasses relative to forbs. Further, Scheu (2003) presented data showing that perennial ryegrass (*Lolium perenne*) benefited more from the presence of earthworms than did white clover (*Trifolium repens*) in pot experiments. The mechanism for such effects is not fully understood, but it has been suggested (Scheu 2003; Wurst et al. 2005), and recently shown (Eisenhauer and Scheu 2008), that earthworms enhance the competitive ability of grasses against legumes by increasing the supply of nitrogen for grasses. The promotion of graminoids relative to other plant functional groups might also result from grasses being more associated with earthworm casts, where enhanced nutrient availability stimulates their growth (Zaller and Arnone 1999). Further, grasses may benefit from earthworms causing a more patchy distribution of soil organic matter; grasses tend to be better able to forage resource patches through root proliferation than are forbs and legumes (Scheu 2003; Wurst et al. 2003). Earthworms might also influence plant community dynamics by affecting seed transport and germination, for instance through preferential ingestion of seeds of certain plant species and sizes, and deposition of them in nutrient-rich casts (Piearce et al. 1994; Thompson et al. 1994; Scheu 2003). However, very little is known about the significance of these processes in natural plant communities.

The dominant ecosystem engineers of tropical soils are termites, which can reach densities as high as 7000 m^2 and constitute some 95% of belowground insect biomass (Aber et al. 2000). Termites feed mainly on decaying organic matter and roots, and are best known for their large effects on soil nutrient cycling and the physical architecture and hydrology of soils, through their mound-building and gallery-excavating activities (Aber et al. 2000; Hyodo et al. 2000). Most termites depend on symbiotic gut bacteria for energy metabolism, whereas others such as the 'fungus growers' (which will be discussed below in the context of carbon cycling) lack these gut flora and rely on externally cultivated fungi for their nutrition (Aber et al. 2000). Some termites also contain populations of flagellate protozoa in their guts that have the ability to degrade cellulose and other plant polysaccharides, whereas others have gut bacteria which fix atmospheric nitrogen and can potentially act as an ecosystem nitrogen input (Breznak and Brune 1994; Aber et al. 2000). Further, termite mounds serve as foci for nutrient redistribution in many landscapes (Spain and McIvor 1988; Aber et al. 2000). The effects of termites on vegetation dynamics are most noticeable in 'termite savanna' ecosystems, which is a term coined for vegetation formations in Africa and South America, consisting of grasslands and discrete islands of woodland based on large termite mounds (Harris 1964). In these ecosystems, termites can greatly influence vegetation succession. When mounds become physically stable, either through colony death or because the mound is so large that only part of it is colonized by termites, a vegetation succession occurs whereby grasses colonize, and are then replaced by shrubs and finally tall trees. When favourable conditions permit, these islands of woodland increase in size until they merge with each other and produce a closed canopy forest (Harris 1964).

Ants similarly influence vegetation patchiness and dynamics by altering soil physical properties and through the formation of nests. The vast majority of ants build nests within soil, where they concentrate organic material and create hotspots or 'oases' of high nutrient availability which serve to stimulate plant growth and create vegetation mosaics (Hölldobler and Wilson 1990). For instance, leaf-cutter ants, which are a principal herbivore of neotropical regions, build large nests which extend deep into the soil (Hölldobler and Wilson 1990). These nests create patches of high nutrient availability which promote plant growth, thereby forming vegetation patches that are distinct from non-nest areas and areas where nests have been abandoned (Garrettson et al. 1998; Verchot et al. 2003; Moutinho et al. 2003). These types of positive effect of ant nests on vegetation have been shown to be reversible by natural fire, particularly in neotropical regions (Sousa-Souto et al. 2008). However, burning in the neotropics can also facilitate establishment of leaf-cutter ants because they prefer open terrain (Vasconcelos and Cherret 1995). This greater activity of leaf-cutter ants can in turn encourage greater plant nutrient availability on nutrient depleted soils, thereby accelerating vegetation recovery after fire (Sousa-Souto et al. 2007). Another recently discovered mechanism by which leaf-cutter ants influence nutrient cycling in neotropical ecosystems is via their engagement in mutualistic associations with nitrogen-fixing bacteria, which supplement the nitrogen budget of the their fungal gardens that they cultivate for food. Using acetylene reduction and stable isotope experiments, Pinto-Tomás et al. (2009) showed that nitrogen fixation occurred in the fungus gardens of a wide range of leaf-cutter ant species, and that this fixed nitrogen was incorporated into ant biomass (Fig. 2.14). Moreover, they found that symbiotic nitrogen-fixing bacteria were consistently isolated in fungus gardens of leaf-cutter ant colonies collected from across several sites in Argentina, Costa Rica, and Panama. Through this mechanism, the authors estimated that a single mature leaf-cutter ant colony could contribute as much as 1.8 kg of fixed nitrogen per year into neotropical ecosystems. Given the dominance of leaf-cutter ants, this could represent major source of nitrogen in neotropical ecosystems (Pinto-Tomás et al. 2009).

While most ants build nests in soil, a small group of ants, mostly of the genus *Formica*, build large parts of their nests on the soil surface using organic materials such as needles, twigs, and bark collected from the surrounding area (Jurgensen et al. 2008). These surface mound-building ants are common in many temperate and boreal grasslands and forests in Europe and North America (Hölldobler and Wilson 1990) where they can have substantial effects on vegetation development and ecosystem processes. These effects can be direct, for instance through enrichment of organic matter and nutrients in mounds relative to adjacent soil (Jurgensen et al. 2008). They can also be indirect, through predation on decomposer fauna or other predators such

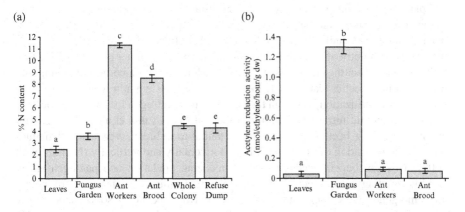

Fig. 2.14 Evidence for nitrogen fixation in the fungus gardens of leaf cutter ants: (a) nitrogen content of the different components of five leaf cutting ant colonies and (b) nitrogen-fixation activity measured by acetylene reduction for different components of 10 leaf-cutting ant colonies. All results are shown as means ± SEM. Means labelled with different letters (a–e) are statistically different ($P < 0.05$). From Pinto Tomás et al. (2009), with permission from the American Association for the Advancement of Science.

as spiders (Laakso and Setälä 1997; 1998; Hawes et al. 2002), thereby potentially triggering trophic cascades that influence decomposition processes and nutrient availability. Mound-building ants can also indirectly influence soil nutrient supply to plants either by tending and protecting aphids which feed on phloem sugars from leaves and twigs in tree canopies (Stadler and Dixon 2005; Styrsky and Eubanks 2007), or by attacking leaf-defoliating insects (Warrington and Whittaker 1985). Both of these activities have the potential to alter the quantity and quality of carbon inputs to soil from litter and leachates of needles and leaves (Stadler et al. 2001), and thereby influence soil organisms and the processes of nutrient cycling that they drive. Further, because ant mounds facilitate colonization and growth by some plant species above others, they can exert important effects on vegetation composition and succession. For example, Blanka et al. (2009) showed that mounds formed by the ant *Lasius flavus* in Slovakian mountain grasslands supported colonization of spruce (*Picea abies*) forest, accelerating succession of grassland to spruce forest. The role of such aboveground trophic interactions as drivers of soil and vegetation dynamics will be considered in more detail in Chapter 4.

Ecosystem processes and vegetation dynamics can also be affected by dung beetles, a widely distributed group of insects which are most abundant in tropical forests and savannas (Hanski and Cambefort 1991). As will be discussed in Chapter 4, significant quantities of nutrients are recycled in terrestrial ecosystems via vertebrate faeces, and the principal role of dung beetles is to transfer into and relocate this material within the soil, thereby instigating a series of changes that affect soil nutrient cycling and plant growth (Nichols et al. 2008). For instance, several studies show that dung beetles stimulate soil microbial activity and rates of nitrogen mineralization (Yokoyama and Kai 1993; Yokoyama et al. 1991; Yamada et al. 2007), and increase soil pH, cation exchange capacity, and concentrations of nutrients (e.g. phosphorus, nitrogen, potassium, calcium, and magnesium) (Yamada et al. 2007). As a consequence, numerous studies have reported beneficial effects of dung beetle activity on the growth and nutrient content of agricultural crops, and on net primary production (NPP) in natural settings (Nichols et al. 2008). Moreover, by incorporating dung into the soil, dung beetles can also help to reduce ecosystem nitrogen loss through ammonium volatilization, although they can stimulate nitrogen loss by denitrification through promoting nitrate availability in the soil (Yokoyama et al. 1991). As with earthworms, dung beetles also have the potential to influence vegetation dynamics through the relocation of plant seeds, both horizontally and vertically (Nichols et al. 2008). This can benefit seed survival by reducing predation and pathogen attack and by directing seed dispersal to more favourable microclimates for germination and emergence (Andresen and Levey 2004). Dung beetles also influence plant communities indirectly through predation of leaf-cutting ants (e.g. Vasconcelos et al. 2006), which as discussed above strongly impact on ecosystem processes and vegetation dynamics in neotropical regions. Before leaving the topic of dung beetles, it is important to note that most studies on their influence on soil processes and vegetation have been done in grasslands and agricultural settings, often under artificial conditions and with single plant and dung beetle species. Their role in regulating ecosystem processes and vegetation dynamics in natural ecosystems is hence poorly understood and represents a significant gap in understanding (Nichols et al. 2008).

Soil macrofauna can substantially modify the physical structure of soil, and in many cases significantly improve soil porosity and the movement or water and nutrients through the soil (Bardgett et al. 2001a). For instance, it is well documented that earthworms increase the density of macropores and channels in soil (Knight et al. 1992; Lavelle et al. 1997), thereby enhancing water infiltration and the movement of both water and soluble nutrients through soil and to inter-connected waterways (Sharply et al. 1979; Bardgett et al. 2001a). Similarly, nests of soil-dwelling ants provide extensive macroporosity to soil, which affects the rates of infiltration of both water and soluble nutrients (Eldridge 1993). While such improvement in soil porosity and water movement caused by earthworm burrowing can significantly enhance plant growth, increased rates of infiltration (especially when caused by deep-burrowing species) can also cause increased leaching, and hence loss of nutrients from soil. For example, the inoculation of grain-crop soils with earthworms was found to result in a 4–12-fold increase in leachate volumes and a 10-fold increase in the amount of dissolved nitrogen contained within them (Subler et al. 1997). Further, the inoculation of limed coniferous forest soils with earthworms was found to cause a 50-fold increase in the concentration of nitrate and cations in soil solution, therefore greatly enhancing the potential for nutrient loss for these soils (Robinson et al. 1992; 1996). Although not quantified, these types of effects of soil animals on water infiltration and leaching are likely to have strong influences on nutrient availability to plants, and also on the transfer of water and nutrients to ground-waters and adjacent ecosystems, with potential consequences for water quality (Bardgett et al. 2001a).

2.5 Soil biotic interactions, carbon dynamics, and global change

So far, we have considered the role of soil biotic interactions as regulators of nutrient cycling and plant community dynamics at local scales. However, soil biotic interactions also play a key role in regulating carbon dynamics in terrestrial ecosystems, with potential global consequences for land–atmosphere exchanges of carbon, and carbon-cycle feedbacks that could amplify climate change. Soils play a major role in climate feedbacks because they release and absorb greenhouse gases, such as carbon dioxide and methane, while storing large quantities of carbon and acting as a significant global carbon sink (Schimel et al. 1994; Heimann and Reichstein 2008; Chapin et al. 2009) (Fig. 2.15). Indeed, it has been estimated that soils contain some 80% of the Earth's terrestrial carbon stock and as much as 90% of the carbon pool of grasslands, deserts, tundra, wetlands, and croplands (IPCC 2007). The flux of below-ground carbon dioxide to the atmosphere due to respiration is an order of magnitude larger than emissions of carbon dioxide through anthropogenic sources (Raich and Potter 1995; IPCC 2007). Further, there is considerable concern that global warming

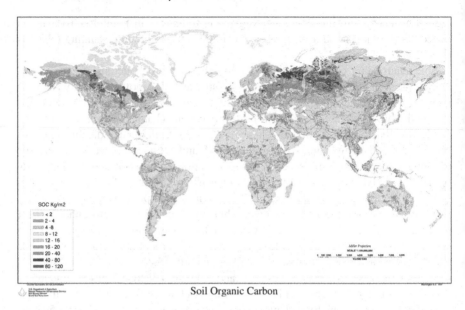

Soil Organic Carbon

Fig. 2.15 Soil organic carbon map showing the distribution of the soil organic carbon to 1 m depth. From FAO-UNESCO, Soil Map of the World, digitized by ESRI. Soil climate map, USDA-NRCS, Soil Survey Division, World Soil Resources, Washington DC.

will increase this liberation of carbon dioxide from soil to atmosphere due to enhanced microbial breakdown of soil organic matter (Jenkinson et al. 1991; Davidson and Janssens 2006). Such acceleration in carbon loss could significantly exacerbate the soil carbon-cycle feedback if predicted climate change scenarios are correct (Cox et al. 2000; Friedlingstein et al. 2006). In this section, we discuss the role of soil biotic interactions in regulating soil carbon flux to the atmosphere, and consider how such activities contribute to climate change via carbon-cycle feedbacks.

2.5.1 Soil biotic interactions and ecosystem carbon exchange

The vast majority of plant production (80–90%) enters the soil food web as exudates, dead leaves, roots, wood, or animal wastes. Although abiotic factors, especially moisture and temperature, act as primary determinants of the rate of decomposition of this organic matter, soil microbes are directly responsible for its breakdown because they are often the only organisms in soil that produce enzymes capable of degrading recalcitrant plant-derived compounds such as lignin and cellulose. As a consequence, a large proportion ($\approx$50%) of soil respiration can be attributed to the activity of heterotrophic microbes, the remainder being due to plant roots (autotrophic respiration) and associated mycorrhizal fungi (Högberg and Read 2006). While a large amount of respiration can be attributed directly to heterotrophic microbes, there is ample evidence that biotic interactions between microbes and animals, including collembolans, mites, enchytraeids, isopods, and earthworms, indirectly stimulate rates of organic matter decomposition and respiration through a variety of mechanisms (Mikola et al. 2002; Wardle 2002; Bardgett 2005; Cole et al. 2006). First, they fragment plant material, which increases the surface area available for microbial colonization (Swift et al. 1979; Seastedt 1984; Wardle 2002; Bardgett 2005). Second, they partially digest dead plant matter and return the rest of it to the soil as faecal pellets which have a higher surface-to-volume ratio and provide a more favourable environment for microbes than does the original material consumed; this in turn enhances rates of decomposition and nutrient release (Webb 1977; Teuben and Verhoef 1992; Zimmer and Topp 2002; Zaady et al. 2003). Third, they bring decomposer microbes and organic matter into more direct contact. For example, the guts of termites and earthworms provide a microenvironment in which microbes form symbiotic relationships with their host through producing extracellular enzymes that break down recalcitrant organic compounds, thereby making nutrients more accessible to the host (Zimmer and Topp 1998; Slaytor 2000; Dillon and Dillon 2004). Some fauna, such as ants, also concentrate carbon in their nests, which consequently have higher rates of respiration than adjacent areas (Risch et al. 2005; Ohashi et al. 2007), although it is not known how much this contributes to total soil carbon dioxide emissions (Jurgensen et al. 2008).

The positive effect of soil macrofauna on decomposition through bringing microbes and organic matter into closer contact does not end in the gut. Microbes from the gut are also released in faecal material where they continue to decompose egested organic matter (Frouz et al. 2002; 2003). Further, some fauna such as termites and leaf-cutting ants play an important role in bringing plant matter into contact with decomposers through their external symbioses with fungal saprotrophs. For example, termite populations and fungal combs (i.e. 'gardens') have been found to account for between 5 and 39% of the annual respired litter carbon in tropical savannah and dry tropical forest (Yamada et al. 2005). Similarly, earthworms are known to drag litter from the soil surface into burrows, thereby increasing its availability to soil microbes and enhancing its decomposition (Tiunov and Scheu 1999). As discussed above, dung beetles also play a crucial role in enhancing rates of decomposition and nutrient availability both by stimulating microbial activity and by relocating mammal faeces and carrion into soil (Yokoyama et al. 1991; Nichols et al. 2008). This may be especially important in tropical forests where they are capable of transferring all deposited faeces into soil within hours of deposition (Arrow 1931; Slade et al. 2007), resulting in enhanced rates of organic matter decomposition and soil respiration (Stevenson and Dindal 1987). Likewise, earthworms consume vast amounts of both plant litter and soil mineral particles, which are mixed together in the earthworm gut and egested as surface and subsurface casts. Such casts, which can be produced at annual rates ranging from 1 to 500 t ha^{-1}, contain greater numbers of microbes and have higher enzyme activities than does surrounding soil (Edwards and Bohlen 1996). As a consequence, the presence of earthworms is well known to enhance rates of decomposition and carbon mineralization (Cortez et al. 1989), and rates of nitrogen and phosphorus mineralization (Scheu 1987; Lavelle and Martin 1992; Sharpley and Syers 1976).

Another mechanism by which animals influence soil respiration and carbon dioxide release is through their feeding activities. As discussed previously, the primary food source of many soil animals are microorganisms. Faunal grazing on microbes can markedly affect the growth, activity, and composition of microbial communities with important consequences for decomposition and soil respiration. For instance, low and intermediate levels of grazing by collembolans have been shown to stimulate the growth, respiration, and enzyme production of decomposer fungi (Hanlon and Anderson 1979; Bengtsson and Rundgren 1983; Hedlund et al. 1991; Bardgett et al. 1993c), in some cases leading to enhanced litter mass loss (Cragg and Bardgett 2001). Also, grazing by nematodes, protozoa, and mites has been shown to influence the activity and growth of soil bacteria and fungi (Dyer et al. 1992; Vreeken-Buijs et al. 1997; Hedlund and Öhrn 2000; Bonkowski 2004), which is in turn likely to alter rates of organic matter decomposition. Further, selective feeding by collembolans has been shown to alter fungal community structure, with consequences for ecosystem level processes such as plant litter decomposition (Newell 1984a, 1984b). Collectively, these studies point to biotic interactions between microbial-feeding fauna and microbes as important drivers of organic matter decomposition and carbon exchange in terrestrial ecosystems.

Most of the aforementioned studies have considered the influence of single species or functional groups of soil organisms on aspects of the carbon cycle. As discussed above, there is potential for trophic cascades within the soil food web to influence organic matter decomposition. However, studies that have explored how variations in the diversity and composition of the soil community influence decomposition processes and respiration mostly point to the important role of species traits as drivers of these processes. For instance, Cragg and Bardgett (2001) used three species of collembolans to create model communities of one, two, or three species in all possible combinations, and found that rates of soil respiration were better explained by the presence versus absence of one species (*Folsomia candida*) than by species richness *per se*. Likewise, Heemsbergen et al. (2004) found that species richness of experimental soil macrofauna communities (ranging from one to eight species) was a poor predictor of the rates of litter mass loss and soil respiration. However, they also found that the functional dissimilarity of species explained much of the variation in these processes (Fig. 2.16). The positive effect of functional dissimilarity on decomposition processes that they observed was attributed to facilitative interactions between component species, which were greatest when component species had different functional roles (Heemsbergen et al. 2004). These findings suggest that to predict the consequences of species loss on decomposition processes first requires an understanding of how individual species contribute to multiple species interactions. In cases where diversity effects on decomposition processes have been found, they tend to occur at the lower end of the diversity gradient (Liiri et al. 2002; Setälä and McLean 2004; Tiunov and Scheu 2005), pointing to a high degree of functional redundancy within soil communities (Liiri et al. 2002; Setälä and McLean 2004). However, as discussed earlier in the chapter some species are more redundant than others, and some are functionally irreplaceable (Laakso and Setälä 1999a,b).

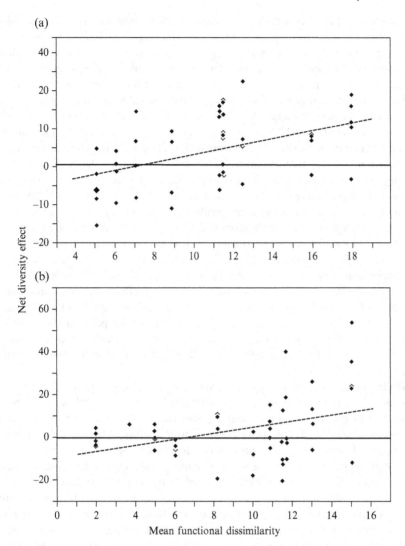

Fig. 2.16 Influence of macro-detritivore diversity in experimental communities on soil respiration (a) and leaf-litter mass loss (b) in relation to mean functional dissimilarity, defined as the degree to which component species are functionally different in those processes determined from their performance in monoculture. Each series of dots represents a treatment ($n = 5$ replicates per treatment; some dots overlap). A significant positive regression between the mean functional dissimilarity of the communities and the net diversity effect for soil respiration and leaf-litter mass loss indicates that positive net diversity effects are more pronounced in communities consisting of functionally dissimilar species. Functional dissimilarity was related to neither species number nor taxonomic group number. From Heemsbergen et al. (2004), with permission from the American Association for the Advancement of Science.

It is important to recognize that the role of soil biotic interactions relative to abiotic factors that regulate decomposition processes and carbon exchanges varies tremendously across biomes. For instance, in a comparative study across ecosystems, Gonzalez and Seastedt (2001) found that the role of soil fauna in governing decomposition was disproportionally greater in tropical wet forests than in tropical dry or subalpine forests. Likewise, Wall et al. (2008) conducted a global decomposition experiment to assess the importance of soil animals in carbon mineralization across 30 sites distributed from 43°S to 68°N on six continents, and found that soil animals increase decomposition rates in temperate and wet tropical climates, but have neutral effects where temperature or moisture constrains biological activity. Overall, these findings were taken to suggest that faunal influences on decomposition processes are dependent on prevailing climatic conditions, and are therefore of most relevance at the regional scale when attempts are made to predict the effects of global change scenarios on carbon dynamics (Wall et al. 2008) (Fig. 2.17). In contrast, Powers et al. (2009) found for 23 tropical forest sites that variation in annual rainfall among sites from 760 to 5797 mm per year had no effect on whether mesofaunal exclusion altered decomposition, even though precipitation was strongly related to decomposition rate. The relationship between macroclimate and faunal effects on decomposition is complicated further by the fact that the complexity of soil communities also varies greatly across biomes, and it has been suggested that decomposition processes may be especially susceptible to changes in soil diversity in species poor soils of extreme environments such as hot and cold deserts (Freckman and Mankau 1986; Freckman and Virginia 1997; Wall 2007).

While soil respiration is typically separated into autotrophic and heterotrophic components, it is in reality driven by a continuum from roots with their autotroph-dependent mycorrhizal fungi, through to other rhizosphere microorganisms that are supported largely by recent photosynthates, to the classical heterotrophs discussed above that decompose larger macromolecules in soil organic matter (Fig. 2.18) (Högberg and Read 2006). There is now mounting evidence that the autotroph-linked component of this continuum, which is fueled by recent photosynthate, is as important a driver of carbon exchange as are the much slower fluxes of carbon arising from the decomposition of plant litter (Högberg and Read 2006). There are two broad lines of evidence that support this view. First, as discussed further in Chapter 3, field experiments using physiological manipulations (e.g. tree-girdling) and canopy-labelling techniques show that as much as half of the soil respiratory carbon release from soil is derived from recent photosynthate fixed over the previous hours or days (Craine et al. 1999; Högberg et al. 2001; Steinmann et al. 2004; Pollierer et al. 2007). Second, mycorrhizal fungi which colonize around 80% of land plants and support vast extraradical mycelial systems provide the largest sink for these photosynthates (Johnson et al. 2002; Leake et al. 2004). The onward transfer of this recent photosynthate from hyphae to the soil occurs by respiration and exudation from hyphal tips as labile organic compounds such as sugars, low-molecular-weight carboxylic acids, and amino acids (Johnson et al. 2002; Jones et al. 2004), and by grazing of hyphae by fungal-feeding animals. For instance, Johnson et al. (2005) showed that the addition of

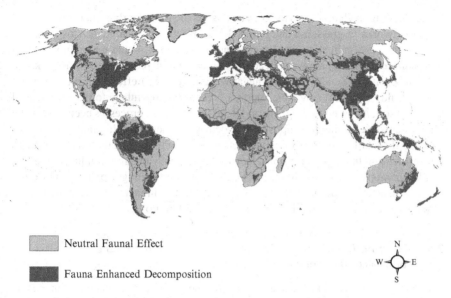

Neutral Faunal Effect

Fauna Enhanced Decomposition

Fig. 2.17 Map showing climatic regions where soil animals were found to enhance decomposition rates, adapted from Wall et al. (2008). Results from temperate and wet tropical climatic regions (light gray) show fauna to increase decomposition rates, but to have neutral effects in other regions (dark gray). This suggests that in future scenarios of climate change for regions predicted to be warmer and wetter, enhanced decomposition rates may result from greater effects of soil animals and other biota. Results are based on the Global Litter Invertebrate Decomposition Experiment (GLIDE). From Wall et al. (2008), with permission from Wiley-Blackwell.

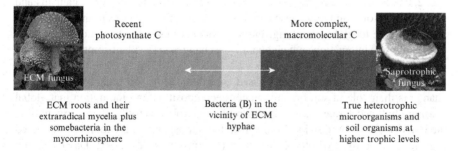

Recent photosynthate C

More complex, macromolecular C

ECM fungus

Saprotrophic fungus

ECM roots and their extraradical mycelia plus somebacteria in the mycorrhizosphere

Bacteria (B) in the vicinity of ECM hyphae

True heterotrophic microorganisms and soil organisms at higher trophic levels

Fig. 2.18 The soil autotroph–heterotroph respiratory continuum. Movement to the right means a decreasing use of recent photosynthate, but increasing heterotrophy and metabolism of carbon of increasing 'age' (i.e. time since incorporation of carbon during photosynthesis). ECM, ectomycorrhizal. From Högberg and Read (2006), with permission from Elsevier.

fungal-feeding collembolans to grassland soils disrupted arbuscular mycorrhizal fungal networks in grassland and reduced ^{13}C-enrichment of mycorrhizosphere respiration by 32%, indicating that the presence of some soil fauna may inhibit carbon fluxes through certain pathways. In addition, grazing of mycorrhizae by soil fauna can result in positive (Setälä 1995) or negative (Finlay 1985) effects on plant growth, which could in turn influence the supply of recent photosynthate to the mycorrhizal fungi and soil. As noted by Högberg and Read (2006), these processes occur within the same microscale as the decomposition of detritus by heterotrophic microorganisms, creating difficulties in discriminating between the activity of autotroph-dependent mycorrhiza and that of heterotrophs. However, these new insights emphasize the role of plant canopy processes, living roots, their symbiotic fungal partners and other closely associated microbes, and soil fauna, as determinants of soil activity and respiratory release of carbon (Högberg and Read 2006; Pollierer et al. 2007).

2.5.2 Contribution of soil biotic interactions to climate change via carbon-cycle feedbacks

Ultimately, the net effect of climate change on ecosystem carbon budgets depends on the balance between photosynthesis and both autotrophic and heterotrophic respiration. Although our knowledge of the assimilatory component (i.e. photosynthesis) of the carbon cycle and its response to climate change is well advanced, there are major gaps in our understanding of the response of soil respiration (Trumbore 2006). Climate change has both direct and indirect effects on the activities of soil organisms that return greenhouse gases to the atmosphere and contribute to global warming. Direct effects include the influence of temperature, changing precipitation, and extreme climatic events on soil organisms and greenhouse gas production. Meanwhile, indirect effects result from climate-driven changes in plant productivity and species composition which alter soil physicochemical conditions, the supply of carbon to soil and the structure and activity of microbial communities involved in decomposition processes and carbon release from soil (Bardgett et al. 2008; Fig. 2.19). Here, we consider direct effects of climate change on soil organisms and biotic interactions, and how these can feed back greenhouse gases to the atmosphere and contribute to global warming. Indirect effects of climate change on soil organisms and biotic interactions which operate via changes in the plant community will be discussed in Chapter 3.

One of the most commonly discussed contributions of soil organisms to climate change is their role in soil organic matter decomposition and the notion that global warming will accelerate rates of heterotrophic microbial activity, thereby increasing the transfer of carbon dioxide from soil to the atmosphere and exports of dissolved organic carbon by hydrologic leaching (Jenkinson et al. 1991; Davidson and Janssens, 2006). The concern here is that, because rates of soil respiration are more sensitive than is primary production to temperature (Jenkinson et al. 1991; Schimel et al. 1994), it is thought that climate warming will increase the net transfer of carbon from soil to atmosphere, thus creating a positive feedback on climate change

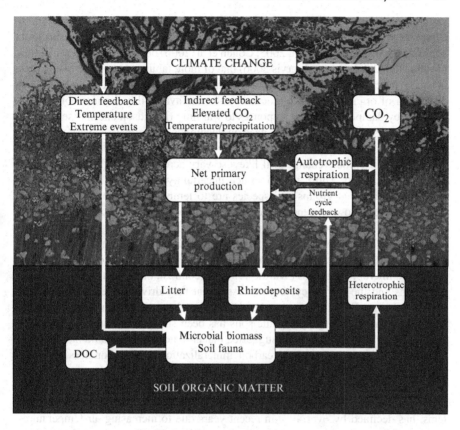

Fig. 2.19 Direct and indirect effects of climate change on soil microbial communities and routes of feedback to global warming through carbon dioxide production. Direct effects include the influence of temperature, changing precipitation, and extreme climatic events on soil microbes and greenhouse gas production. Indirect effects result from climate-driven changes in plant productivity and vegetation structure which alter soil physicochemical conditions, the supply of carbon to soil and the structure and activity of microbial communities involved in decomposition processes and carbon release from soil. DOC, dissolved organic carbon. From Bardgett et al. (2008), with permission from Macmillan Publishers Ltd.

(Cox et al. 2000). Although it is well known that temperature is an important determinant of rates of organic-matter decomposition, the nature of the relationship between temperature and heterotrophic respiration, and its potential to feed back to climate change, are far from clear (Davidson and Janssens 2006; Trumbore 2006).

There are several reasons for this uncertainty. First, soil organic matter is inherently complex and the temperature dependence of decomposition of soil carbon compounds of differing chemical composition and substrate quality vary substantially (Davidson and Janssens 2006). For instance, some studies indicate that the temperature sensitivity of litter decomposition increases as the quality of organic carbon consumed by microbes declines (Fierer et al. 2005; Conant et al. 2008). Conversely, other studies report that the temperature sensitivity of more recalcitrant substrates is similar (Fang et al. 2005; Conen et al. 2006) or less than (Luo et al. 2001; Melillo et al. 2002; Rey and Jarvis 2006) that of more labile substrates. Second, there is much potential for environmental constraints, such as physical and chemical protection of organic matter, to decrease substrate availability for microbial attack, and thereby dampen microbial responses to warming (Davidson and Janssens 2006). As already discussed, faunal influences on decomposition processes vary regionally depending on prevailing climatic conditions, suggesting that the sensitivity of decomposition to climate change will likewise vary across biomes (Wall et al. 2008; Powers et al. 2009). Third, there is much uncertainty about how reactive different microbial and faunal groups and species are to temperature change. For instance, several studies show a general lack of effects of elevated temperature on soil communities (Hodkinson et al. 1996; Kandeler et al. 1998; Bardgett et al. 1999b), and literature syntheses report only weak relationships between mean annual temperatures and densities of soil faunal groups (Petersen and Luxton 1982; Wardle 2002) and microbial biomass (Wardle 1992). In contrast, other studies show that soil organisms and the carbon-cycling processes that they drive are responsive to temperature change. For instance, the abundance of enchytraeid worms which dominate the fauna of acid and highly organic soils has been shown to be strongly related to temperature. As such, it has been suggested that climate warming could increase their abundance leading to enhanced carbon mineralization and carbon loss from soil (Briones et al. 1998; Cole et al. 2002a, 2002b). Likewise, the abundance of soil nematodes in the Dry Valleys of Antarctica, including the dominant species *Scottnema lindsayae* which plays a significant role in soil carbon cycling in these ecosystems, has declined by over 40% in recent years due to increasing air temperatures, with possible implications for carbon dynamics (Fig. 2.20) (Barrett et al. 2004). Also, even subtle warming (by $\approx 1°C$) has been found to cause a sustained increase in ecosystem respiration rates in a sub-Arctic peatland, particularly in the subsurface layers (25–50 cm depth), indicative of a large and long-lasting positive feedback of carbon stored in northern peatlands to the global climate system (Dorrepaal et al. 2009).

Further evidence of effects of climate change on decomposer communities arises from studies on mushroom phenology. A recent analysis by Gange et al. (2007) related temporal shifts in autumnal fruiting patterns of macrofungi in southern England to shifts in climate and found that the average first-fruiting date of 315 species is now earlier, whereas last fruiting date is now later, than was the case 56 years ago. Their study also found that many species are now fruiting twice a year, indicative of increased mycelial activity and possibly greater decomposition rates in ecosystems. Similarity, a recent analysis of herbarium records over the period 1940–2006 in Norway found that the time of fruiting of mushrooms has changed considerably over recent years, but in this case with an average delay in fruiting since 1980 of 12.9 days and with changes differing strongly according to taxa (Kauserud et al. 2008). These changes in autumnal mushroom phenology coincide with the

(a)   (b)

Fig. 2.20 Antarctic Dry Valleys (a) and the dominant species of nematode *Scottnema lindsayae* (b), which plays a significant role in soil carbon cycling in these ecosystems. Images by Diana Wall (a) and M. Mundo (b).

extension of the growing season caused by global climate change and are likely to continue under current climate change scenario (Kauserud et al. 2008). Finally, it is uncertain as to whether short-term increases in carbon mineralization, which are commonly observed in warming experiments in the field (Luo et al., 2001; Melillo et al., 2002; Bradford et al. 2008), will be sustained due to depletion of substrate availability and acclimation of soil communities to higher temperatures (Kirschbaum 2004; Bradford et al. 2008; but see Hartley et al. 2008). This level of uncertainty about the response of soil organisms to climate change extends to unreliable model predictions of soil carbon feedbacks to climate change (Kirschbaum 2006), and resolving this issue represents a major research challenge for the future.

Climate-change-driven increases in the frequency of extreme weather events, such as droughting and freezing, may have an even greater effect on soil organisms and their activities than will overall changes in temperature and precipitation. For instance, it is well known that both droughting and freezing have substantial direct effects on microbial physiology and the composition of the soil microbial community, with important consequences for ecosystem-level carbon dynamics (Schimel et al. 2007). Further, the occurrence of soil surface ice layers, which are predicted to become more frequent with climate change as a result of diminished snow cover, are known to have strong negative effects on soil animals (Coulson et al. 2000) with likely consequences for carbon and nutrient dynamics. However, the effects of stressors resulting from climate change on soil biotic communities and the consequences for carbon exchange are likely to vary substantially across ecosystems. For instance, increased frequency and intensity of drought in drier ecosystems should result in moisture-limiting conditions for microbial activity, creating a negative feedback on microbial decomposition and soil carbon loss from microbial respiration (Nardo et al. 2004; Henry et al. 2005). Further, microbial responses to drying and rewetting are likely to be less pronounced in soils that are frequently exposed to natural drying/rewetting cycles (Birch 1958; Fierer and Schimel 2002). In contrast, increased drought and drying in wetlands and peatlands, which will lower the water table and introduce oxygen into previously anaerobic soil, will in turn create more favourable conditions for microbial activity (Freeman et al. 2004a). Also, increasing oxygen levels in previously anaerobic peatland soils has been shown to enhance the activity of phenol oxidases which play a pivotal role in the breakdown of recalcitrant organic matter (Freeman et al. 2004a; Zibilske and Bradford 2007). Because peatlands and wetlands represent among the largest stocks of terrestrial carbon globally (Ward et al. 2007), such enhanced breakdown of recalcitrant organic matter under drying could have major implications for the global carbon cycle (Freeman et al. 2004a). Before leaving this topic, it is important to note that methanogenic pathways are also affected by increased oxygen availability associated with drought, in that methane emissions are reduced by toxic effects of oxygen on methanogens (Roulet and Moore 1995; Freeman et al. 2002). Also, drought can have marked effects on nitrous oxide emission from soils, a potent greenhouse gas that is increasing in atmospheric concentrations at the rate of 0.2–0.3% per year (Houghton et al. 1996). However, responses depend on the severity of drought, in that modest summer drought is likely to have limited effect on soil nitrous oxide emissions, whereas more extreme drought can greatly increase them (Dowrick et al. 1999).

Another climate-change-related factor that is likely to have a profound influence on soil communities and decomposition in Arctic and alpine regions is reduced snow cover and the thawing of permafrost. It has been estimated that 25% of Earth's permafrost could thaw by 2100 due to climate warming, exposing considerable amounts of otherwise protected organic matter to microbial decomposition (Anisimov et al. 1999), thus creating a positive feedback on climate change (Davidson and Janssens 2006; Heimann and Reichstein 2008). Consistent with this, Schuur et al. (2009) found that permafrost thaw over decadal timescales in an Alaskan tundra landscape has caused significant losses of soil carbon, despite increased plant growth and ecosystem carbon input. Because snow is an important insulator of winter soil biological processes, predicted reductions in snow cover in alpine and Arctic regions should also increase soil freezing, with consequences for root mortality, nutrient cycling, and microbial-driven soil processes (Groffman et al. 2001; Bardgett et al. 2005). Strong microbial responses to freeze–thaw cycles have been shown in several studies, leading to increased microbial activity and greenhouse gas emission (Christensen and Tiedje 1990; Sharma et al. 2006), altered microbial substrate use (Schimel and Mikan 2005) and the expression of denitrifying genes leading to the release of nitrous oxide gas (Sharma et al., 2006). However, a recent synthesis of literature on this topic concluded that while freeze–thaw events might induce gaseous and/or solute losses of nitrogen from soils that are relevant at an annual time scale, they may have little effect on (or even reduce) soil carbon losses relative to unfrozen conditions in the longer term (Matzner and Borken 2008). Further, recent studies in subalpine forest in Colorado indicate that reduced snow cover can suppress rates of soil respiration through impairing a unique and highly temperature-sensitive soil microbial community that occurs beneath snow (Monson et al. 2006); this may significantly affect winter soil microbial activity, carbon storage, and carbon dioxide efflux in alpine and Arctic regions.

2.5.3 Multiple global change drivers and soil biotic interactions

The majority of studies to date that have explored effects of climate change on biological systems and soil organisms have considered single factors, such as elevated atmospheric carbon dioxide concentration, warming, and drought. However, there is much potential for interactions between these factors to have additive or antagonistic effects on soil organisms and the activities that they drive (Shaw et al. 2002; Mikkelsen et al. 2008; Bardgett et al. 2008; Tylianakis et al. 2008). Very little is known about the influence of multiple and interacting climate drivers on soil organisms and their activities, although some studies do point to strong non-additive belowground effects of these drivers, with feedback consequences for carbon exchange. For instance, microbial decomposition of peat was found to be significantly greater when subject to both elevated temperature and atmospheric carbon dioxide than when these factors were each elevated singly (Fenner et al. 2007a, 2007b). This created an even stronger positive feedback on carbon loss from soil as dissolved organic carbon in drainage water and respiration (Freeman et al. 2004b). Added to this complexity is our knowledge that other organisms and trophic groups that influence soil microbes directly, such as microbial-feeding fauna, will also respond to multiple climate change factors (Wardle 2002; Bardgett 2005; Tylianakis et al. 2008). This complexity further hampers our ability to predict effects of multiple climate change drivers on soil biological communities and carbon-exchange feedbacks.

In addition to multiple climate change drivers, soil organisms and their activities are also affected substantially by other global change phenomena, such as nitrogen deposition, invasion of new species, and land-use change. Perhaps the strongest driver is land-use change (see Sala et al. 2000), and it is widely documented that changes in the intensity of land use or the conversion of natural vegetation to agriculture or forestry can have substantial, and often strongly negative and irreversible, effects on soil biological communities and their activities (Brussaard et al. 1997; Wardle 2002; Bardgett 2005). One pattern that commonly emerges in the context of land-use change is that intensification of farming, including increased tillage, fertilizer use, and grazing, is typically associated with an increased role of the bacterial-based energy channel relative to the fungal-based channel (Hendrix et al. 1986; Wardle 2002; Bardgett 2005). As discussed above, this increased bacterial role is associated with faster, leakier nutrient cycling and more losses of nutrients and carbon in water and greenhouse gases to the atmosphere (Wardle et al. 2004a; Van der Heijden et al. 2008). In contrast, low-intensity management systems often encourage fungal-based soil food webs that are more similar to those of natural systems, and tend to be associated with more efficient nutrient cycling (Bardgett and McAlister 1999; de Vries et al. 2006; Gordon et al. 2008) and enhanced soil carbon sequestration (Six et al. 2006; De Deyn et al. 2008).

Soil biological communities are also strongly affected by nitrogen enrichment, which is of high relevance because anthropogenic activities have substantially increased global rates of nitrogen fixation and deposition (Vitousek et al. 1997a; Holland et al. 1999; Bobbink and Lamers 2002; Galloway et al. 2008). For instance, it is well known that nitrogen enrichment can have direct and differential impacts on extracellular enzymes involved in decomposition processes. This typically involves stimulation of the synthesis of cellulases which degrade labile, high-cellulose litter, but suppression of the synthesis of ligninolytic enzymes by white rot fungi which decompose recalcitrant, high-lignin litter (Carreiro et al. 2000; Frey et al. 2004; Waldrop et al. 2004; Allison et al. 2008). Also, nitrogen deposition is known to influence the abundance and diversity of different components of the soil microbial community. These include bacteria, saprophytic fungi (Donnison et al. 2000; Bardgett et al., 2006; Allison et al. 2008), mycorrhizal fungi (Egerton-Warburton and Allen 2000; Frey et al. 2004), and soil fauna (Scheu and Schaefer 1998; Ettema et al. 1999), which are also affected by climate change and are well known to have substantial effects on decomposition processes and ecosystem-level carbon exchange. A recent meta-analysis on this topic revealed that soil microbial biomass declined 15% on average under nitrogen fertilization, but that declines in abundance of microbes and fungi were more evident in studies of longer durations and with higher total amounts of nitrogen added (Treseder 2008). Moreover, that study showed negative responses of microbial biomass to nitrogen fertilization to be significantly correlated with declines in soil carbon dioxide emissions, indicating that moderate declines in microbial biomass under nitrogen fertilization may also have consequences for carbon fluxes. However, another meta-analysis of 109 studies across the globe revealed that nitrogen enrichment had no significant effect on net ecosystem carbon dioxide exchange in non-forest natural ecosystems, but did increase methane and nitrous oxide emissions by 97 and 216%, respectively (Liu and Greaver 2009). It was suggested, therefore, that any potential positive effects of nitrogen enrichment on the global terrestrial carbon sink should be offset by the stimulation of methane and nitrous oxide emissions, which are more potent greenhouse gases than is carbon dioxide (Liu and Greaver 2009). Importantly, and as will be considered in Chapter 3, nitrogen deposition and other global changes can also influence soil microbes and decomposition processes indirectly through altering vegetation composition and productivity, and by alleviating progressive nitrogen limitation of plant growth, which typically occurs under elevated atmospheric carbon dioxide (Finzi et al. 2002; Luo et al. 2004; de Graaff et al. 2006).

In general, very little is known about the combined effects of global changes on soil biological communities and their activities, but they clearly have the potential to amplify, suppress or perhaps even neutralize climate-change-driven effects on soil microbes and their feedback to carbon exchange (Bardgett et al. 2008). A recent synthesis (Tylianakis et al. 2008) of data from 688 published studies on the effects of global change on biotic interactions in terrestrial ecosystems (including those that occur in the decomposer food web) highlighted that there is substantial variability among studies in both the magnitude and direction of effects of any given global change driver on any given type of biotic interaction. Further, that analysis highlighted that the unanticipated effects of multiple drivers acting simultaneously create major challenges in predicting future responses to global environmental change. Experimental studies that simultaneously vary two or more global change drivers within the same experiment therefore have considerable promise for improving our understanding of how interactions involving soil communities and their effects on ecosystem-level processes may respond to current global change scenarios.

2.6 Conclusions

In this chapter, we have explored how the activities and interactions of belowground communities regulate belowground processes, and the consequences of this for the productivity, diversity, and composition of plant communities. We have also discussed the role of soil biotic interactions in a global change context, focusing on their influence on ecosystem carbon dynamics and contribution to climate change through carbon-cycle feedbacks. While it has long been recognized that soil microorganisms play a pivotal role in the breakdown of organic matter, and liberation of plant growth-limiting nutrients, the last decade has seen several significant advances in this area. For instance, recent recognition of the importance of direct uptake of organic nitrogen by plants has led to a radical rethink of terrestrial nitrogen cycling and in particular of the microbial processes that control plant nitrogen availability (Schimel and Bennett 2004; Jones et al. 2005). We also have greater awareness that the cycling of labile nitrogen pools relies on intimate, temporal coupling between plants and microbes and their seasonal resource demands (Jaeger et al. 1999; Bardgett et al. 2005). Further, we now recognize that partitioning of nitrogen (and possibly phosphorus) on the basis of chemical form can provide a mechanism for plants to efficiently partition a limited soil nutrient pool, thereby facilitating species coexistence and the maintenance of plant diversity (McKane et al. 2002; Kahmen et al. 2006; but see Harrison et al. 2007). Recent years have also witnessed a growing understanding of the mechanisms by which mycorrhizal fungi contribute to plant community dynamics, for instance by promoting more efficient exploitation and equitable distribution of nutrients within the plant community (Van der Heijden et al. 1998; Maherali and Klironomos 2007), and of the evolutionary mechanisms that explain why certain types of ecosystems lack symbiotic nitrogen-fixers and therefore experience greater nitrogen limitation (Menge et al. 2008). Despite these advances, the contribution of soil microbes to ecosystem functioning is still poorly understood, largely because over 95% of microbes are unable to be cultured on conventional media (Van der Heijden et al. 2008). Recent advances in ecological genomics and the use of microarrays to detect key genes responsible for key ecosystem processes (Van Straalen and Roelofs 2006;

Zak et al. 2006; He et al. 2007) offer potential to overcome this problem and improve understanding of the roles that soil microbes play in regulating nutrient and plant community dynamics (Van der Heijden et al. 2008).

While plant nutrient availability is strongly regulated by microbial activities, the soil also hosts an abundant and highly complex community of invertebrates which derive most of their nutrition and carbon either directly from roots and exudates, or indirectly from feeding on microbes (Pollierer et al. 2007). There is ample evidence that faunal consumers of microbes and their predators strongly regulate nutrient and plant community dynamics by altering the balance between microbial immobilization and mineralization, and hence nutrient availability (Wardle 2002; Bardgett 2005). Moreover, there is greater recent recognition that larger fauna such as ecosystem engineers can substantially modify soil processes and the temporal and spatial dynamics of plant communities through several mechanisms including alteration of the physical structure of soil, transportation of faecal material and propagules, and influencing the movement of water and soluble nutrients (Lavelle et al. 1997; Bardgett et al. 2001a). Another important development is the recognition that nutrient availability to plants can also be controlled several trophic steps away from the plant root through top-down control (Moore et al. 2003). This indicates that an understanding of how belowground communities influence aboveground community properties requires explicit consideration of the role of predation as a control point in belowground processes (Moore et al. 2003). A further development in this area is the recognition that compositional changes in the soil food web, including broad shifts between fungal and bacterial-based energy channels, have important consequences for ecosystem processes of decomposition and nutrient cycling (Wardle et al. 2004a; Van der Heijden et al. 2008). However, it is also evident that understanding the functional consequences of changes in soil food web composition also requires recognition of how individual species contribute to multiple species interactions (Heemsbergen et al. 2004) and of environmental context. This latter point is especially important given that the role of soil biotic interactions relative to abiotic factors in regulating ecosystem processes varies tremendously under different environmental conditions and across biomes (Gonzalez and Seastedt 2001; Wall et al. 2008).

A topic that has attracted much recent attention is the role that soil organisms and biotic interactions play in regulating carbon dynamics in terrestrial ecosystems, especially in the context of land–atmosphere exchanges of carbon and carbon-cycle feedbacks that could amplify climate change. One key development is the increased recognition that recent plant photosynthate acts as a major driver of soil biological activity and carbon flux in terrestrial ecosystems, accounting for as much as half of the respiratory carbon release from soil (Högberg et al. 2001; Steinmann et al. 2004; Pollierer et al. 2007). However, much remains to be learned about the mechanisms that govern respiratory fluxes from soil, especially in the context of climate change which can affect soil organisms and their activities directly via changes in temperature and precipitation, and indirectly via changes in vegetation and photosynthate supply (Bardgett et al. 2008). For instance, although it is well established that temperature acts as an important determinant of the rate of organic matter decomposition, and recent evidence points to warming and permafrost thaw causing sustained loss of carbon from tundra soils (Schuur et al. 2009; Dorrepaal et al. 2009), the nature of the relationship between temperature and heterotrophic microbial respiration and its potential to feed back to climate change is unclear (Davidson and Janssens 2006; Trumbore 2006). Also, while there is mounting evidence that decomposer organisms and their activities are strongly affected by extreme events associated with climate change (Schimel et al. 2007), the consequences of this for decomposition processes and carbon exchange remain little understood. This level of uncertainty about the response of soil organisms and their activities to climate change extends to unreliable model predictions of soil carbon feedbacks (Kirschbaum 2006), and resolving this issue represents a major research challenge for the future.

Finally, we highlight that global change drivers other than climate change also have strong effects on soil organisms and their activities, and that these effects often have far-reaching consequences for ecosystem-level properties, including soil carbon sequestration. However, it is becoming increasingly apparent that our ability to predict future responses to global change, and the potential for climate-change mitigation through carbon sequestration, requires a greater understanding of the simultaneous effects of multiple global change drivers on soil biological communities (Tylianakis et al. 2008; Bardgett et al. 2008). As highlighted in this chapter, very little is known about this, although there is clearly much potential for interactions between global change drivers to amplify, suppress, or even neutralize climate change driven effects on soil microbes and their feedback to carbon exchange (Bardgett et al. 2008). As will also be argued in Chapter 3, the unanticipated effect on soil biological communities of multiple drivers acting simultaneously represents a major research challenge for the future.

3

Plant community influences on the soil community and plant–soil feedbacks

3.1 Introduction

The community structure of plants and that of soil organisms are coupled to one another to varying degrees. In Chapter 2, we described how communities of soil organisms could influence the productivity and composition of plant communities through a variety of mechanisms. In turn, plants serve as determinants of soil communities. This may happen through either the indirect or direct pathways highlighted in Chapter 2 (Fig. 2.1): the indirect pathway involves plants influencing organisms in the decomposer food web by determining the quantity and quality of litter that enters the soil, whereas the direct pathway involves the characteristics of live roots determining the accessibility of root-associated organisms to the resources produced by the roots. While the indirect pathway usually involves a relatively low level of specificity between plants and soil organisms, there is increasing recognition that the direct pathway often involves a high level of specificity among aboveground and belowground taxa (Wardle et al. 2004a). Further, there is growing recognition that the effects of plants and soil organisms on one another results in important feedback mechanisms between the aboveground and belowground biota (Van der Putten et al. 1993; Bever et al. 1997; Wolfe and Klironomos 2005; Kulmatiski et al. 2008).

Since the early 1990s, the issue of how plant species may drive ecosystem processes, including those driven by soil organisms, has attracted considerable attention from ecologists (e.g. Hobbie 1992; Lawton 1994). However, it is important to recognize that this issue has a strong historical basis. For example, the mull and mor theory of Müller (1884) explicitly recognizes the role that plant species, and variation within plant species, can have in determining the community composition of soil fauna, as well as feedbacks through effects on nutrient supply from the soil. Subsequently, Handley (1954, 1961) showed through a series of experiments in the UK that *Calluna vulgaris* differed from coexisting plant species because it could produce tannin–protein complexes that inhibited mineralization of nitrogen, and therefore the nutrition of coexisting plant species. Furthermore, the effect of plant species on those soil organisms intimately associated with plant roots has long been recognized by agronomists. For example, crop rotations have long been used by farmers to prevent specific crop species from accumulating soil-borne pathogens that in turn would impair productivity of that crop through negative feedback. Further, as noted in Chapter 2, farmers have also recognized for centuries that utilizing plant species mixtures that include legumes which support nitrogen-fixing bacteria can greatly enhance soil fertility and crop nutrition.

To understand how plant communities influence soil communities, it is first necessary to understand how and why different species in a plant community vary in their belowground effects. Plant communities usually consist of several coexisting species that differ in their ecophysiological characteristics, and the literature is replete with examples of how different plant species vary in their effects on soil communities and processes. As such, comparisons of highly contrasting species in the same plant community provide compelling evidence that interspecific variation in litter quality is a powerful determinant of plant litter decomposition and the availability of nutrients required for plant growth (e.g. Berendse 1998; Bowman et al. 2004; Santiago 2007). Similarly, coexisting species can differ greatly in their rhizosphere communities of both microbes (Aberdeen 1956; Grayston et al. 1998) and microbe-feeding fauna (De Deyn et al. 2004; Viketoft 2008). The functional characteristics of plant species also determine the communities of those organisms that interact more directly with plant roots, such as root pathogens (Korthals et al. 2001) and mycorrhizal fungi (Cornelissen et al. 2001a). While the observation that plant species differ in the communities of soil organisms that they support is perhaps unsurprising, understanding the mechanistic basis of these species effects and their implications for ecosystem processes is essential for understanding the functioning of terrestrial ecosystems.

The goal of this chapter is to provide an outline of how plant community characteristics influence the soil biological community and thereby affect ecosystem processes and properties. In doing this, we first outline how variation among and within plant species influences soil communities and thereby the biogeochemical processes that they drive. We also consider the overarching effect of plant attributes or traits in governing these effects. Based on this framework, we then consider aboveground–belowground feedbacks in the context of vegetation succession and ecosystem development, and finally in the context of shifts in vegetation composition brought about by human-induced global change. In Chapter 2, we considered the effects of the belowground subsystem on the aboveground subsystem; our ultimate goal in this chapter is to complement Chapter 2 by highlighting how aboveground and belowground communities can operate in tandem to drive community and ecosystem properties.

3.2 How plants affect the belowground subsystem

3.2.1 Differential effects of different plant species

Within the soil food web, any given group of organisms is likely to be strongly affected by differences among plant species only if it is driven primarily by bottom-up control (i.e. resource quantity and quality) rather than by top-down control (i.e. regulation by its predators) or by abiotic factors. As discussed in Chapter 2, the importance of top-down and bottom-up regulation in the soil food web differs among major food web components, as well as within major groups under differing environmental conditions (Bengtsson *et al.* 1995; Mikola and Setälä 1998b; Wardle 2002; Moore et al. 2003). As a consequence, variable effects of plant species on soil taxa and food-web groups may be expected.

One mechanism through which different plant species may vary in their effects on soil food web biota is through differences in net primary production (NPP); that is, the quantity and quality of resources that they return to the soil as plant litter and rhizodeposits. However, effects of NPP on components of the soil food web are far from consistent, and different studies that have investigated how components of the soil food web vary across gradients of NPP have found somewhat contrasting results: responses can be positive, neutral, or even negative depending on environmental context (Fig. 3.1). In this light, it is perhaps unsurprising that experimental studies which compare planted monocultures of several plant species do not necessarily find the most productive plant species to support the greatest density or biomass of soil organisms. For example, in monoculture plots of each of eight grass and forb species, De Deyn et al. (2004) found little relationship between plant shoot mass and abundance of soil nematodes, including those involved in the decomposer food web as well as those that function as root herbivores. Similarly, Hooper and Vitousek (1997, 1998) did not find the most productive monocultures of plant functional groups in serpentine grassland to necessarily support higher levels of soil microbial biomass. Further, Wardle et al. (2003d) found for monocultures of nine herbaceous plant species that while the biomass of the basal trophic level of the soil food web (i.e. the microflora) was generally greatest for the most productive plant species, this pattern was not generally apparent for the higher consumer trophic levels.

Bottom-up effects of plant species differences on soil food-web groups and the belowground processes that they drive are, however, apparent from the large number of studies that have considered interspecific differences in litter quality. There is a vast literature showing that differences in litter chemical constituents between different plant species (e.g. nitrogen, phosphorus, polyphenolics, lignin, soluble carbon, calcium) drives litter decomposition rates, and therefore the activity of decomposer organisms (Swift et al. 1979; Berg and McClaugherty 2003). This is consistent with studies showing that species that produce high-quality (i.e. rapidly decomposing) litter support greater densities or biomasses of litter-decomposer microflora and fauna (e.g. Parmelee et al. 1989; Hansen 2000), and greater litter-consumption rates by saprophagous fauna such as arthropods (Nicolai 1988) and earthworms (Hobbie et al. 2006). Litter-quality differences among plant species may also serve as powerful determinants of the community structure of both decomposer microflora (Widden and Hsu 1987) and fauna (Wardle et al. 2006) inhabiting the litter. Further, there is some evidence that some plant species can preferentially select for decomposer taxa that enhance the decomposition of their own litter (e.g. Hansen 1999; Negrete-Yankelevich et al. 2008; Vivanco and Austin 2008; Ayres et al. 2009), indicative of feedback mechanisms that will be discussed later in this chapter.

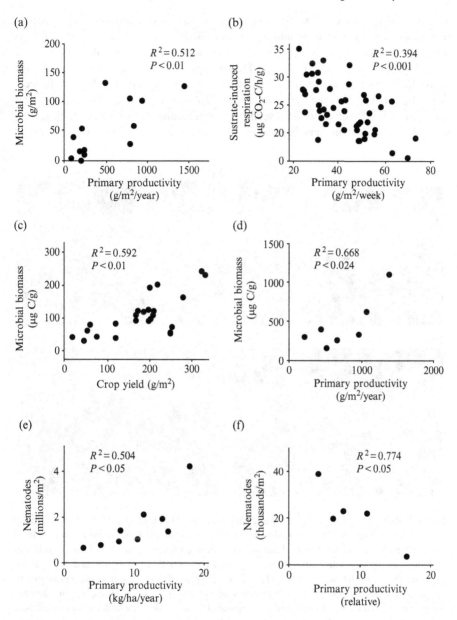

Fig. 3.1 Biomass or populations of consumer organisms in the decomposer food web in response to gradients of NPP. Consumer organisms considered are: (a) soil microbial biomass in late successional North American forests (Zak et al. 1994); (b) substrate-induced respiration (relative measure of microbial biomass) in a New Zealand grassland (Wardle et al. 1995); (c) soil microbial biomass in cropping fields in Alabama, USA (Insam et al. 1991); (d) soil microbial biomass in coniferous forests in Oregon, USA (Myrold et al. 1989); (e) soil nematode populations in a New Zealand grassland (Yeates 1979); (f) soil nematode populations in Japanese forests (Kitazawa in Yeates 1979). Derived from Wardle (2002).

Plant species that vary greatly in their litter quality also often differ in the rates of processes that occur in their underlying soil (Binkley and Giardina 1998). This is especially apparent for comparisons of coexisting plant species that have vastly differing litter properties (Fig. 3.2). For example, in Dutch moist heathlands, the grass *Molinia caerulea* produces litter of a much higher quality (i.e. higher nitrogen, lower lignin) than does the coexisting dwarf shrub *Erica tetralix*. Consequently, rates of microbially driven processes, such as nitrogen mineralization, are much greater in soil under *Molinia* than beneath *Erica* (Berendse 1998). Likewise, *Empetrum*

Fig. 3.2 Pairs of co-occurring plant species that have vastly contrasting effects on belowground organisms and processes. (a) *Empetrum hermaphroditum* and (b) *Deschampsia flexuosa* from the northern Swedish boreal zone. *Empetrum* is a slow-growing species with long-lived leaves that produces poor-quality litter containing high concentrations of lignin and the stilbene phenolic batatasin III which adversely affects a range of aboveground and belowground processes and organisms. In contrast the grass *Deschampsia* is a fast-growing species that has a rapid turnover of tissue and that produces high-quality poorly defended litter with low levels of lignin and phenolics (see Nilsson et al. 2002). (c) *Erica tetralix* with *Narthecium ossifragum* and (d) *Molinia caerulea* in Dutch moist heathlands. *Erica* grows slowly and produces well-defended poor-quality litter with a high lignin concentration that only slowly releases mineral nitrogen, while *Molinia* produces high-quality litter and promotes nitrogen mineralization (see Berendse 1998). Disturbance and addition of fertilizers can result in a switch from domination by *Empetrum* to domination by *Deschampsia* in northern Sweden, and from *Erica* to *Molinia* in Dutch heathlands. Photo credits: (a, b) M.-C. Nilsson; (c, d) J. Janssen.

hermaphroditum in the northern Swedish boreal forest produces litter of poor quality with high concentrations of the stilbene phenolic batatasin III, which impairs soil microbial biomass and activity, and litter-decomposition rates relative to litter from coexisting plant species such as *Vaccinium myrtillus* and *Deschampsia flexuosa* (Nilsson and Wardle 2005). Further, in alpine grasslands in Colorado, the forb *Geum* (formerly *Acomastylis*) *rossii* produces shoot and root litter with a high concentration of labile polyphenolics, while the coexisting and fast-growing grass *Deschampsia caespitosa* does not. These phenolics appear to be readily utilized by the soil microflora, leading to greater microbial biomass and promotion of microbial immobilization of nitrogen, which reduces nitrogen availability to *D. caespitosa*, thereby reducing its growth (Bowman et al. 2004; Meier et al. 2008).

Plant species also differ greatly in the resources that they contain in their roots and release through the rhizosphere. It is well established that plant species vary greatly in their effects on those soil organisms that are intimately associated with their roots, and that there is a high degree of specificity between plant species and taxa of ectomycorrhizal fungi (Smith and Read 1997), arbuscular mycorrhizal fungi (Van der Heijden et al. 1998a), and root pathogens and herbivores (Yeates 1979; Korthals et al. 2001). However, as mentioned in Chapter 2, plants can also exert important effects on decomposer organisms through recently fixed carbon released into the soil from the plant root system. This has long been known to be important in plant communities dominated by fast-growing herbaceous species such as grasslands. For instance, root systems of different herbaceous plant species are known to vary greatly in the densities and community composition of the bacteria, fungi, and microfauna that they support (Rovira et al. 1974; Grayston et al. 1998; Bardgett et al. 1999c), and live roots of different plant species can differ greatly in their effects on microbial activity and decomposition processes (Dormaar 1990). As shown by Van der Krift et al. (2002), live roots of four grass species differed greatly in their effects on decomposition of dead root material, with live roots of *Festuca ovina* having especially strong positive effects, probably because this species exuded high amounts of compounds that stimulate microbial activity. Likewise, the above-mentioned slow-growing alpine forb *G. rossii* has recently been shown to produce phenolic-rich rhizodeposits which strongly influence microbial community dynamics and soil nutrient availability, with consequences for the growth of the coexisting grass *D. caespitosa* (Meier et al. 2008, 2009).

As noted in Chapter 2, there is also recent evidence from experiments in forests that have utilized C-13 enrichment labelling (Pollierer et al. 2007; Högberg et al. 2008) and tree girdling (Högberg et al. 2001) to show that tree roots release substantial quantities of recently fixed carbon to the belowground subsystem, and that this can serve as a major driver of soil respiration, mycorrhizal fungal production (Högberg et al. 2001), saprophytic microflora (Högberg et al. 2008), and soil invertebrate communities (Pollierer et al. 2007). Although to date studies of this type have been performed on single tree species, it is plausible that these short-term allocation patterns may vary greatly among coexisting tree species, thus serving as an important mechanism by which differences among tree species effects are manifested belowground. Testing of this hypothesis, however, remains to be performed.

3.2.2 Effects of within-species variation

Although most work on how different plants affect belowground organisms and processes has been at the across-species level, there is increasing recognition that variability at the within-species level may also be an important ecosystem driver. Many studies have shown that measures of leaf litter quality, decomposition, and nutrient mineralization can vary within species across environmental gradients (e.g. Crews et al. 1995; Northup et al. 1995) or among individuals or groups of individuals of a single species (e.g. Madritch and Hunter 2002). Such studies highlight the magnitude and ecological significance of phenotypic variability that can exist at the within-species level. However, it is only relatively recently that the importance of within-species variability at the genetic level has become appreciated as an important ecological driver (Bailey et al. 2009).

A small but growing number of studies have characterized plant genetic variability within species or within hybrid swarms, and related this variability to effects of the plant on belowground organisms or processes (Fig. 3.3). For example, Treseder and Vitousek (2001) grew seedlings of *Metrosideros polymorpha* collected from each of three sites in Hawai'i varying in nutrient limitation in a common garden experiment. They found that plants from the three sites were genetically distinct, and that they differed in nutrient-resorption potential, concentrations of litter lignin and nitrogen, and litter decomposability. Schweitzer et al. (2004) similarly utilized a common garden experiment to show that concentrations of condensed leaf tannins varied greatly among hybrids of *Populus fremontii* × *Populus angustifolia*, and that this was in turn closely related to variability among the hybrids in litter decomposability and rates of soil nitrogen mineralization. Subsequent work on that experiment revealed strong linkages between plant genotype and both microbial biomass and microbial community structure (Schweitzer et al. 2008). Studies by Classen et al. (2007) on *Pinus edulis* revealed that strains that differed in their level of herbivore resistance also differed in the rate of shoot and root litter decomposition and nutrient release. Similarly, Silfver et al. (2007) utilized a common garden experiment to show that *Betula pendula* genotypes which differed in their herbivore resistance also differed in their foliar nitrogen and protein concentrations, as well as in their litter decomposability. With regard to decomposer biota, Crutsinger et al. (2008) found that plots planted with different genotypes of *Solidago altissima* sometimes also supported different densities of litter-dwelling microarthropods. Collectively, these studies point to wide ranging effects of genotypes on the belowground subsystem, although the magnitude of these within-species effects relative to that for across-species effects remains mostly unexplored.

Those studies that have explored belowground consequences of within-species genetic variability serve to highlight the linkage between community genetics and ecosystem processes. As such they have contributed to the development of the concept of the 'extended phenotype' (*sensu* Whitham et al. 2003) in which plant genetics at the within-species level drives community properties and ecosystem functioning. Recent work has also highlighted that the potential of plant genes to

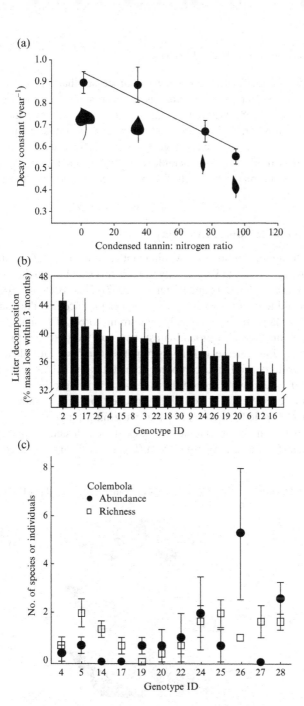

Fig. 3.3 Examples of the magnitude of effects of genetic variability on belowground processes and organisms. (a) Litter decay rates and condensed-tannin-to-nitrogen ratios for *Populus fremontii*, *Populus angustifolia*, and their hybrids. From Schweitzer et al. (2004), with permission from Wiley Blackwell; (b) mean mass loss (± 1 SE) of decomposing litter for each of 19 genotypes of *Betula pendula* in a common garden experiment. From Silfver et al. (2007), with permission from Springer Science+Business Media; (c) litter-dwelling collembolan abundance and species richness (mean ± 1 SE) in plots planted with each of 12 genotypes of *Solidago altissima*. From Crutsinger et al. (2008), with permission from Springer Science+Business Media.

influence soil microbial communities and the processes that they drive is able to be inherited by subsequent plant generations, indicative of 'community heritability' (Schweitzer et al. 2008; Whitham et al. 2003, 2008). So far, the evolutionary implications of this heritability are unclear, and it is not known whether plant genotypes derive a selective advantage through it. Nevertheless, the concept of community heritability has considerable potential for the integration of genetic characteristics at the plant level with feedbacks between the producer and decomposer subsystems.

3.2.3 Spatial and temporal variability

There is increasing recognition that, within plant communities, soil organisms and the processes that they drive can vary greatly in both space and time (Ettema and Wardle 2002; Bardgett et al. 2005; Berg and Bengtsson 2007). This is determined in a large part by the spatial and temporal variability of resources produced by the plant, especially when the soil biota is strongly driven by bottom-up forces. The spatial and temporal effect of plants on the belowground subsystem has the potential to operate as an important driver of the release and supply of nutrients required for plant growth.

Spatial patterning of soil biota can occur both vertically and horizontally (Berg and Bengtsson 2007); here we focus on the latter as this is more likely to be driven by plant species effects. Spatially explicit approaches for studying soil organisms and processes have revealed patterns of spatial patterning ranging from millimetres to hundreds of metres, meaning that patterning exists in a nested manner across a wide range of scales (Ettema and Wardle 2002) (Fig. 3.4). Plants influence these spatial

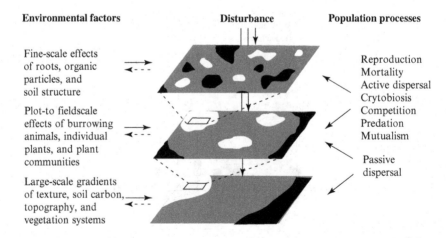

Fig. 3.4 Determinants of spatial heterogeneity of soil organisms. Spatial heterogeneity of soil organisms and processes is distributed on nested scales, and is driven by both biotic and abiotic factors including the effects of plant species. From Ettema and Wardle (2002), with permission from Elsevier.

patterns at several of these scales, suggesting that differences among plant species are important drivers of the spatial structure of the belowground subsystem. For example, a spatially explicit approach was used by Grundmann and Debouzie (2000) to show that bacteria capable of oxidizing ammonium and nitrate were aggregated over the order of millimetres, apparently reflecting the spatial distribution of fine roots. Meanwhile, Wachinger et al. (2000) found that spatial heterogeneity of methanogenic bacteria was related to small-scale patchiness of plant litter. At larger scales, spatial patterning of soil organisms is well known to be reflective of the spatial patterning of plants (Klironomos et al. 1999; Wardle 2002).

The role of plant species effects in driving soil processes is most apparent when plant species of contrasting ecology occur together in the landscape, but are spatially segregated from one another. In these cases each plant can directly influence the soil under it, and when horizontal transport rates of plant-derived resources are low this creates a spatial mosaic of soil biota and the processes that they drive. This is especially apparent when individuals of woody plant species (trees and shrubs) are scattered through the landscape, yielding island-like effects in which soil organism densities and soil processes are greatly altered relative to the rest of the landscape. Examples of this phenomenon include individual shrubs in arid ecosystems that create 'islands of fertility' (e.g. Schlesinger and Pilmanis 1998), and 'tree islands' following krummholz formation in alpine tundra above the treeline (Seastedt and Adams 2001). Coexisting tree species that produce resources of greatly contrasting quality also influence the spatial patterning of soil organisms and processes. For example, spatial distribution of the soil microbial community in a Swedish boreal forest was found by Saetre and Bååth (2000) to be determined by the spatial arrangement of the two dominant tree species, the deciduous *Betuna pendula* and the evergreen *Picea abies*. This type of phenomenon may also occur among genotypes within tree species, and Madritch et al. (2009) showed that coexisting clones of *Populus tremuloides* in forests in Wisconsin caused distinct spatial mosaics of soil processes relevant to carbon and nitrogen cycling. Finally, contrasting types of vegetation in the landscape can also differ in their effects on the spatial heterogeneity of belowground properties. For example, Kleb and Wilson (1997) found that woody vegetation promoted spatial heterogeneity of soil carbon and nitrogen relative to grassland vegetation, apparently due to woody species producing a coarser root system. However, a recent synthesis of global literature (Pärtel et al. 2008) showed that this pattern was apparent only in temperate and not in tropical ecosystems.

Belowground biota are also affected by plants at a hierarchy of temporal scales (Bardgett et al. 2005), due in a large part to temporal variability of resource input to the soil (Wardle 2002). Different plant species and functional groups differ greatly in the temporal variability of their resource production. For example, at a regional spatial scale, Knapp and Smith (2001) found large differences in inter-year variability of NPP across 11 sites in North America, with vegetation dominated by herbaceous species showing greater variability than that dominated by woody species. On a more local spatial scale, Wardle et al. (1999) found for a small plot experiment in a grazed grassland that different plant functional groups differed in the temporal variability of their productivity, and that this in turn affected the temporal variability of soil microbe and nematode densities. These plant-driven effects on belowground temporal variability are likely to serve as important temporal regulators of key ecosystem processes and the supply of plant-available nutrients from soil.

Plant species within a community that provide resources in pulses are likely to exert particularly important temporal effects on belowground organisms and processes (Wardle 2002; Yang et al. 2008). For example, climatically driven pulsed phenomena such as mast seeding can result in large quantities of reproductive material that decomposes and serves as a potential resource for decomposers. As such, Zackrisson et al. (1999) showed that mast seeding in boreal forest by *Picea abies* provided a pulse of readily available nitrogen that stimulated the growth of tree seedlings, presumably through stimulating soil microbial activity, and similar effects may also result from plant species that produce pulsed supplies of pollen (see Greenfield 1999). Further, different coexisting plant species are also likely to differ greatly in relation to seasonal production of resources; for example, deciduous tree species usually provide more distinct pulses of litter fall than do evergreen species. In the shorter term, pulses of root exudates are powerful drivers of root-associated microbes, and extrinsic factors that encourage pulses of root exudates, such as foliar and root herbivory, in turn induce pulsed responses of the soil biota (Bardgett and Wardle 2003; see also Chapter 4). However, despite the likely importance of pulsed supply of plant-derived resources in an aboveground–belowground context, this issue to date has been explored in comparatively few studies.

3.2.4 Multiple species effects

Plant species do not occur singly, but rather as members of multiple species communities. The issue that therefore arises is whether each species in that community exerts effects on the belowground subsystem that are independent of those of the other species. Since the mid-1990s this topic has been formulated as the so-called diversity-function issue which seeks to address whether increasing plant species richness influences ecological processes and properties (see review by Hooper et al. 2005), but it nevertheless has a long historical basis. For example, agronomists have long been interested in the potential benefits of multiple species cropping and intercropping (Vandermeer 1990), and Odum's theory of ecosystem succession (Odum 1969) speculated on whether species diversity altered the physical stability of ecosystems and was a necessity for 'long life of the ecosystem'.

The issue of how combinations of multiple plant species affect belowground organisms and processes has been widely studied, and not just in recent years. For example, Christie (1974, 1978) performed several conceptually simple experiments to show that when pairs of plant species were grown together they could stimulate microbial densities on each other's roots relative to when they were grown separately (Fig. 3.5). This means that mixtures of plant species could potentially stimulate microbial biomass relative to corresponding plant monocultures. Further, in mixed stands of tree species,

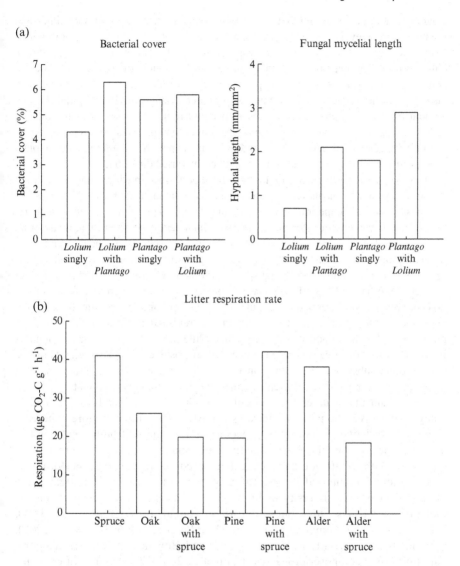

Fig. 3.5 Early studies that provide evidence that plant species can have different effects on soil biota and processes when grown in mixtures than when grown alone. (a) Cover by bacteria and length of fungal mycelium on roots of each of two herbaceous plant species when grown in mixture and grown singly. *Lolium* supports significantly more of both bacteria and fungi when grown with *Plantago* than when grown alone ($P < 0.025$), while *Plantago* supports significantly more fungi when grown with *Lolium* than when grown alone ($P < 0.01$). Derived from Christie et al. (1974). (b) Litter-layer respiration rates in monocultures and two-species mixtures of different tree species. Respiration in the spruce+pine mixture is greater than expected based on the component monocultures while that for the spruce+alder and spruce+oak mixtures is less than expected ($P < 0.001$ in all cases). Derived from Chapman et al. (1988).

Chapman et al. (1988) found that pairs of tree species grown together sometimes had greatly different levels of litter respiration and densities of enchytraeids and earthworms in the litter layer relative to when they were grown singly. This mixture effect could be either positive or negative depending on the combination of tree species present (Fig. 3.5). Other studies considering pairwise effects of plant species on soil processes also show varied effects. For example, Wardle and Nicholson (1996) found that the amount of soil microbial biomass present per unit root could either be promoted or reduced by growing plant species in mixtures rather than monocultures, depending on the plant species considered. Such studies point to pairwise combinations of plant species strongly influencing the soil biota, albeit in varied directions.

Over the past 15 years many studies have been performed in which some measure of the diversity of live plants (e.g. species richness, functional group richness, species evenness) has been experimentally varied and various community or ecosystem response variables subsequently measured (see Balvanera et al. 2006; Cardinale et al. 2006). Several of these have focused on the belowground subsystem, and the majority of these find soil processes and/or densities of soil organisms to be either weakly affected or unresponsive to the numbers of plant species or functional groups present (see Hooper et al. 2005). A handful of studies do claim to present evidence for strong effects of plant diversity on soil processes and biota, but such results appear to be specific for particular experiments and approaches. This issue will be revisited in Chapter 5, but for the purposes of this discussion it is apparent that while increasing plant species richness from one to two or three species may exert either positive or negative effects on the belowground subsystem depending on the species present, there is little consistent evidence of a strong effect of increasing plant species richness beyond that.

One variant of the diversity manipulation approach that has been adopted in many studies over the past two decades is the so-called litter-mixture approach. Here, litters of two or more plant species are decomposed both singly and in multiple species combinations, and the performance of each mixture is then compared with what would be expected based on the average performance of the litters decomposed singly. Litter-mixture studies have investigated the effects of increasing diversity of litter types on a range of response variables, including litter-mass loss rates, detritivore populations (Blair et al. 1990; Hansen 2000) and community structure (Wardle et al. 2006; Ball et al. 2009), nutrient dynamics (Ball et al. 2008; Meier and Bowman 2008), and temporal variability of decomposer processes (Keith et al. 2008). The balance of evidence points to mixtures of plant species having idiosyncratic effects on decomposer processes and organisms, but with effects that are more often positive than negative (see reviews by Gartner and Cardon 2004; Hättenschwiler et al. 2005). While most studies have considered mixtures of two or three plant species, those studies that have utilized mixtures of multiple plant species find little evidence of increasing litter diversity beyond three species exerting important ecological effects (e.g. Wardle et al. 1997a; Bardgett and Shine 1999; Perez Harguindeguy et al. 2008). Factors that regulate the effects of litter mixing are discussed later in this chapter.

3.3 Overriding effects of plant traits

3.3.1 Contrasting plant species and trait axes

How plant species affect ecosystem properties is determined to a large extent by their ecophysiological attributes or functional traits. It has long been recognized that plant species (or groups of plant species) differ greatly in their traits, with traits of species adapted for high resource availability differing from those of species adapted for low resource availability (Grime 1977; Coley et al. 1985). For example, conifers which are usually adapted for low-nutrient conditions characteristically have low rates of tissue production, long-lived leaves, lower leaf nutrient concentrations, lower specific leaf areas, and lower maximum photosynthetic rates, compared to angiosperm species (Cornelissen et al. 1996; Aerts and Chapin 2000). Comparative studies involving multiple species also show consistent relationships among these traits (e.g. Poorter and Remkes 1990), and provide evidence for an axis of evolutionary specialization amongst species which reflects a trade-off between the acquisition of resources and conservation of resources (Grime et al. 1997). Analysis of global data-sets containing plant trait (Díaz et al. 2004) or leaf trait (Wright et al. 2004) data confirm the consistency of this trade-off across contrasting floras and across major phylogenetic groups of vascular plants. As such, Díaz et al. (2004) showed that the primary axis of specialization was determined by traits relating to resource capture, usage, and release, with the secondary axis being more related to traits reflecting plant size.

Plant trait combinations are important drivers of litter quality, and plant species that are more adapted to nutrient-poor conditions and resource conservation generally produce litter with lower concentrations of water soluble compounds, and higher concentrations of structural carbohydrates such as cellulose and lignin and defense compounds such as polyphenolics, than species adapted to high-resource conditions. These properties of plant species adapted to poor resource availability make their leaf litter less desirable for decomposer microbes and fauna; as a consequence, litter from these species decomposes more slowly than species adapted to high-resource conditions. As such, litter from coniferous species generally decomposes more slowly than that from woody angiosperm species, which in turn breaks down more slowly than that from herbaceous species (Enríquez et al. 1993; Cornelissen 1996). Comparative studies of multiple plant species also shows litter mass loss to be related to a range of plant traits, including plant growth rate (Cornelissen and Thompson 1997; Wardle et al. 1998a), tissue strength (Cornelissen and Thompson 1997), and specific leaf area (Santiago 2007; Kurokawa and Nakashizuka 2008). These studies collectively point to suites of leaf traits as an important determinant of litter breakdown rates across species.

Recent studies have also focused on how plant traits might affect decomposability of plant tissues other than leaves. For example, there is some evidence that root traits may be correlated with leaf traits across species (Craine et al. 2002; Tjoelker et al. 2005; Freschet et al. 2010), suggesting that decomposability of root litter may be driven by similar suites of traits that drive decomposition of aboveground litter

(Wardle et al. 1998a). However, there is also evidence that plants may combine leaf traits typical for fast-growing 'competitor' species with root traits typical for slower-growing 'stress tolerator' species, or the other way round, and such reciprocal trait combinations could potentially stabilize their co-existence (Personeni and Loiseau 2004). In forested ecosystems, tree species produce most of their biomass and litter as wood. Little is understood about how decomposability of woody litter across plant species is related to leaf traits, or how decomposability of wood litter is related to the decomposability of leaf litter. However, a recent global meta-analysis (Weedon et al. 2009) revealed that gymnosperm wood litter decomposes more slowly than that from angiosperms, and that wood traits such as concentrations of carbon and nitrogen and the carbon-to-nitrogen ratio of wood were all related to wood decomposability, at least for angiosperms. Further, there is recent evidence (Cornwell et al. 2009) that wood traits determine not just the decomposability of wood, but also what proportion of it enters the decomposer pathway as opposed to being consumed by fire or invertebrates.

While plant traits are well known to influence decomposition of plant litter, less is known about the linkage between traits and soil organisms, or whether this has multitrophic belowground consequences. Linkages between traits and belowground biota are perhaps best understood for those soil organisms directly linked to plant roots, notably mycorrhizal fungi. A synthesis of data for 83 British plant species showed that mycorrhizal status is linked to plant traits such as relative growth rate and foliar nutrient concentrations, as well as to the decomposability of leaf litter (Cornelissen et al. 2001a) (Fig. 3.6). Specifically, ericoid mycorrhizae and to a lesser extent ectomycorrhizae tend to be associated with plant species that have traits best suited for acidic and resource-poor conditions, while arbuscular mycorrhizae are associated with species that have traits more suited for base-rich and nitrogen-rich but often phosphorus-poor conditions (Read 1991; Cornelissen et al. 2001a). Associations of plant traits with soil organisms have been seldom explored explicitly for the decomposer subsystem. However, Wardle et al. (1998a) found across 20 herbaceous plant species that the effect of plant species on microbial biomass (the primary consumer trophic level of the soil food web) was significantly related to particular root traits, such as specific root length, as well as some morphological aboveground traits. This is suggestive of multitrophic consequences of plant traits. Across plant species, consumption of dead leaves by litter-feeding invertebrates such as earthworms appears to be related to key traits such as leaf nutrient content (Hendriksen 1990). It has also been proposed that plant species which have traits associated with more fertile environments are more likely to support bacterial-based rather than fungal-based soil food webs (see Chapter 2), and to promote domination of the soil meso- and macrofauna by earthworms and enchytraeids rather than by microarthropods (Wardle et al. 2004a). Although there is evidence across habitats that this is likely to be the case, direct experimental evidence under controlled conditions, for example from common garden experiments that link plant traits to the abundance, activity, and composition of decomposer food webs, is scarce.

While plant traits have long been recognized as important ecosystem drivers, recent studies have highlighted that coexisting species may vary tremendously in their traits,

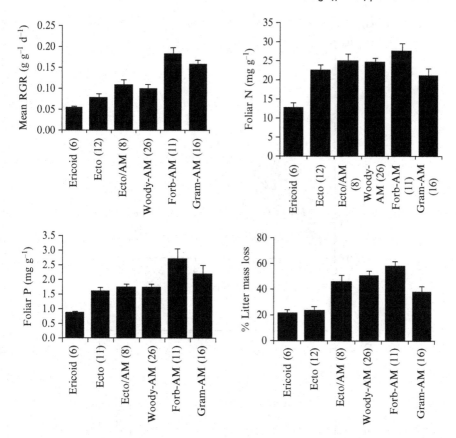

Fig. 3.6 Mean values (+1 SE) for plant traits of six plant functional groups based on mycorrhizal association type and life form. Numbers in parentheses refer to the number of species. AM, arbuscular mycorrhizal fungi; RGR, relative growth rate in a standard growth-chamber environment; Gram, graminoid. Mass-loss data are for a 20 week simultaneous multispecies incubation in a litter bed. From Cornelissen et al. (2001a), with permission from Springer Science+Business Media.

even over small spatial scales. For example, Hättenschwiler et al. (2008) showed that 45 coexisting Amazonian tree species varied three-fold in their litter nitrogen concentrations and seven-fold in their foliar phosphorus concentrations (Fig. 3.7). Further, Richardson et al. (2008) collected foliar and litter samples from a 100 ha area of temperate New Zealand rainforest representing a range of slopes and aspects, and found over a 10-fold range in leaf nitrogen and phosphorus concentrations; this range represented a large portion of the range reported globally for these nutrients (Fig. 3.7). The ranges in leaf and litter nutrient concentrations found at local scales are likely to have important knock-on effects to the decomposer subsystem. For instance, Cornwell et al. (2008) synthesized data from 66 common garden decomposition experiments, each involving a range of litter types decomposed in the same environment, and found

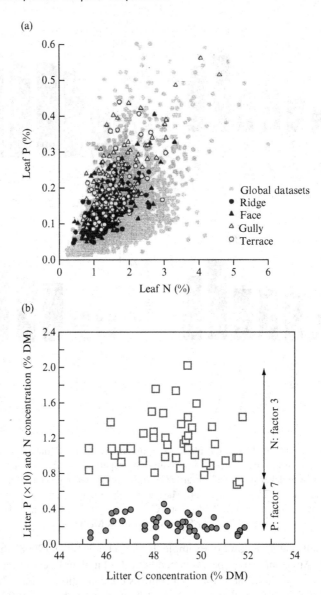

Fig. 3.7 Substantial variation in leaf and litter nitrogen and phosphorus concentrations over relatively small spatial scales. (a) Range of leaf nitrogen and phosphorus concentrations for four different landforms in a 100 ha area of rainforest in New Zealand, relative to that for global-scale data-sets. From Richardson et al. (2008), with permission from Wiley-Blackwell. (b) Litter phosphorus (P; circles) and nitrogen (N; squares) in relation to carbon (C) concentration for 45 coexisting Amazonian rainforest tree species; each species is represented by a different point. The phosphorus and nitrogen concentrations vary by factors of three and seven respectively. DM, dry matter. From Hättenschwiler et al. (2008), with permission from Wiley-Blackwell.

that rates of litter decomposition among species varied more than 10-fold within experiments when the middle 90% of species (i.e. between the 5th and 95th percentiles) were considered. This compared with a 5.5–5.9-fold variation in decomposition of common substrates in different biomes of greatly contrasting climate (see Berg et al. 1993; Parton et al. 2007) and points to the considerable importance of interspecific variation in plant traits (notably those that drive litter quality) relative to the direct effects of climate as a driver of plant litter decomposition.

Most studies on the effects of plant traits on the belowground subsystem have considered only higher plant species, and few have considered other plant groups such as ferns and bryophytes despite their likely role in driving nutrient cycling (e.g. DeLuca et al. 2002b; Turetsky 2003) and regeneration of dominant tree species (e.g. Coomes et al. 2005; Nilsson and Wardle 2005) in many forested ecosystems worldwide (Fig. 3.8). However, Amantangelo and Vitousek (2008) found polyploid and non-polyploid ferns in Hawai'ian rainforest to differ both with each other and with coexisting vascular plant species in their leaf-litter chemistry. Further, Wardle et al. (2002) found that litter from ferns in New Zealand rainforests had higher concentrations of lignin and fibre and lower rates of decomposition than coexisting angiosperm species, and that lignin concentration was an important predictor of differences in decomposition rate among fern species. With regard to mosses, Wardle et al. (2003b) found litter from feather mosses in boreal forest to decompose much more slowly than that from any of the coexisting vascular plant species. Similarly, Dorrepaal et al. (2005) and Lang et al. (2009) found for the sub-Arctic tundra that litter decomposition for moss species (notably *Sphagnum* spp.) generally decomposed more slowly than that for vascular species. Lang et al. (2009) also found large differences among moss and liverwort taxa in terms of both litter quality and decomposition rate. The available evidence suggest that ferns and mosses may differ greatly in their traits from that of higher vascular plants, and that this may have important ecosystem-level consequences whenever these plants are abundant. A likely consequence is the accumulation in the ecosystem of slowly decomposing fern and bryophyte litter relative to faster-decomposing litter provided by angiosperm species, as is evident for instance in high-latitude peatlands.

The vast majority of studies on the ecological impact of plant traits have focused on across-species variability and neglected within-species variability, often by using a single data point for each species. However, foliar and litter traits, notably nutrient concentrations, may also vary several-fold within species (Richardson et al. 2005, 2008), and large differences exist for litter decomposition rates within species both across environmental gradients (Crews et al. 1995; Wardle et al. 2009a) and among genotypes (Schweitzer et al. 2004; also see earlier discussion). It also appears likely, therefore, that variation in key traits within a given species can affect the decomposer subsystem in the same way as does variation in traits amongst species (Schweitzer et al. 2004). As such, assignment of single trait values on a per-species basis, such as is commonly done in large traits databases, has obvious limitations especially for species that have considerable plasticity. Much remains to be understood about the importance of within-species variability, and the relative role of within- and across-species variability in key traits, as drivers of the belowground subsystem and ecosystem properties.

(a)

(b)

Fig. 3.8 Few studies have considered the role of plant traits in affecting the decomposer subsystem for lower plant species, despite their important role in driving ecosystem processes in many ecosystems, such as ferns (e.g. *Blechnum discolor*) in New Zealand rain forest (a) and feather mosses (e.g. *Hylocomium splendens*) in boreal forest (b). However, litter from these components often decomposes much more slowly than that from the higher plants with which they coexist, and as such may have an important role in driving belowground organisms and processes. Images by D. Wardle (a) and A. Lagerström (b).

3.3.2 Trait dominance, trait dissimilarity, and multiple species effects

Plant species in any community differ greatly in their contribution to total biomass or NPP. The mass ratio hypothesis proposed by Grime (1998) predicts that in a multiple-species community the relative effect of each species on ecosystem properties should be proportional to its relative contribution to the total NPP of the ecosystem. As such, it serves as a useful framework for predicting the ecosystem-level consequences of plant traits on ecosystem processes at a community-level scale, because it suggests that the effects of the functional traits of each species in a given community on ecosystem properties should be related to its contribution to the total productivity or biomass of that community.

Some recent studies have used variants of this approach to relate ecosystem process rates to community-weighted plant trait values, both above- and below-ground. The procedure is conceptually straightforward: key functional traits are measured for each species in the community and these traits are then weighted for each species in terms of its proportional contribution to total community biomass. These weighted trait values are combined to provide a community-aggregated value for each trait, and ecosystem processes can then be related to these community-aggregated trait values at the whole-community level. For example, Vile et al. (2006) found that intersite variation in ecosystem NPP in a post-agricultural succession in France could be predicted by using community-weighted values of a key functional trait, namely relative growth rate. Similarly, community-weighted trait values for functional leaf traits, for example specific leaf area, leaf dry weight matter content, and leaf nitrogen concentration, have been shown to serve as effective predictors of total plot-level litter decomposition rate across plots of primarily herbaceous vegetation in France (Garnier et al. 2004), Sweden (Quested et al. 2007), and throughout Europe and Israel (Fortunel et al. 2009) (Fig. 3.9). Such studies point to the tractability of scaling up from traits of individual leaves to process rates at the whole-ecosystem scale.

Despite its obvious intuitive appeal, there are also two types of situation in which the mass ratio hypothesis and community-aggregated trait values may be unable to reliably predict community and ecosystem processes. The first situation involves plant species that have disproportionate effects on ecosystem processes relative to their contribution to total community biomass. For example, Peltzer et al. (2009) used plant removal experiments in a primary successional New Zealand floodplain to show that low-biomass invasive species had disproportionately large effects on soil microbial and nematode communities in relation to their contribution to total plant community biomass (Fig. 3.10). Similarly, Wardle and Zackrisson (2005) used removal experiments on islands with boreal forest to show that dwarf shrubs, which represent only a small proportion of the total plant biomass present, had much greater effects on plant litter decomposition, soil nutrients, and microbial biomass than did the trees and mosses. The second situation occurs when the dominant plant species (or those that drive ecosystem processes) have strong interactive (non-additive) effects on the ecosystem process in question. There are numerous examples of this phenomenon

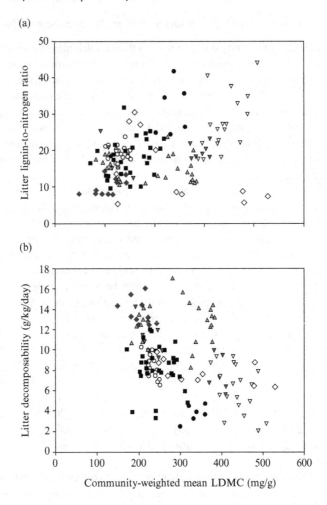

Fig. 3.9 Relationship of (a) litter lignin-to-nitrogen ratio and (b) litter decomposability at the whole-plot scale with a community-aggregated value of a key leaf trait; that is, leaf dry matter content (LDMC). Each point represents a different plot, and different symbols represent different locations throughout Europe and Israel. From Fortunel et al. (2009), with permission from the Ecological Society of America.

from litter-mixing studies, in which plant litter decomposition in multiple-species litter mixtures can differ greatly from what would be expected based on the decomposition of the components considered singly (Gartner and Cardon 2004; Hättenschwiler et al. 2005; Ball et al. 2008; but see Hoorens et al. 2003). In this light, when aggregate trait values have been used to predict plant-litter decomposition rates at the community level, a considerable proportion of the total variation remains unaccounted for (see Fortunel et al. 2009; Fig. 3.9). In any case, the mass ratio hypothesis provides a useful null model for predicting ecosystem processes in multiple-species ecosystems,

Fig. 3.10 Differences in plant community structure at the time of harvest after 4 years of plant-removal (exclusion) treatments applied to a primary floodplain succession in the eastern South Island of New Zealand: (a) all species present (i.e. no plants excluded); (b) both dominant shrub species (the native nitrogen fixer *Coriaria arborea* and the invasive *Buddleja davidii*) present but all other non-native species excluded; (c) *Buddleja* excluded; (d) *Coriaria* excluded; (e) *Buddleja* and *Coriaria* excluded; and (f) *Buddleja*, *Coriaria*, and all other non-natives excluded. The experiment showed that *Coriaria* and *Buddleja* together comprised 97% of aboveground biomass at the end of the experiment whereas other non-native species comprised 3%; despite this the other non-natives had greater effects on soil nutrients and belowground biota than did the two dominant shrub species, in contrast to the predictions of the mass ratio hypothesis. From Peltzer et al. (2009), with permission from Wiley-Blackwell.

and exploration of situations in which it does not hold may yield useful insights about the non-additive nature of multiple species effects in driving ecosystem processes.

3.3.3 Ecosystem stoichiometery

One set of plant traits that has attracted considerable recent attention involves the stoichiometric ratios of major elements, notably carbon, nitrogen, and phosphorus (Koerselman and Meuleman 1996; Elser et al. 2000; Güsewell et al. 2003). At the global scale, there is some evidence that leaf nitrogen-to-phosphorus ratios of forest plants decrease with increasing latitude, indicative of increasing limitation by nitrogen relative to phosphorus (McGroddy et al. 2004). However, there is also considerable variability in nitrogen-to-phosphorus ratios within biomes, as a result of environmental variation in soil fertility, topography, and geology (Güsewell et al. 2003; Parfitt et al. 2005). For example, Townsend et al. (2007) found a 12-fold range in the nitrogen-to-phosphorus ratio of rainforest trees in Costa Rica and Brazil. Similarly, Högberg (1992) provided data showing large differences in foliar nitrogen-to-phosphorus ratios for three functional groups of trees in savanna in Zambia and Tanzania, with nitrogen-fixing arbuscular mycorrhizal species having higher ratios than other arbuscular mycorrhizal species, and with both groups having higher ratios than ectomycorrhizal species. Important differences even exist among coexisting plant species; for example, Hättenschwiler et al. (2008) found foliar ratios of carbon to nitrogen, carbon to phosphorus, and nitrogen to phosphorus to all vary more than three-fold among 45 coexisting tree species in Amazonia. Similarly, large differences were found in the nitrogen-to-phosphorus ratio among coexisting species as well as within species (as a result of fertilizer treatments) in the mountains of the north-west Caucasus in Russia (Soudzilovskaia et al. 2007).

The belowground subsystem is, however, only indirectly influenced by foliar nutrient characteristics and more directly governed by litter nutrient properties. Therefore, the amount of nitrogen and phosphorus resorbed by the plant from foliage prior to litter fall is also an important determinant of the nutrient ratios of the resources entering the decomposer subsystem. As such, litter often contains much lower concentrations of nitrogen and phosphorus than does foliage from the same species (Killingbeck 1996; McGroddy et al. 2004). The resorption potential by plants of nitrogen and phosphorus, expressed in terms of concentration or pool sizes in foliage relative to that of litter (Killingbeck 1996), also shows a several-fold difference across environmental gradients both among and within species (Kobe et al. 2005; Richardson et al. 2005), as well as among coexisting plant species (Hättenschwiler et al. 2008; Wardle et al. 2009b). As such, and from a stoichiometric perspective, the decomposer subsystem should be influenced by the combined effects of two sets of live plant traits, i.e. the nutrient ratios of the leaves themselves, and the plant's resorption proficiency of these nutrients prior to litter fall.

There are several ways by which stoichiometric ratios of plant litter may in turn impact on the decomposer subsystem. First, they can potentially affect elemental ratios in tissues of decomposer organisms. A global literature synthesis by Cleveland and Liptzin (2007) showed that ratios of carbon to nitrogen to phosphorus in the soil microbial biomass were relatively constrained, even across large spatial scales and types of ecosystems. However, some variation was present, which reflected the ratios of these elements in the soil, and presumably in the litter input from the dominant plant species. Further, Martinson et al. (2008) synthesized the limited amount of data available on nutrient ratios for decomposer arthropods and found that there was some variation in ratios of these elements across orders. Second, these ratios in turn affect the activity of decomposer organisms, in terms of production of enzymes by the decomposers (Sinsabaugh et al. 2008), and the enzymatic release of carbon and nutrients from plant litter during decomposition. The regulation of litter decomposition and nitrogen release by litter carbon-to-nitrogen ratios has been long recognized (Swift et al. 1979), and there is increasing evidence that decomposer-driven processes can be regulated by the availability of both nitrogen and phosphorus (Hobbie and Vitousek 2000; Kaspari et al. 2008) as well as by the ratio of nitrogen to phosphorus in litter (Wardle et al. 2002; Zhou et al. 2008). Therefore, as decomposition proceeds, ratios of carbon to nitrogen to phosphorus of litter converge to values closer to that of decomposer organism tissues. Third, the relative availability of carbon, nitrogen, and phosphorus in the soil and in plant litter can serve as a powerful driver of the decomposer microbe and invertebrate abundance and community structure (Wardle et al. 2004b; Doblas-Miranda et al. 2008), the relative densities of bacteria and fungi (Güsewell and Gessner 2009), and the body size distribution of soil animals (Mulder and Elser 2009). These effects on decomposer organisms are in turn likely to have important consequences for the processes of carbon and nutrient cycling that they drive, and ultimately the functioning of the ecosystem.

3.4 Plant–soil feedbacks

It is apparent from the material presented so far in this chapter and in Chapter 2 that plant species can select for particular soil communities and processes, and that soil communities are in turn important drivers of plant growth and community composition. The consequence of this is that important feedbacks can exist between plant and soil communities, in which individual plant species select for soil communities which can then alter the success of both that plant species and of coexisting plant species (Fig. 3.11). Many studies have explicitly studied this type of feedback, especially over the past 15 years (e.g. Bever 1994; Van der Putten et al. 1993; Klironomos 2002; Bezemer et al. 2006; Kardol et al. 2007). Although plant–soil feedbacks can be inferred from a variety of approaches, a popular approach has involved the use of 'feedback' experiments, of which there are several variants, but the basic approach involves two steps. First, soil (or soil inoculum) is collected that has been cultivated by monocultures of known plant species, and which therefore contains the soil communities associated with those species. Second, each plant species is then grown in soil that has been cultivated both by its own species and by other species.

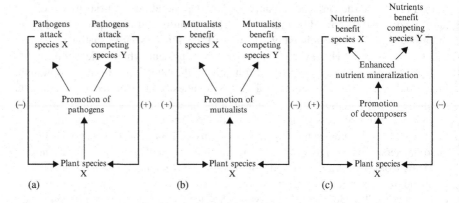

Fig. 3.11 Depiction of mechanisms by which a hypothetical plant species X can exert either positive (+) or negative (−) feedbacks with the soil community that it promotes, depending on whether the mechanism mainly affects species X directly or exerts a stronger effect on hypothetical competing plant species Y. (a) Pathogens, (b) mutualists, (c) decomposers.

This allows determination of the response of any given plant species to its own soil community relative to that associated with other species. A positive feedback for a given species exists when it performs better in its own soil than in soil from the other species; a negative feedback exists when it performs worse. A third step can subsequently be applied to identify the mechanistic basis of the feedback, through the use of soil analyses, and soil sterilization and inoculation experiments (Kulmatiski and Kardol 2008).

Feedbacks between plant and soil communities can involve both the direct and indirect pathways identified in Fig. 2.1 (Wardle et al. 2004a). Soil organisms that interact with plants via the direct pathway include both those that have mutualistic and antagonistic associations with the plants. Plant mutualists such as symbiotic nitrogen-fixing bacteria and mycorrhizal fungi usually demonstrate a positive feedback with their host plant species, and there is a vast literature on the benefits to plant species from these associations, as discussed in Chapter 2. As such, many plant species select for nitrogen-fixing bacteria that greatly enhance their success under nitrogen-poor conditions. Examples include *Lupinus* spp. on newly formed surfaces following volcanic activity on Mount St Helens (Morris and Wood 1989), *Alnus* spp. on freshly deposited gravels on floodplains (Walker 1989), and *Carmichaelia odorata* colonization on newly formed alluvial river terraces (Bellingham et al. 2001). Similarly, plant species selection for a compatible mycorrhizal fungal community in turn has very well-known benefits for the plant through increasing its access to soil nutrients and thus improving its nutrition (Smith and Read 1997). The positive effect of plant species on their mycorrhizal associates also exerts important effects at the community level. For example, there is evidence that mycorrhizal fungi often promotes coexistence of plants and plant species, apparently through reducing the intensity of competition (Grime et al. 1987; Hartnett and Wilson 1999; but see Connell and Lowman 1989). Similarly, several studies have suggested that mycorrhizal hyphal links among coexisting plants can be important in assisting the establishment of new seedlings into communities, although the evidence for this is mixed (Perry et al. 1989; Kytöviita et al. 2003; Selosse et al. 2006). While positive feedbacks between plants and their mycorrhizal associates are undoubtedly common, negative feedbacks are also theoretically possible, especially when a plant species favours mycorrhizal fungal associates which then benefit competing plant species (Bever 2003), or in fertile habitats where some plant species might even be adversely affected by mycorrhizal fungi (Francis and Read 1995).

Soil pathogens and root herbivores in contrast usually demonstrate negative feedbacks with plants. This is very well known for agroecosystems, and crop rotations are routinely performed to reduce accumulation of soil-borne pathogens under particular crop species. As mentioned in Chapter 2, negative feedbacks involving pathogens have also become increasingly studied in other types of ecosystems and shown to have important consequences for plant growth in several studies (Kulmatiski et al. 2008). For example, in a classical study in Dutch fore dunes, Van der Putten et al. (1993) used soil feedback experiments to show that accumulation of pathogenic nematodes under the pioneer plant species *Ammophila arenaria* contributed to a decline of that species and enhancement of coexisting plant species, thereby promoting vegetation succession. Similarly, studies on *Prunus serotina* in North America (Packer and Clay 2000, 2003; Reinhart et al. 2005) provide evidence for accumulation of oomycete pathogens under mature *Prunus* trees that greatly enhance *Prunus* seedling mortality and reduce seedling growth. Sterilization of soil from under the trees almost entirely eliminates seedling mortality (Packer and Clay 2000). Several other studies using soil feedback experiments in native ecosystems also provide evidence for negative feedbacks which are likely to result in a large part from soil pathogen accumulation (e.g. Bever 1994; Klironomos 2002; Kardol et al. 2006, 2007; Van der Putten et al. 2007; Petermann et al. 2008). However, plant–soil feedbacks involving root pathogens need not always be negative, and the possibility theoretically exists for plant species to benefit from promoting pathogens that reduce the fitness of competing plant species (Bever 2003).

As discussed in Chapter 2, decomposer organisms interact with plants via the indirect pathway (Fig. 2.1) whereby plants provide resources for the decomposers as plant litter and rhizosphere materials, and the decomposers in turn release nutrients from these materials for plant growth. The relationship between producers and decomposers is frequently mutualistic and therefore leads to a positive feedback. For example, as discussed earlier in this chapter, the quantity and quality of organic materials that plants input to the soil is often a powerful determinant of the abundance and activity of decomposer microflora and soil fauna. Further, several studies have shown that the composition of the decomposer community and structure of the soil food web in turn influences plant growth rates and tissue nutrient concentrations (Chapter 2; Wardle 2002), potentially altering the quantity and quality of the resources that the plant produces. For example, Setälä and Huhta (1991) found that addition of soil fauna to microcosms planted with seedlings of *Betula pendula* caused an increase in shoot nitrogen concentration of more than 2.3-fold; the improved quality of litter subsequently produced by these seedlings would likely to be in turn beneficial for the soil fauna. While positive feedbacks of this type are likely to be important whenever a particular plant species dominates the soil under it, negative feedbacks are also possible, especially when the plant promotes belowground conditions that then promote competing plant species. For example, in moist heath lands in The Netherlands, the dwarf shrub *Erica tetralix* dominates during early succession when nutrient availability is low, but promotes accumulation of organic matter and faster net nitrogen mineralization rates, in turn encouraging replacement of *E. tetralix* by the grass *Molinia caerulea* which has a higher potential growth rate (Berendse 1998). This, and the rapid rate of mineralization of nitrogen from the new litter produced by *M. caerulea*, sets a feedback in motion that disadvantages *E. tetralix* and encourages further growth of *M. caerulea* (Fig. 3.2).

With regard to the indirect pathway, an intriguing possibility is that plant communities or species may select for decomposer communities that preferentially mineralize their own litter (Wardle 2002). Despite the tractability of studying this question, it has attracted surprisingly little attention to date. At the across-ecosystem scale, Hunt et al. (1988) collected litter from the dominant plant species of an alpine meadow, a pine forest, and a prairie in Colorado and decomposed all three litters in all three habitats. In all cases, litter decomposed faster than expected in its own habitat relative to when placed in either of the other habitats, suggesting that the vegetation in each habitat selects for decomposers that break down its own litter (Fig. 3.12). Conversely, Wardle et al. (2003a) collected litter of dominant species from each of 30 islands in northern Sweden that differed greatly in vegetation composition, and decomposed each litter on its own island and on other islands; here, placing litter in its site of origin had no effect on decomposition. At the within-ecosystem scale, Vivanco and Austin (2008) collected litter from each of three *Nothofagus* species in Patagonia, Argentina, and decomposed each litter under each of the three species. In all cases they found that litter decomposed fastest when placed under its own species, suggesting a capacity for even closely related coexisting tree species to select for decomposers that preferentially break down their own litter (Fig. 3.12). Similarly, Strickland et al. (2009) found in an incubation study that recalcitrant tree litter decomposed faster in the presence of microbial communities from forest ecosystems than from herbaceous ecosystems, and Ayres et al. (2009) found that litters of three dominant tree species of high-elevation forest in Colorado, USA, decomposed faster when they were incubated with their own soil biota. Conversely, Ayres et al. (2006) performed a similar experiment involving three tree species in northern England and failed to find such an effect. When a plant species or community of plants selects for a soil community that preferentially breaks down its own litter, it may potentially derive a competitive advantage through having preferential access to nutrients released from that litter. However, explicit tests of this have yet to be performed.

As is apparent through Fig. 3.11 and the above discussion, there are several mechanisms through which plant–soil feedbacks can occur, and these can be either positive or negative. This leads to the question of what the net effect of plant–soil feedbacks is on plant growth, and this issue remains poorly understood. In a recent

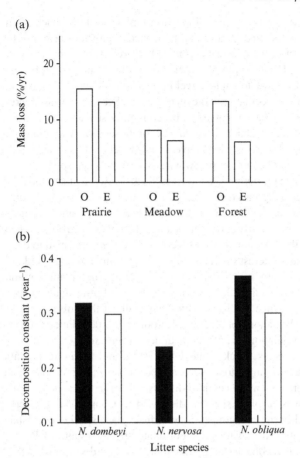

Fig. 3.12 Evidence that plant communities and species can select for decomposer communities that preferentially decompose their own litter. (a) Results of a litter reciprocal transplant experiment in which litter was taken from the dominant plant species in a prairie, a meadow, and a pine forest, with each litter then being decomposed in all three habitats. For each habitat, O is the decomposition rate observed when litter collected from that habitat was decomposed in the same habitat, and E is the expected decomposition rate for litter from that habitat based on its decomposition rate in the other two habitats. Overall the O and E values differ at $P = 0.001$. Based on analysis of data presented by Hunt et al. (1988). (b) Litter mass loss (expressed as decomposition constant) of litter of each of three *Nothofagus* species decomposed under its own species (black bars) and under the other species (white bars). The overall effect of location on decomposition is significant at $P = 0.001$. From Vivanco and Austin (2008), with permission from Wiley-Blackwell.

meta-analysis of 45 published studies, Kulmatiski et al. (2008) showed that net plant–soil feedback effects were more often negative than positive, and suggested that the high abundance of negative feedbacks could have important consequences for structuring plant communities. However, such analyses need to be treated with caution for variety of reasons. First, most plant–soil feedback studies have been performed in glasshouse studies that may encourage short-term proliferation of plant antagonists relative to those that promote plant growth, especially when compared with field situations. For example, short-term glasshouse studies may be unable to provide conditions conducive to rapid development of soil macrofaunal communities and mycorrhizal networks that promote plant growth. Second, the vast majority of feedback studies done to date have involved herbaceous early-successional plant species, and it is likely that positive feedbacks involving decomposer biota would be more important in later successional systems which have a well-developed litter layer and are dominated by woody plants. The indirect pathway between plants and soil biota (Fig. 2.1) is likely to be important in reinforcing positive feedbacks between plants and decomposer biota, especially if plant species can select for biota that preferentially decompose their own litter. Much remains unknown about the net direction of plant–soil feedback effects on plant growth, but it is likely that positive feedbacks are far more important than suggested by the current balance of feedback studies performed to date.

Despite recent interest in plant–soil feedback, much remains unknown about how these feedbacks are in turn important in structuring plant communities in real ecosystems. Plant–soil feedbacks involving soil-borne antagonists may result in enhanced coexistence of plant species especially if they disadvantage the dominant plant species (Bonanomi et al. 2005; Kardol et al. 2006; Petermann et al. 2008). For example, accumulation of pathogens under *Prunus serotina* leads to loss of dominance by that species (Packer and Clay 2000) and therefore potentially a greater abundance of its competitors. Such a mechanism may promote plant diversity, as well as accelerate plant succession (discussed later in this chapter) and encourage the success of invasive plant species (discussed in Chapter 5). Conversely, feedbacks involving soil-borne antagonists may reduce plant species coexistence when they disproportionately disadvantage subordinate species (Van der Putten 2005). Similarly, plant–soil feedbacks involving mutualists or decomposers that benefit plant growth could conceivably reduce species coexistence when dominant species are favoured, or increase coexistence when subordinate species are favoured (Van der Putten 2005). Indeed, there is theoretical evidence that plant–soil feedbacks involving decomposition and mineralization processes could maintain plant species coexistence and diversity, especially if each plant species in the community regulates the supply of nutrients from the soil in its immediate vicinity (Huston and De Angelis 1994).

3.5 Succession and disturbance

The influence of plant species on belowground biota and processes, and the role of plant–soil feedbacks, is highly apparent from studies on vegetation succession. Here, plant-species composition changes over time on either newly formed surfaces (primary succession) or following disturbance events (secondary succession). The extent and rate of this change is determined in a large part by others, notably facilitation (where one or several species improve conditions for the establishment of one or

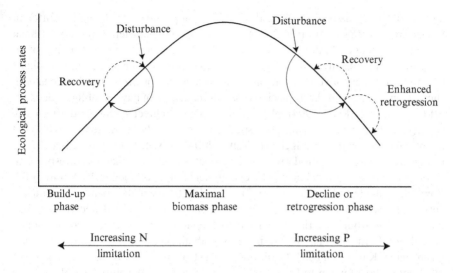

Fig. 3.13 The build-up and decline or retrogressive phases of succession. The solid curves represent primary succession while the dashed curves represent secondary succession. Adapted from Walker et al. (2001), with permission from Wiley-Blackwell.

more species), competitive, or interference interactions, and tolerance by species of the effects of other species (see Connell and Slatyer 1977; Pickett et al. 1987; Walker and del Moral 2003). These interactions and species-replacement patterns are often driven by changes in soil resource availability which itself changes during succession. Succession and ecosystem development consists of the build-up phase during which plant biomass accumulates, the maximal biomass phase, and (given sufficient time) the retrogressive or decline phase (Walker et al. 2001) (Fig. 3.13). The build-up phase is characterized by the increased availability of nutrients (often nitrogen) while the decline phase is characterized by declining availability of nutrients, especially phosphorus (Vitousek 2004; Wardle et al. 2004b).

3.5.1 The build-up phase of succession

Upon the formation of new surfaces during primary succession, there is colonization by both plant and soil communities. Although soil communities can establish independently of plants (Bardgett et al. 2007a; Nemergut et al. 2007; Schmidt et al. 2008), they mainly develop in association with the plant community. This applies to soil biotic components that interact with plants both via the direct and indirect pathways (Fig. 2.1). With regard to the direct pathway, communities of colonizing plant species often include those that form symbiotic associations with nitrogen-fixing bacteria, and which are well known to contribute greatly to the accumulation of ecosystem nitrogen during the build-up phase (Morris and Wood 1989; Bellingham et al. 2001). During this phase the mycorrhizal fungal community also develops, with colonizing plant species preferentially selecting for strains of fungi that are characteristic of pioneer communities (Ishida et al. 2008) and with different plant species favouring different fungal species (Pueschel et al. 2008). Over time the character of the mycorrhizal community changes, in that while during early succession a proportion of the colonizing plants are non-mycorrhizal, in mid-succession the dominant herbaceous plants tend to have a facultative requirement for arbuscular mycorrhizal fungi, while later in succession the tree and shrub species which dominate often have an obligate need for ectomycorrhizae (Read 1994). With regard to the indirect pathway, during early succession different colonizing plant species are frequently associated with different microbial communities (Bardgett and Walker 2004). As succession proceeds and there is a greater amount of plant-derived organic matter in the soil, there is corresponding colonization of soil invertebrates and development of the soil food web (Wardle 2002; Bardgett et al. 2005; Neutel et al. 2007). Such development and changes in components of the soil food web during vegetation succession are well documented and include shifts from bacterial-dominated to fungal-dominated food webs (Ohtonen et al. 1999; Bardgett et al. 2007a), increasing abundance of detritivorous meso- and macrofauna (Kaufmann 2001; Doblas-Miranda et al. 2008), and enhanced densities of invertebrate taxa known to be *K*- rather than *r*-selected (Wasilweska 1994). These shifts are also accompanied by increasing food-chain length and complexity of the soil food web (Wardle 2002; Hodkinson et al. 2004; Neutel et al. 2007).

Concomitant with changes in belowground communities that occur during vegetation succession are changes in the rates of ecosystem processes driven by the soil biota. Build-up of nitrogen during the initial stages of succession often results in the microbial community shifting from being nitrogen-limited to carbon-limited, resulting in a reduced loss of carbon through respiration per unit amount of microbial biomass (Insam and Haselwandter 1989). This contributes to the rapid accumulation of organic matter during primary succession, and which can be especially apparent when plant species capable of significant nitrogen fixation are dominating (Halvorson et al. 1991). As nitrogen availability increases and soil food webs develop, there is frequently a greater rate of decomposition of plant litter (e.g. Crews et al. 1995) and supply rates of nutrients, notably nitrogen, from the soil (Chapin et al. 1994). These changes in soil processes can be driven by shifts in plant community composition succession, especially when early successional species have large differences in functional traits to species that dominate later (Berendse 1998; Grime 2001). However, as succession continues and the maximal biomass phase is approached, a greater proportion of the ecosystem's available nutrients become immobilized in biological tissues. This contributes to a decline in plant litter quality, activities of soil organisms, and rates of mineralization and supply of available nutrients from the soil (Brais et al. 1995; Hättenschwiler and Vitousek 2000; DeLuca et al. 2002a). At this stage, nutrients are cycled more conservatively, and both aboveground and belowground communities can be colimited by nitrogen and phosphorus (Vitousek 2004). The eventual consequence of this limitation by nutrients is ecosystem retrogression, which is discussed in section 3.5.2.

Plant species and their functional attributes may also be important drivers of the disturbance regime that the ecosystem is subjected to (McIntyre et al. 1999), and disturbances during the build-up and maximal biomass phases of succession frequently reverses successional development (Fig. 3.13). Major disturbances have important effects on both the quantity and quality of the resources input to the belowground subsystem, and this can influence both the soil biota and the processes that they drive. For example, the formation of canopy gaps, a common consequence of forest disturbance, can alter (and often impair) soil microbial biomass levels and rates of litter decomposition and nutrient mineralization, presumably through a reduction of plant-derived inputs (Bauhus and Barthel 1995; Zhang and Zak 1998; Sariyildiz 2008). Forest damage through storms, notably hurricanes, can potentially have far-ranging effects on belowground processes and organisms, through causing immediate inputs of tree green-fall and mortality of fine roots (Herbert et al. 1999) and longer-term reductions in the quantity and quality of litter inputs during stand recovery (Hunter and Forkner 1999; Lugo 2008). Further, fire is a major contributor to ecosystem disturbance regimes worldwide, and can exert a multitude of effects on belowground organisms and processes (reviewed by Certini 2005). The most obvious and widespread of these effects is to impair soil biota through causing loss of organic matter (and, in forests, the humus layer), leading to a loss of habitat and resources for soil organisms (Certini 2005). However, positive effects can also arise, for example through fire encouraging domination by plant species that produce litter of superior quality for decomposers (Nilsson and Wardle 2005), and the production of charcoal that can greatly promote both soil microbial growth (Zackrisson et al. 1996) and the belowground processes that they drive (Wardle et al. 2008b).

3.5.2 Ecosystem retrogression

In the long-term absence of catastrophic disturbance, characteristically in the order of millennia, a decline or retrogressive phase often occurs which is associated with a distinct reduction of plant biomass (Crews et al. 1995; Richardson et al. 2004; Wardle et al. 2004b). This decline is accompanied by reductions in NPP, and shifts in the respiratory and photosynthetic characteristics of the dominant vegetation (Turnbull et al. 2005; Whitehead et al. 2005). Ecosystem retrogression has been characterized for several long-term chronosequences around the world, including those initiated by volcanic activity in Hawai'i (Crews et al. 1995; Vitousek 2004), isostatic land uplift from the ocean in New Zealand (Ward 1988; Coomes et al. 2005), retreat of glacial ice in New Zealand (Walker and Syers 1976; Richardson et al. 2004), wildfire in northern Sweden (Wardle et al. 1997b, 2003a), and aeolian sand movement in Queensland, Australia (Thompson 1981; Walker et al. 2001) (Fig. 3.14). This decline phase is linked to long-term reductions in the availability of nutrients, notably phosphorus (Walker and Syers 1976; Vitousek 2004), although increased soil wetness through reduced soil drainage in old-growth forest could also sometimes contribute (Coomes et al. 2005). As such, it has been suggested that during retrogression, phosphorus becomes more limiting relative to nitrogen. This is because phosphorus is non-renewable and eventually becomes depleted from the ecosystem or bound in increasingly unavailable forms while nitrogen can be renewed by biological nitrogen fixation (Wardle et al. 2004b). In this light, ecologically

(a) (b)

(c) (d)

Fig. 3.14 Reduction in forest plant biomass resulting from nutrient limitation due to ecosystem retrogression (see Wardle et al. 2004b). (a, b) Forest at the maximal biomass phase (a) and long-term retrogressive phase (b) for the Waitutu chronosequence in southern New Zealand (Ward 1988), a sequence spanning 600 000 years following uplift of land from the ocean; (c, d) forest at the maximal biomass phase (c) and long-term retrogressive phase (d) for the Cooloola sequence in Queensland, Australia, a sequence spanning 600 000 years following the formation of sand dunes through aeolian sand movement. Images (a, b) by D.A. Wardle and (c, d) by R.D. Bardgett.

significant rates of non-symbiotic biological nitrogen fixation have been found for ecosystems undergoing retrogression (Lagerström et al. 2007; Menge and Hedin 2009; Gundale et al. 2010). Consistent with this, there is evidence of a consistent increase in nitrogen-to-phosphorus ratios of forest humus (and often of litter and foliage) when long-term chronosequences enter a retrogressive phase, with ratios tending towards values that are indicative of ecosystem limitation by phosphorus rather than nitrogen (Wardle et al. 2004b) (Fig. 3.15).

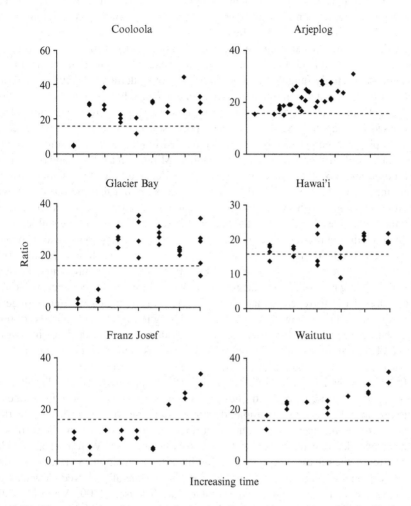

Fig. 3.15 Nitrogen-to-phosphorus ratios for humus substrate (or uppermost mineral soil substrate in the case of Cooloola) of each of six widely contrasting long-term chronosequences that have undergone ecosystem retrogression, in relation to increasing time since the catastrophic disturbance that initiated the chronosequence (expressed as a rank scale). The chronosequences are: Cooloola, Queensland, Australia (caused by aeolian formation of sand dunes; spans 600 000 years); Glacier Bay, Alaska (caused by retreat of glacial ice; spans 14 000 years); Franz Josef, New Zealand (caused by retreat of glacial ice; spans 120 000 years); Arjeplog island sequence, northern Sweden (caused by time since last fire, spans 6000 years); Hawai'i island sequence (surfaces formed by volcanic activity, spans 4 100 000 years); and Waitutu, southern New Zealand (surfaces formed by uplift from the ocean, spans 600 000 years). The Redfield ratio (nitrogen:phosphorus = 16), above which phosphorus is believed to become limiting relative to nitrogen (Redfield 1958), is shown for comparative purposes in each panel as a dashed line. From Wardle et al. (2004b), with permission from the American Association for the Advancement of Science.

The reduced availability of nutrients and plant productivity in ecosystems that have undergone retrogression has knock-on consequences for the key components of the soil food web. For example, Wardle et al. (2004b) found while comparing six long-term chronosequences worldwide that retrogression had generally negative consequences for the soil microbial biomass, and often caused the microbial community to become increasingly fungal- (as opposed to bacterial-) dominated. This is consistent with fungal-dominated communities being better adapted for nutrient-poor conditions and maintaining nutrient cycles that are more conservative in retaining nutrients, as discussed in Chapter 2. Studies on decomposer soil fauna also reveal evidence for multitrophic consequences of ecosystem retrogression. For example, studies on both the Franz Josef Glacier chronosequence (Doblas-Miranda et al. 2008) and the Waitutu chronosequence (Williamson et al. 2005) of New Zealand found that all major trophic groupings of nematodes declined during retrogression, but that the ratio of fungal-feeding nematodes to bacterial-feeding nematodes increased. This further suggests that the fungal-based, rather than the bacterial-based, component of the soil food web is favoured during retrogression, consistent with recent studies pointing to increasing phosphorus limitation favouring domination by fungi (Güsewell and Gessner 2009) and their consumers (Mulder and Elser 2009). Furthermore, measurements of a range of soil-dwelling macroinvertebrate groups across the Franz Josef sequence by Doblas-Miranda et al. (2008) revealed that while different taxa dominated at different stages of the sequence, all taxa showed strong declines during retrogression regardless of body size or trophic position. It is therefore apparent that the strong declines in nutrient availability that occur in the order of millennia as retrogression proceeds exert pervasive and strong bottom-up controls of the entire soil food web.

Diminished nutrient availability and impaired decomposer activity during retrogression in turn leads to reduced rates of decomposer processes and release rates of nutrients from the soil. For example, decomposition rates of standardized substrates across chronosequences show reduced rates of decomposition when placed in sites that have undergone retrogression (e.g. Crews et al. 1995; Wardle et al. 2003a). Furthermore, the release of nutrients during litter decomposition and rates of nutrient fluxes driven by decomposer organisms are both generally slowed down as the retrogressive phase is approached (Hobbie and Vitousek 2000; Vitousek 2004; Wardle et al. 2004b). Retardation of decomposer processes during retrogression is further enforced by the response of the dominant plant species in the community to retrogression. During retrogression, plants resorb a greater proportion of their nutrients from their foliage prior to litter fall (Richardson et al. 2005) and the quality of litter entering the decomposer subsystem is therefore poorer (Crews et al. 1995; Wardle et al. 2009a). This litter quality reduction occurs both across species (i.e. species that dominate in retrogressive stages produce poorer-quality litter) and within species (i.e. species that dominate across multiple chronosequence stages produce poorer-quality litter during retrogression) (Richardson et al. 2005; Wardle et al. 2009a). As such, reciprocal litter-transplant experiments, in which litters from both non-retrogressive and retrogressive chronosequence sites are each decomposed in both types of site, show that decomposer processes are impaired by retrogressive stands both through supporting a less-active microbial community and through dominant species producing a lower quality of litter (Crews et al. 1995; Wardle et al. 2003a). Impairment of decomposer processes and litter quality during retrogression may be partly reversible by fertilization of the nutrient that is most limiting during retrogression; that is, phosphorus (Hobbie and Vitousek 2000).

3.5.3 Succession and plant–soil feedbacks

The plant–soil feedback mechanisms described earlier in this chapter (Fig. 3.11) can serve as powerful determinants of species replacement and therefore vegetation succession. Both positive and negative feedbacks can either maintain a given species and thus retard succession, or encourage succession by encouraging replacement of a given species by new species. Feedbacks during succession involve both the direct and indirect pathways (Fig. 2.1), although the direct pathway has been by far the most frequently studied.

Plant mutualists that exert a positive feedback with their host plant species are well known as important drivers of succession. For example, many early successional plant species form mutualistic associations with symbiotic nitrogen-fixing bacteria, and while in the short term this benefits the host plant, in the longer term this leads to improved soil nitrogen availability and therefore facilitation of later successional plant species that then dominate. This is well known to contribute to species replacement, with the earlier successional species being suppressed and excluded (Chapin et al. 1994; Fastie 1995; Bellingham et al. 2001). Mutualisms between plant species and mycorrhizal fungi may either enhance or retard succession. On one hand, pioneer plant species may promote build-up of fungal species that can then serve as mutualists with other later-successional plant species, leading to facilitation of these new species and ultimately species replacement (Nara and Hogetsu 2004; Nara 2006). On the other hand, mycorrhizal networks hosted by dominant plant species can also encourage recruitment of seedlings of the same species (Dickie et al. 2002; Teste and Simard 2008), potentially allowing that species to maintain dominance and thus impair species replacement.

There is increasing recognition that negative plant–soil feedbacks can encourage successional replacement of plant species, and several studies have tested this using the feedback experimental approach described earlier in this chapter. Such studies often find a given plant species to promote its soil-borne pathogens, leading to accelerated succession by later successional species that are less susceptible to the pathogens (Van der Putten et al. 1993; Kulmatiski et al. 2008). For example, Kardol et al. (2006) used a feedback experiment involving plant species that dominate at different stages of grassland succession to show that early-successional plant species showed a negative plant–soil feedback while mid-successional species showed a neutral feedback and late-successional species showed a positive feedback (Fig. 3.16). This is indicative of a capability of earlier successional species to influence the soil biota in such a way as to hasten successional replacement of species. Further, De Deyn et al. (2003) used an experimental approach to show that

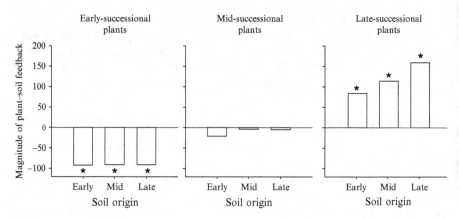

Fig. 3.16 Results of a soil feedback experiment in which 12 grassland plant species (four species characteristic of each of early, middle, and late successional stages) were all grown in sterilized soils collected from early-, mid-, and late-successional fields (first growth period), the vegetation then harvested, then the soil replanted again with the same 12 species (second growth period). For each bar, the soil feedback effect is the percentage change in plant growth at the second growth period relative to that for the first growth period. * indicates that feedback effect is significantly different to 0 at $P = 0.05$. Derived from Kardol et al. (2006).

soil faunal communities extracted from soils differing in successional stage consistently selectively impaired early successional plant species, thus encouraging species characteristic of later successional stages. They attributed these effects to root-feeding parasitic nematodes in the soil, suggesting that soil-borne antagonists can be important in encouraging the replacement of early successional plant species. Most studies of this type have involved grassland successions, and whether such mechanisms can also speed up succession in other ecosystem types such as forests remains largely unexplored.

Fewer studies have considered the role of feedbacks between plants and the decomposer subsystem (i.e. the indirect pathway in Fig. 2.1) in influencing succession. This is despite the ubiquity of these feedbacks in terrestrial ecosystems, and their likely importance in influencing plant species replacement through affecting the availability of soil nutrients. These feedbacks are driven in the first instance by plant litter. For example, the study by Berendse (1998) discussed earlier provides evidence that the dwarf shrub *Erica tetralix* promotes organic-matter build-up and increased rates of nitrogen mineralization; this encourages domination by the grass *Molinia caerulea* that in turn replaces *E. tetralix*. Conversely, if later successional plant species produce litter that impairs decomposer activity, this could conceivably impair species replacement. For example, maintenance of *Pinus muricata* dominance in nutrient-poor sites in California could in part result from the trees producing polyphenol-rich litter that is unfavourable for decomposer biota (or at least for those decomposers that rapidly mineralize litter nutrients), leading to reduced mineralization of nitrogen in the litter and therefore the availability of nitrogen to competing plant species (Northup et al. 1995). Despite the well-known importance of plant–decomposer feedbacks in determining plant growth, the consequences of these for plant succession remain little understood.

3.6 Indirect belowground effects of global change via vegetation

The Earth and its ecosystems are undergoing substantial and increasing modification as a result of human-induced global change (Millennium Ecosystem Assessment 2005). The principal global change drivers, namely climate change (including atmospheric carbon dioxide enrichment), increased anthropogenic inputs of nitrogen, invasion of exotic species, and land-use change, all act simultaneously to greatly modify both the composition and functioning of terrestrial ecosystems. As discussed in Chapter 2, the response of belowground communities and the processes and feedbacks that they drive to global change phenomena operate via both direct and indirect mechanisms. Direct effects of global change on soil biological communities and the processes that they drive were discussed in Chapter 2, largely in the context of carbon dynamics and climate change. Here, we consider how effects of global change phenomena on soil communities and the ecosystem-level functions and feedbacks that they drive are affected indirectly by changes in the productivity and shifts in composition of plant communities. Indeed, we argue that an understanding of the effects of these global change phenomena on terrestrial ecosystems, and also their potential mitigation for instance through soil carbon sequestration, requires explicit consideration of linkages between aboveground and belowground subsystems (Wolters et al. 2000; Wardle et al. 2004a; Tylianakis et al. 2008; Bardgett et al. 2008; Kardol et al. 2010). We illustrate this using climate change and nitrogen enrichment as examples, whereas effects of invasive species will be considered in Chapter 5. We do not provide an exhaustive review of the indirect impacts of these global change phenomena on soil biological properties and ecosystem-level feedbacks, which is beyond the scope of this book; rather, we use selected examples to illustrate key mechanisms at play.

3.6.1 Indirect belowground effects of climate change

Rising atmospheric concentrations of carbon dioxide represents the clearest and most documented signal of human alteration of the Earth-system (IPCC 2007); since pre-industrial years (pre-1800), atmospheric carbon dioxide concentrations have risen by some 30% from 280 to 365 ppm, due mostly to the burning of fossil fuels and clearance of natural lands for agriculture, and, depending on future trends, they could reach 540–970 ppm by the end of this century. This rise in atmospheric carbon dioxide has already substantially affected the world's climate, and the consensus is that it will force further change in coming years, leading to a rise in average temperatures and a greater occurrence of extreme weather events (IPCC 2007).

These changes in climate have the potential to have marked indirect effects on soil biological communities and their activity, and therefore feedback to climate change, by influencing plant growth and vegetation composition, and therefore the allocation of carbon belowground. Such plant-mediated indirect effects of climate change on belowground communities can operate via a variety of mechanisms, but can broadly be separated into two (Bardgett et al. 2008; Kardol et al. 2010). First, rising atmospheric concentrations of carbon dioxide can indirectly impact on soil microbes and their predators via increased plant photosynthesis and transfer of photosynthate carbon to fine roots and soil biota, and through altering plant litter quality. Second, in the longer term indirect effects of climate change on soil organisms and their activity operate via shifts in the functional composition and diversity of vegetation. We now discuss these mechanisms in turn.

It is well established that elevated carbon dioxide increases plant photosynthesis and growth, especially under nutrient-rich conditions (Curtis and Wang 1998; De Graaff et al. 2006), and this in turn increases the flux of carbon to roots, their symbionts, and heterotrophic microbes via root exudation of easily degradable sugars, organic acids, and amino acids (Zak et al. 1993; Diaz et al. 1993; Iversen et al. 2008). The consequences of increased carbon flux from roots to soil for microbial communities and carbon exchange are difficult to predict, because these effects are influenced by a range of factors including plant species identity and traits, soil food web interactions, and soil fertility. However, we can identify some possible outcomes for soil organisms and carbon exchange. For instance, increased belowground allocation of carbon to roots, and its transfer from roots to soil, can stimulate microbial abundance and activity, enhancing the mineralization of both recent and old soil organic carbon (i.e. priming), and hence carbon loss from soil (Körner and Arnone 1992; Zak et al. 1993; Freeman et al. 2004; Heath et al. 2005; Fontaine and Barot 2005; Kuzyakov 2006; Dijkstra and Cheng 2007; Jackson et al. 2009). Indeed, several free-air carbon dioxide enrichment (FACE) experiments have shown that elevated carbon dioxide can lead to substantial increases in root biomass and soil respiration (e.g. Körner and Arnone 1992; Hungate et al. 1997; Norby et al. 2004; Pritchard et al. 2008; Jackson et al. 2009), and that, in general, belowground responses to elevated carbon dioxide are often greater than aboveground responses in the same systems (Jackson et al. 2009). As highlighted in the previous chapter, mycorrhizal fungi also act as a significant sink for recent plant photosynthate (Högberg and Read 2006), and increased supply of photosynthate carbon to these fungi under elevated carbon dioxide can stimulate their growth (Rillig et al. 2000), especially under conditions of low nutrient availability (Klironomos et al. 1997; Staddon et al. 2004). This can in turn feed back positively on plant growth and hence carbon assimilation. Changes in the growth of mycorrhizal fungi in response to elevated carbon dioxide concentrations can also potentially influence soil and ecosystem-level carbon dynamics by controlling the release of carbon to the soil microbial community (Högberg and Read 2006), and by enhancing the stabilization of soil organic carbon through promotion of soil aggregation (Rillig and Mummey 2006; Six et al. 2006; De Deyn et al. 2008; Wilson et al. 2009).

Conversely, the stimulation of microbial biomass as a result of enhanced root carbon supply can lead to immobilization of soil nitrogen, thereby limiting nitrogen availability to plants (Diaz et al. 1993; De Graaff et al. 2007), creating a negative feedback that constrains future increases in plant growth and carbon transfer to soil. Also, increased plant-microbial competition for nitrogen, which as discussed in Chapter 2 can lead to reduced soil nitrogen availability and microbial activity, can suppress microbial decomposition, leading to increased ecosystem carbon accumulation (Hu et al. 2001). A key mechanism here is likely to be reduced substrate quality, in that elevated carbon dioxide concentrations reduce nitrogen concentrations of plant tissue, resulting in larger carbon-to-nitrogen ratios of litter inputs entering soil (Curtis and Wang 1998; Coûteaux et al. 1999). This in turn leads to nitrogen limitation of the microbial biomass, resulting in nitrogen immobilization and a reduction in the supply of nitrogen to plants, thereby constraining the fertilizing effects of elevated carbon dioxide on plants (Curtis and Wang 1998; Coûteaux et al. 1999). Interestingly, a recent meta-analysis by De Graaff et al. (2006) of data from 117 studies revealed that on average elevated carbon dioxide stimulated gross nitrogen immobilization and microbial nitrogen by 22 and 5.8% respectively, and increased microbial biomass and soil respiration by 7.1 and 17.7%, respectively. However, this meta-analysis revealed that elevated carbon dioxide stimulated overall plant biomass substantially, especially when additional nitrogen was added, thereby outweighing carbon loss through microbial respiration and causing a net increase in soil carbon of 1.2% per year. Other studies have shown that in unfertilized ecosystems microbial nitrogen immobilization enhances acclimation of plant growth to elevated carbon dioxide in the long-term (Finzi et al. 2002; De Graaff et al. 2006). As such, increased soil carbon input and sequestration under elevated carbon dioxide can only be sustained in the long term when additional nutrients are supplied, for instance from fertilizers or atmospheric nitrogen pollution (Finzi et al. 2002; De Graaff et al. 2006). The decline in nitrogen availability under elevated carbon dioxide is central to the progressive nitrogen limitation concept, whereby in the absence of new nitrogen inputs (e.g. from fertilizers or nitrogen pollution) the availability of nitrogen declines over time at elevated carbon dioxide relative to its availability at ambient carbon dioxide levels (Luo et al. 2004).

In the longer term (decades to centuries), climate change indirectly influences the belowground subsystem via shifts in the functional composition and diversity of vegetation. It is well established that global warming and altered precipitation regimes can alter the distribution of plant species and functional groups at both local and global scales (Prentice et al. 1992; Woodward et al. 2004). For instance, recent changes in precipitation patterns have markedly affected vegetation composition in tropical rainforest (Engelbrecht et al. 2007) and African savanna (Sankaran et al. 2005), and warming is leading to rapid replacement of Canadian tundra by boreal forest (Danby and Hik 2007), declines in cryptogam abundance (mosses and lichens) (Cornelissen et al. 2001b; Walker et al. 2006), rapid losses of plant diversity in salt marshes (Geden and Bertness 2009), pan-Arctic shrub encroachment in Arctic tundra (Sturm et al. 2001; Epstein et al. 2004; Tape et al. 2006; Wookey et al. 2009), and widespread increases in tree mortality rates in the western USA (Van Mantgem et al. 2009). Likewise, elevated atmospheric carbon dioxide has been shown to cause shifts in the vegetation composition of grassland (Körner et al. 1997; Niklaus et al. 2001), in some cases promoting the growth of legumes with consequences for soil nitrogen cycling (Newton et al. 1995; Hanley et al. 2004; Ross et al. 2004). Such shifts in vegetation composition influence the uptake of carbon dioxide by photo-synthesis (Reich et al. 2001; Ward et al. 2009) and hence the transfer of photo-synthate to soil, and also modify the soil physical environment for example by changes in root architecture and rooting depth (Jackson et al. 1996). However, the main route by which climate-driven shifts in vegetation composition influence soil organisms over longer timescales is via changes in the quality and quantity of organic matter entering the soil as plant litter, which as discussed previously in this chapter act as primary drivers of the biological properties of soil.

One well-documented example of the belowground consequences of shifts in vegetation composition resulting from global warming comes from the Arctic (reviewed by Wookey et al. 2009) (Fig. 3.17). It is well known that high-latitude and alpine ecosystems will be especially responsive to rapid climatic warming in coming decades (Overpeck et al. 1997; Serreze et al. 2000; Euskirchen et al. 2006), and a consequence of this is the expansion of dwarf-shrubs, such as *Betula nana* (Hobbie et al. 1999; Sturm *et al.* 2001; Epstein *et al.* 2004; Jónsdóttir et al. 2005; Tape *et al.* 2006). Dwarf-shrubs produce leaf litter and woody material of much poorer quality, and hence lower decomposability, than the graminoids and forbs that they replace (Cornelissen 1996; Quested et al. 2003; Dorrepaal et al. 2005). The conse-quences of this for carbon-cycle feedbacks in the Arctic are uncertain, but it was recently suggested by Cornelissen et al. (2007) that warming-induced expansion of shrubs with recalcitrant leaf litter across cold biomes would constitute a negative feedback to global warming, potentially countering the direct warming enhancement of decomposition and hence carbon loss from Arctic soils. This conclusion was based on the results of a large-scale quantitative analysis of the major climate-change-related drivers of litter decomposition rates in cold northern biomes worldwide, which showed that both plant growth form and direct climate effects were the principal drivers of litter decomposition rate (Fig. 3.18). Specifically, Cornelissen et al. (2007) found that litter of the predominant plant species in 33 experiments in Arctic and alpine sites in the Northern Hemisphere decomposed on average 42% faster in warmer than colder soils (representing 3.7°C higher soil temperatures sustained over a longer growing season), and on average grass and sedge litters were decomposed 40% faster than shrub litter (Fig. 3.19). Interestingly, these finding are consistent with those of the previously mentioned study of Cornwell et al. (2008) who analysed mass-loss data for 818 plant species from 66 decomposition experiments on six continents and found that the magnitude of species-driven differences in decomposition are much larger than previously thought and greater than climate-driven variation.

In drawing conclusions on how increased shrub expansion will affect large-scale soil carbon stocks in the Arctic, other feedbacks on the carbon cycle resulting from climate warming and increasing shrub encroachment need to be considered

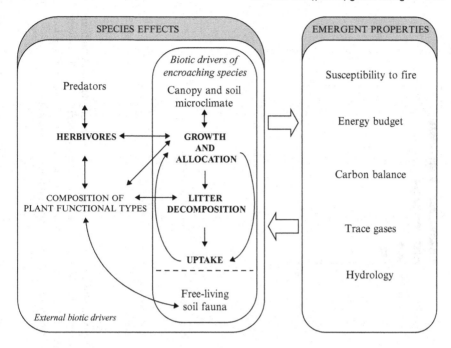

Fig. 3.17 Conceptual framework for considering the internal linkages and feedbacks associated with a shift in plant functional types (PFTs) on ecosystem-level properties in the Arctic, from Wookey et al. (2009). Changes in PFTs (left-hand box, Species Effects) will feed forward to affect consumers (both herbivores and decomposers) through changes in the quality, quantity, and timing of food availability, as well as changing habitat physical characteristics, for example shelter for small herbivores (e.g. invertebrates, small rodents, and birds). Where feedbacks occur between two components of the ecosystem they are indicated by bidirectional arrows. In the decomposition subsystem feedbacks are exerted through potential shifts in the magnitude, timing, and location (e.g. in the rhizosphere versus at the soil surface or the interface of dead roots) of nutrient mineralization, alterations in mycorrhizal associations, and competition between decomposers and autotrophs for nutrients. Changes in the composition of PFTs, and in plant growth and allocation patterns, interface directly with the Earth-system by influencing ecological properties and processes that affect energy and water exchange, soil fertility and physical structure, or susceptibility to fire, while whole-ecosystem metabolism across trophic levels affects carbon fluxes (both gaseous and aqueous). From Wookey et al. (2009), with permission from Wiley-Blackwell.

(Cornelissen et al. 2007). First, it is unclear how the negative effect of shrub expansion on decomposition compares in magnitude with changes in decomposition driven by changes in snow cover, which could be either positive or negative (Chapin et al. 2005; Sturm et al. 2005; Weintraub and Schimel 2005; Wookey et al. 2009). Second, because shrubs tend to have shallower root systems than do graminoids, decomposition of their root litter might be enhanced by warming of upper soil horizons, thereby enhancing soil carbon loss (Mack et al. 2004). Third, the build-up of litter resulting from increased productivity and associated litter production of shrubs (Shaver et al. 2001), combined

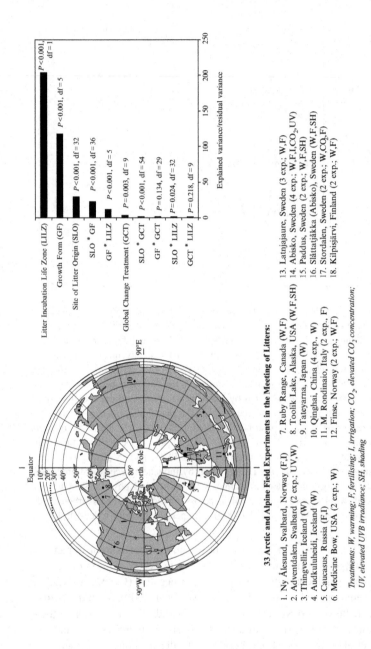

33 Arctic and Alpine Field Experiments in the Meeting of Litters:

1. Ny Ålesund, Svalbard, Norway (F,I)
2. Adventdalen, Svalbard (2 exp.; UV,W)
3. Thingvellir, Iceland (W)
4. Audkuluheidi, Iceland (W)
5. Caucasus, Russia (F,I)
6. Medicine Bow, USA (2 exp.; W)
7. Ruby Range, Canada (W,F)
8. Toolik Lake, Alaska, USA (W,F,SH)
9. Tateyarna, Japan (W)
10. Qinghai, China (4 exp., W)
11. M. Rondinaio, Italy (2 exp., F)
12. Finse, Norway (2 exp.; W,F)
13. Latnjajaure, Sweden (3 exp.; W,F)
14. Abisko, Sweden (4 exp.; W,F,I,CO$_2$, UV)
15. Paddus, Sweden (2 exp.; W,F,SH)
16. Slåttatjåkka (Abisko), Sweden (W,F,SH)
17. Stordalen, Sweden (2 exp.; W,CO$_2$,F)
18. Kilpisjärvi, Finland (2 exp.; W,F)

Treatments: W, warming; F, fertilising; I, irrigation; CO$_2$, elevated CO$_2$ concentration; UV, elevated UVB irradiance; SH, shading

Fig. 3.18 Analysis of the key biotic and abiotic drivers of percentage mass loss of leaf litters (right), based on decomposition of litter collected from the predominant plant species in 33 global change-mimicking experiments in Arctic and alpine sites in the Northern Hemisphere. All leaf litters were incubated in a common litter bed at two contrasting litter incubation life zones (LILZs). One LILZ was at high altitude and the other was at low altitude: the 'meeting of litters' experiment (left). From Cornelissen et al. (2007), with permission from Wiley-Blackwell.

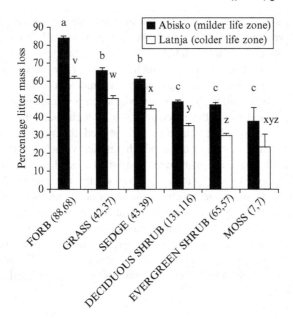

Fig. 3.19 Mean mass-loss data (%) of leaf litters from predominant plants of six growth forms collected from ambient plots in circumarctic-alpine sites (Fig. 3.18), after 2 years of outdoor incubation in two contrasting litter incubation life zones (LILZs). Numbers in parentheses refer to replication in the milder LILZ (Abisko) and the colder LILZ (Latnjajaure), respectively. ANOVA output: growth form (GF): $F = 84.7$, $P < 0.001$; LILZ $F = 50.0$, $P < 0.001$; GF by LILZ interaction $F = 0.98$, $P = 0.43$. Standard errors are shown one-sided. Within each LILZ, growth forms with the same letter are not significantly different in *post-hoc* Games–Howell tests. From Cornelissen et al. (2007), with permission from Wiley-Blackwell.

with their lower rates of decomposition, may increase fuel loads and the flammability of the litter layer, thereby promoting fire-induced carbon loss to the atmosphere. The magnitude of this increase, however, will depend on trends in precipitation and soil hydrology (Chapin and Starfield 1997; Hobbie et al. 2001; Cornelissen et al. 2007). Fourth, Arctic plant functional groups differ in their association with microbial communities (Wallenstein et al. 2007; Wookey et al. 2009) and mycorrhizal status (Read et al. 2004), so climate-driven shifts in vegetation composition might affect the capacity of microbes to decompose plant litter. This could in turn alter nutrient competition between plants and soil microbes with possible consequences for ecosystem nutrient cycling and soil carbon exchange (Bardgett et al. 2008). Finally, as noted by Cornelissen et al. (2007) the relative strengths of these various feedbacks due to growth form shifts need to be considered in the context of longer-term warming effects on deep soil carbon dynamics (Mack et al. 2004) and of positive feedbacks on carbon cycling from disturbances such as land use and herbivory (Wookey et al. 2009).

The above example from the Arctic provides evidence of the strong effect of interactions between plant traits and the belowground subsystem on ecosystem carbon stocks under climate change. As recently reviewed by De Deyn et al. (2008), there is now a growing body of literature showing that soil carbon pools of terrestrial biomass across the globe can be related, in part, to key plant traits and their links with soil biota (Fig. 3.20). These linkages, which operate via a variety of mechanisms at different spatial and temporal scales, govern the assimilation of carbon, its transfer to soil and residence time, and its loss from soil via decomposition processes (Fig. 3.21). While our understanding of the mechanisms by which plant traits influence soil carbon storage is incomplete, these authors argued the need for a mechanistic approach to soil carbon sequestration which implicitly includes the traits of soil biota through their link with plant traits. Importantly, such a trait-based approach could help in the development of strategies to promote carbon sequestration, thereby mitigating increasing atmospheric carbon dioxide concentrations (Lal 2004; P. Smith et al. 2008; Díaz et al. 2009). However, empirical studies are needed to identify general mechanisms by which plant traits and their combinations alter the amount and quality of plant inputs to the soil, and how this can be incorporated into management regimes to promote ecosystem services such as carbon storage (Wardle et al. 2003a; Fornara and Tilman 2008; De Deyn et al. 2008, 2009; Steinbeiss et al. 2008; Jonsson and Wardle 2010).

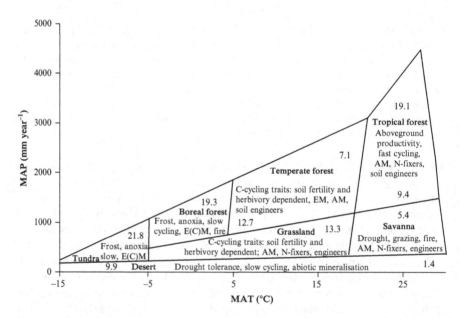

Fig. 3.20 Soil organic carbon pools (kg C m^{-2}; Amundson 2001), and drivers of plant carbon sequestration traits across biomes with characteristic mean annual temperature and precipitation (MAT and MAP). Lower and higher values within biomes represent warm versus cool-temperate forest, respectively, and drier versus wet (peaty) tropical forests; EM, ecto-; ECM, ericoid-; AM, arbuscular mycorrhizal fungi; biome location after Woodward *et al.* (2004). Adapted from De Deyn et al. (2008), with permission from Wiley-Blackwell.

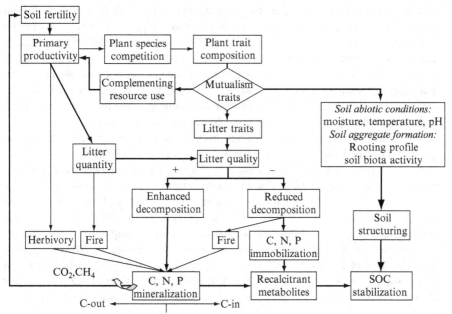

Trait composition enhances soil carbon sequestration when ΔC-out< ΔC-in

Fig. 3.21 Potential effects of plant trait composition on soil carbon sequestration, as manifested through influencing the ratio between carbon gains and losses. SOC, soil organic carbon. Adapted from De Deyn et al. (2008), with permission from Wiley-Blackwell.

3.6.2 Indirect belowground effects of nitrogen deposition

Human activities have lead to major increases in global emissions of nitrogen to the atmosphere, causing dramatic increases in deposition of atmospheric nitrogen to the terrestrial biosphere (Holland et al. 1999; Bobbink and Lamers 2002; Galloway et al. 2004). As with climate change, nitrogen enrichment also has the potential to significantly affect the belowground subsystem and ecosystem-level processes via vegetation change. Because most plant communities are limited by nitrogen (Vitousek and Howarth 1991), the addition of nitrogen substantially alters their productivity and structure, typically favouring fast-growing species that are able to exploit this resource, enhancing NPP. For instance, studies in alpine tundra indicate that nitrogen enrichment causes a shift in community composition in favour of graminoids and sedges over forbs (Bowman et al. 1993, 1995; Soudzilovskaia et al. 2007) and bryophytes (Van der Wal et al. 2003). Similarly, studies of temperate grasslands show that nitrogen enrichment enhances the growth of grasses at the expense of legumes and forbs (Bobbink 1991; Wedin and Tilman 1996; P. Smith et al. 2008), whereas in Arctic tundra increases in woody deciduous shrubs such as *Betula nana* (Bret-Harte et al. 2002; Hobbie et al. 2005) and graminoids (Press et al. 1998; Shaver et al. 1998; Gough et al. 2002) commonly occur.

Nitrogen enrichment increases the amount and quality of litter entering soil, both through promoting the dominance of faster-growing plant species, and through enhancing leaf nitrogen concentrations (Throop and Lerdau 2004), thereby further enhancing the quality of litter inputs to soil. As highlighted previously in this chapter, such changes in organic matter input to soil typically promote the abundance of microbes and animals in soil, and cause a shift to bacterial-based rather than fungal-based soil food webs. This in turn can lead to a strong positive feedback on nitrogen cycling and plant production, where the shift to bacterial-dominance enhances rates of nitrogen mineralization which further promotes soil nitrogen availability and the growth of competitive, fast-growing plant species (Fig. 3.22) (Welker et al. 2001). As a result of this positive feedback on nitrogen cycling, it has been proposed that ecosystem responses to nitrogen enrichment will be non-linear owing to plant species replacements that cause enhanced nitrogen cycling, and will lead to increased ecosystem nitrogen loss as a result of leaching (Welker et al. 2001). However, such positive feedbacks on nitrogen cycling and plant production might also lead to greater performance of invertebrate and vertebrate herbivore populations (Bowman 2000; Van der Wal et al. 2003), which could stabilize or amplify shifts in vegetation composition (Van der Wal et al. 2004; Van der Wal 2006). Also, the potential for positive feedbacks on nitrogen cycling will depend on the capacity of ecosystems to

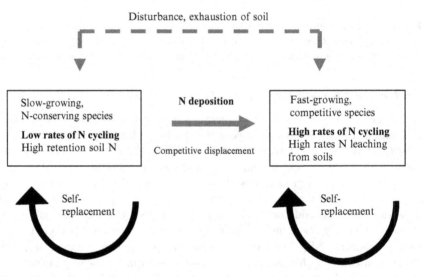

Fig. 3.22 Conceptual model showing the generalized pattern of plant species self-replacement in communities with low and high nitrogen availability. The model shows that in the face of greater nitrogen deposition, nitrogen-conserving species of infertile environments will be replaced by more competitive species, with traits that promote more rapid nitrogen cycling in soil. Under such conditions, the new state would be preserved, even in the absence of additional N inputs, until nitrogen reserves in soil are exhausted when species adapted to low nitrogen conditions will re-establish. From Welker et al. (2001), with permission from Oxford University Press.

act as sinks for atmospheric nitrogen, thereby increasing the resilience of the system to nitrogen enrichment (e.g. Aber et al. 1989; Zogg et al. 2000; Phoenix et al. 2004).

Experimental tests of the response of aboveground–belowground feedbacks to nitrogen deposition are few, and complicated by the existence of strong direct effects of nitrogen on the belowground subsystem, as discussed in Chapter 2. A study of particular relevance is that of Manning et al. (2006), who designed an experiment to tease apart the relative contributions of direct and indirect effects of elevated nitrogen deposition on a suite of ecosystem properties. This study involved the setting up of model plant-based ecosystems, and consisted of two factors, each with two levels, in a factorial design. The first factor comprised low or high N deposition to detect direct effects of nitrogen deposition on ecosystem functions. Meanwhile, the second factor consisted of synthesized plant communities consisting entirely of plant species characteristic of either high- or low-nitrogen-deposition environments. Using this design, they were able to show that the response of model ecosystems to nitrogen deposition was dominated by direct effects upon plant growth and soil biological communities, rather than indirect effects that operated via differences in plant community composition. These effects appeared to be driven by the stimulation of plant growth by nitrogen deposition, which increased litter inputs to soil and stimulated decomposer biomass and activity in high-nitrogen microcosms. Although increased litter inputs were associated with stimulated decomposer activity, this was insufficient to prevent the accumulation of carbon and nitrogen in soil of the high-N-deposition treatment. The authors of this study took care to highlight that caution is needed when extrapolating their results to field conditions. For example, some possible compositional effects of elevated nitrogen, such as regional extinctions and shifts in dominance of major plant functional groups, were not included in their study. However, if similar results were to be obtained in natural ecosystems then this would suggest that the direct effects of nitrogen deposition on ecosystem function could in some circumstances be stronger than expected by indirect effects manifested through changes in plant composition and diversity.

A recent study which sheds further light on the above issue is that of Suding et al. (2008), who carried out a field study on alpine moist-meadow at Niwot Ridge, Colorado, to examine the effects of species loss and nitrogen supply separately and in combination on the vulnerability of the system to accumulated nitrogen. This was tested by using combined treatments of removal of co-dominant plant species [the forb *Geum* (formally *Acomastylis*) *rossii*, and the fast-growing grass *Deschampsia caespitosa*] and manipulation of nitrogen supply over 6 years. They found that *Geum* was associated with soil microbial feedbacks that slowed rates of nitrogen cycling, while *Deschampsia* was associated with microbial feedbacks that increased rates of nitrogen cycling. However, after 4 years of initial resilience to nitrogen supply, *Geum* declined dramatically (by almost 70%) due to increasing nitrogen availability and elimination of the soil microbial feedbacks, which in turn indirectly facilitated *Deschampsia* via competitive release. Because *Deschampsia* was found to exert strong competitive effects on subordinate plant species, it was concluded that this increase in *Deschampsia* abundance might be accompanied by a community-wide drop in diversity. Interestingly, while some evidence of direct effects of enhanced nitrogen supply were detected, such as a change in *Geum* cover, these effects were relatively minor compared to the indirect effects mediated by interactions between biota and resulting changes in nitrogen availability. This finding contrasts with Manning et al. (2006) who found that direct effects of nitrogen addition on ecosystem properties dominated over indirect effects in model systems. Suding et al. (2008) also provide field evidence that plant–soil feedbacks through the microbial community can influence vulnerability to nitrogen enrichment, and that responses to nitrogen enrichment may be non-linear owing to plant species replacements that cause enhanced nitrogen cycling.

Before leaving this topic, it is important to note that there is strong potential for interactive effects between nitrogen enrichment and climate change on the belowground subsystem. As already mentioned, nitrogen enrichment can alleviate progressive nitrogen limitation, which constrains the response of nitrogen-limited ecosystems to elevated carbon dioxide concentrations (Luo et al. 2004). It may also amplify climate-change-driven changes in vegetation composition, thereby increasing the potential for indirect effects of vegetation change on the belowground subsystem (Wookey et al. 2009). Also, as discussed in Chapter 2, nitrogen deposition can have substantial effects on decomposition processes, thereby influencing the response of soil respiration to warming and the capacity of soils to sequester carbon under climate change (Davidson and Janssens 2006; Bardgett et al. 2008), for example by suppressing decomposition and hence soil carbon loss (Craine et al. 2007; Reay et al. 2008). We therefore reiterate the case made in Chapter 2 that there is an urgent need for experimental studies that consider simultaneous responses of aboveground–belowground feedbacks to multiple global change drivers to improve understanding of their effects on ecosystem-level processes under global change scenarios.

3.7 Conclusions

In this chapter we have explored how aboveground communities affect the belowground subsystem, and how these effects in turn feed back aboveground. Some aspects of this topic have attracted considerable attention over the past decade, and there have therefore been several important advances in understanding. While it has been long understood that plant species differ in their effects on the soil community and the processes that they drive, we are now developing an improved mechanistic and predictive understanding about why plant species effects differ, helped in a large part by recognition that fundamental trait differences among species may have major ecological consequences (Grime et al. 1997; Wardle et al. 1998a; Díaz et al. 2004; Wright et al. 2004). Further, recent studies have shown that even within plant communities there is huge trait variation among species (Hättenschwiler et al. 2008; Richardson et al. 2008), and that this trait variation may be much more important than global variation in climate in driving microbial processes such as plant litter decomposition (Cornelissen et al. 2007; Cornwell et al. 2008). Recent studies have also focused on using community-weighted plant trait values to predict aboveground and belowground ecosystem processes, pointing to the tractability in some situations of being able to scale up from traits of individual leaves to processes at the whole ecosystem level (e.g. Quested et al. 2007; Fortunel et al. 2009; but see Wardle and Zackrisson 2005; Peltzer et al. 2009). Another important development is the recognition that differences not just among species, but also within species may be a powerful driver of the belowground system (e.g. Schweitzer et al. 2004; Classen et al. 2007), and the intriguing suggestion has been made that different plant genes or genotypes may be able to select for different soil communities, indicative of 'community heritability' (Whitham et al. 2003, 2008).

Another aspect of plant community effects on the belowground subsystem that has attracted much recent attention is that of plant–soil feedbacks; that is, the notion that plant species may select for soil communities that may either benefit themselves or benefit other species. As such, a substantial number of plant–soil feedback studies have appeared in the recent literature (Kulmatiski et al. 2008). Some recent studies adopting this approach have proven their worth for understanding the role of plant–soil feedbacks in influencing plant diversity (Packer and Clay 2000; Petermann et al. 2008), facilitating vegetation succession (De Deyn et al. 2003; Kardol et al. 2006), and, as discussed in Chapter 5, driving plant invasibility. Much recent work has also focused on improving our understanding of how human-induced global change phenomena such as climate change and nitrogen deposition can influence the belowground subsystem through affecting both the performance of individual plants and the functional composition of the vegetation. Recent studies have highlighted, for example, that global change effects on feedbacks between the plant community and the belowground subsystem have the potential to drive ecosystem processes that influence ecosystem carbon flux (Wardle et al. 2003a; Cornelissen et al. 2007; De Deyn et al. 2008; Chapin et al. 2009). Further, recent manipulative experiments (e.g. Manning et al. 2006; Suding et al. 2008) are greatly enhancing our mechanistic understanding the role of vegetation shifts and dominance in affecting soil communities and the processes that they drive under global change.

Despite significant recent activity in this topic, and increased understanding of it on several fronts, there are many aspects that remain poorly understood and emerge as important topics for future research. For example, while there is much understood about how plant traits may drive aboveground–belowground linkages and feedbacks, our knowledge of how plant traits select for traits of soil organisms and *vice versa* lags well behind. Yet, such information is essential for developing a mechanistic understanding of aboveground–belowground linkages that explicitly incorporates the soil community, and how these linkages drive ecosystem processes. Further, the relative importance of different routes through which plant communities affect the belowground subsystem is only starting to become explored. For example, although the role of plant litter on the belowground subsystem has been thoroughly investigated, we are only beginning to understand the ecological importance of the immediate and much more rapid effects of live plant roots belowground (Pollierer et al. 2007; Högberg et al. 2008; Meier et al. 2009). Further, our understanding of the relative importance of the direct versus indirect pathways (Fig. 2.1) in driving plant–soil feedbacks is still relatively limited, largely because most recent studies on plant–soil feedback have focused on grassland plants and have employed methodologies that are likely to exaggerate the relative importance of short-term feedbacks, especially those that involve pathogen build-up and are often negative (Kulmatiski and Kardol 2008; Kulmatiski et al. 2008). Because of the ease of applying these methods, recent research in this field is very uneven, and more needs to be understood about the role of longer-term feedbacks (many of which are positive), as well as the role of feedbacks in general in non-grassland ecosystems.

Finally, studies on the effects of plant communities on the belowground subsystem and feedbacks aboveground potentially provide us with useful tools for understanding how global change phenomena affect ecosystems and fluxes of carbon between the biosphere and the atmosphere. However, to move towards achieving this goal requires focused effort on two questions that still remain relatively open. First, much research is needed to understand how plant traits and feedbacks between plant and soil communities influence the balance between ecosystem carbon gain and carbon loss, and hence ecosystem carbon sequestration. Such information is essential for predicting which ecosystems will gain or lose carbon under scenarios of global environmental change. Second, we reiterate our point in Chapter 2, and that made by Tylianakis et al. (2008), that the different global change drivers do not operate independently of one another, and that our understanding of global change from an aboveground–belowground perspective will be enhanced by studies that simultaneously consider multiple drivers.

4

Ecosystem-level significance of aboveground consumers

4.1 Introduction

In the last two chapters we explored how biotic interactions between plant and belowground communities act as drivers of terrestrial ecosystem functioning. We also discussed the significance of aboveground–belowground relationships for understanding how global change phenomena affect processes at the ecosystem level, especially in terms of the flux of carbon between the land and the atmosphere. Ecological linkages between plant and belowground communities and their role in regulating ecosystem functioning are also strongly affected by other trophic groups, especially aboveground herbivores that consume plant material and initiate changes in plant community structure and productivity, which propagate belowground (Bardgett et al. 1998b; Bardgett and Wardle 2003; Wardle et al. 2004a). Moreover, top predators can greatly alter the population density of herbivores and thus influence their impacts on ecosystem processes (Post et al 1999; Terborgh et al. 2001, 2006). In particular, effects of carnivores on ecosystem functions can be mediated through changes in plant communities caused by altered herbivore abundance (Wardle 2002; Thébault and Loreau 2003; Dunham 2008), and by shifts in foraging behaviour in response to predation risk (Schmitz et al. 2004). Therefore, understanding the ecological significance of herbivory for ecosystem functioning requires consideration of their role in a multitrophic context, whereby herbivores and carnivores directly and indirectly interact with plants to influence belowground biota and ecosystem processes (Terborgh et al. 2001; Wardle et al. 2005; Dunham 2008; Schmitz 2008a).

Understanding the effects of herbivores and their predators on ecosystem processes, however, is complicated because they involve a variety of mechanisms which operate simultaneously over different temporal and spatial scales (Bardgett and Wardle 2003). For example, the removal of plant shoot tissue by herbivores can induce rapid changes in plant carbon and nitrogen allocation and root exudation patterns, which as discussed in Chapter 3 can influence the growth and activity of rhizosphere organisms that rely on carbon and other nutrients released from plant roots (e.g. Guitian and Bardgett 2000; Hamilton and Frank 2001; Mikola et al. 2009). At the same time, aboveground herbivory can induce a complex suite of chemical changes in foliage, which impact on soil biota and rates of decomposition due to altered litter quality (Bardgett and Wardle 2003). Over longer timescales, selective feeding by herbivores or changes in foraging behaviour in response to predation risk can cause shifts in plant community composition (e.g. Ritchie et al. 1998; Wardle et al. 2001; Fornara and du Toit 2007), changing the amount and quality of litter entering to soil and hence rates of decomposition and nutrient cycling (e.g. Pastor et al. 1993; Wardle et al. 2001; Fornara and du Toit 2008). Added to this, animal wastes (i.e. dung, urine, and insect frass) can create resource-rich patches in the soil surface which influence the growth of plants, soil organisms and decomposition processes (McNaughton et al. 1997a; Bardgett et al. 1998a; Frost and Hunter 2008a; Mikola et al. 2009). Further, disturbance by large herbivores can alter soil physical properties, for instance through trampling and cultivation which can negatively impact on soil organisms and processes of nutrient cycling (King and Hutchinson 1976; King et al. 1976; Mikola et al. 2009). The relative importance of these herbivore-mediated mechanisms relative to landscape factors varies tremendously across ecosystems (Fig. 4.1), thereby causing differences in the scale and direction of herbivore effects (Bardgett and Wardle 2003).

The goal of this chapter is to draw on the concepts developed in Chapters 2 and 3 to outline how aboveground consumers, including herbivores and their predators, act as drivers of ecosystem processes through their influence on soil biota and belowground processes. There has been substantial progress in the last few years in understanding the significance of the effects of aboveground consumers at the ecosystem level. First, we synthesize what has been learned about the various mechanisms by which herbivores can influence soil biota and ecosystem processes at the individual plant and plant community level, and discuss how traits of plant species may have a role in mediating herbivore effects on terrestrial ecosystem function. Second, we discuss the influence of herbivores on belowground communities and ecosystem processes in a multitrophic context, describing recent advances in understanding about how predator–herbivore interactions and trophic cascades influence belowground properties and the functioning of terrestrial ecosystems. Third, we describe recent studies which demonstrate how aboveground biota move resources both within ecosystems and across ecosystem boundaries, and discuss the consequences of this for the functioning of terrestrial ecosystems. Finally, we evaluate how aboveground consumers might be indirectly influenced by global change through alteration of vegetation productivity and composition, and how they might in turn modify responses of terrestrial ecosystems to global change through their influence on plant and soil communities.

4.2 Herbivore-mediated effects on plant–soil feedbacks and ecosystem processes

Herbivores can have a wide range of effects on the functioning of terrestrial ecosystems. Some of these effects are direct, involving the removal and consumption of herbage (which can vary some 100-fold across terrestrial ecosystems), trampling, and the return of animal excreta or insect frass (Floate 1981;

Fig. 4.1 Fenced grazing exclosures erected across a range of ecosystems to study impacts of herbivores on aboveground and belowground properties: (a) exclusion of sheep (*Ovis aries*) grazing from hill grasslands in Llyn Llydaw, Snowdonia, Wales; (b) exclusion of sheep (*O. aries*) from glacial outwash, Iceland; (c) exclusion of sheep (*O. aries*) grazing from montane heath, Scotland; (d) exclusion of European red deer (*Cervus elaphus*) from regenerating birch (*Betula pubescens*) forest, Scotland; (e) exclusion of elk (*Cervus canadensis*), bison (*Bison bison*), and pronghorn antelope (*Antilocarpa americana*) from grasslands at Yellowstone National Park, USA; and (f) exclusion of reindeer (*Rangifer tarandus*) from boreal forest with lichen (*Cladina* spp.) understorey in northern Finland. Photo credits: (a–d) by R.D. Bardgett; (e) by D. Frank; (f) by D.A. Wardle.

McNaughton 1985; McNaughton et al. 1989; Bardgett and Wardle 2003; Frost and Hunter 2008a). However, herbivores also have indirect effects on terrestrial ecosystem functioning by modifying feedbacks between plants and soil biota which influence rates of nutrient and carbon cycling (Bardgett et al. 1998b; Bardgett and Wardle 2003; Wardle et al. 2004a). Such indirect effects of herbivores on plant–soil feedbacks operate over different temporal and spatial scales, ranging from short-term responses at the individual plant level to longer-term responses at the plant community level, and can either accelerate or decelerate rates of nutrient cycling (Ritchie et al. 1998; Bardgett and Wardle 2003; Wardle et al. 2004a) (Fig. 4.2a). For instance, acceleration of nutrient cycles occurs when herbivores promote the supply of labile substrates to soil as faeces and/or root exudates, which stimulates soil decomposer activity, rates of nutrient mineralization, and uptake of nutrients by grazed plants (Bardgett and Wardle 2003). In contrast, deceleration of nutrient cycling occurs when selective feeding on nutrient-rich plant species leads to the dominance of defended plants that produce poor quality litter (Pastor et al. 1993; Ritchie et al. 1998), or when herbivory induces the production of secondary metabolites in foliage which reduce litter quality and decomposability (Rhoades 1985; Findlay et al. 1996) (Fig. 4.2b). In this section, we provide an overview of some of the ways by which aboveground herbivory can positively or negatively influence plant nutrition and productivity via changes in belowground processes of decomposition and nutrient cycling. We illustrate these pathways using specific examples from a range of ecosystems, thereby providing a unifying perspective on herbivore effects on ecosystem processes (Fig. 4.3).

4.2.1 Positive effects of herbivores on belowground properties and ecosystem functioning

Positive effects of foliar herbivory on soil biota and nutrient cycling occur when dominant plant species respond to grazing by exhibiting compensatory growth (Augustine and McNaughton 1998). This pathway is most common in grasslands where optimization of aboveground NPP by grazing mammals is often reported (McNaughton 1985; McNaughton et al. 1997a; De Mazancourt et al. 1999). In these situations, ungulate grazing positively affects the decomposer subsystem by preventing colonization of later successional plants which produce poorer litter quality, as well as through returning carbon and nutrients to the soil in labile forms such as dung and urine, and as enhanced rhizodeposition (McNaughton et al. 1997a, 1997b; Bardgett et al. 1998a; Bardgett and Wardle 2003). Acceleration of nutrient cycling by herbivores has also been reported in other ecosystems, including temperate forests (Frost and Hunter 2008a), Arctic tundra (Van der Wal et al. 2004), and boreal forest (Stark et al. 2000), albeit via different mechanisms. Moreover, the ability of herbivores to accelerate soil nitrogen cycling has led to the suggestion that they can alleviate nutrient deficiency and enhance their own carrying capacity (McNaughton et al. 1997a).

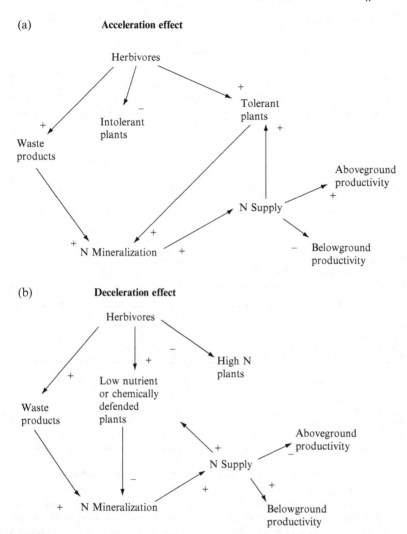

Fig. 4.2 Feedback loops illustrating (a) the acceleration effect and (b) the deceleration effect of herbivores on feedbacks between plant species and nutrient cycling. Arrows indicate net indirect effect of herbivores on the abundance of plants or the rate of the process. From Ritchie et al. (1998), with permission from the Ecological Society of American.

One of the most well-documented mechanisms by which herbivores promote nutrient cycling is via the deposition of animal wastes, including urine, dung, and insect frass. Herbivores consume and digest large quantities of aboveground plant material, and nutrients can be released far more quickly into soil from the resulting faeces than if released from litter. This in turn leads to increased nutrient availability and plant nutrient uptake. In essence, this short-cuts the litter-decomposition

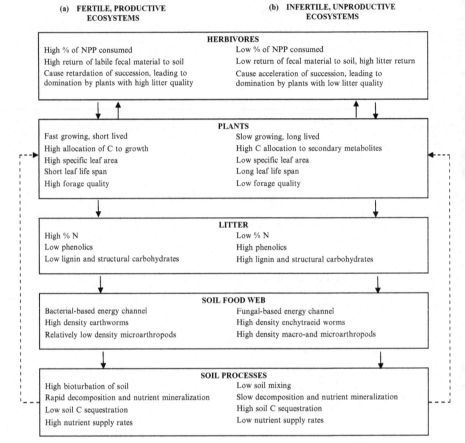

Fig. 4.3 The difference in fundamental plant traits between species which dominate (a) fertile systems that support high herbivory and (b) infertile habitats that support low herbivory functions as a major ecological driver. This occurs through plant traits affecting the quality and quantity of resources that enter the soil, and the key ecological processes in the decomposer subsystem driven by the soil biota. These linkages between belowground and aboveground systems feed back (dashed line) to the plant community positively in fertile conditions (a) and negatively in infertile ecosystems (b). From Wardle et al. (2004a), with permission from the American Association for the Advancement of Science.

pathway, and provides soil organisms with a highly decomposable resource that is rapidly mineralized, thereby liberating nutrients for plant use. There are numerous published studies of mammalian herbivore wastes stimulating soil microbial activity, nutrient cycling, and plant production. For example, in British hill grasslands, urine and dung from grazing cattle and sheep are well known to increase soil microbial activity and nitrogen cycling, leading to greater plant production in grazed than ungrazed areas (Floate 1970a, 1970b; Bardgett et al. 1997, 2001b).

Also, in dry grassland and shrub-grassland at Yellowstone National Park, Frank and McNaughton (1992) detected a positive association between dung deposition and aboveground primary production, and suggested that grazing and productivity are coupled to herbivore-facilitated nutrient cycling in this system. Likewise, Tracy and Frank (1998) found that grazing by elk (*Cervus canadensis*), bison (*Bison bison*), and pronghorn antelope (*Antilocarpa americana*) at Yellowstone National Park caused increased soil microbial biomass and rates of nitrogen mineralization, which was partly attributed to inputs of labile carbon from dung. In the Serengeti National Park, Tanzania, McNaughton et al. (1997a, 1997b) found that urine from gazelles enriched soils with nitrogen from urea, leading to a burst of organic-matter mineralization, and in African savanna ungulate droppings accelerated nitrogen mineralization in heavily grazed sites (Fornara and du Toit 2008). Further, in the high Arctic Van der Wal et al. (2004) found that experimental addition of reindeer (*Rangifer tarandus*) faeces to moss-dominated tundra increased soil microbial biomass and grass growth, which again was attributed to increased nutrient availability (Fig. 4.4). It is important to note that herbivores can also change the spatial distribution of nutrients within ecosystems through their dunging behaviour. For instance, although rabbits deposit dung throughout their range, dung deposition is higher in places where the animals congregate and a large amount of dung is deposited in a small number of latrines (Willot et al. 2000). Consequently, soil nitrogen availability in latrines is very high, despite high rates of nitrogen loss by volatilization, leaching, and denitrification (Willot et al. 2000).

Although less studied, there is accumulating evidence that invertebrate herbivores can also stimulate nutrient cycling and plant production. For instance, in a field experiment, Belovsky and Slade (2000) found that grasshoppers enhanced soil nitrogen cycling and plant abundance in prairie grassland. This was attributed in part to grasshoppers preferentially feeding on plants of lower nitrogen content which decompose slowly, thereby increasing the input to soil of litter from plants of higher nitrogen content which decompose faster and therefore liberate more nutrients for plant use. There is also evidence that waste from invertebrate herbivores (i.e. frass) stimulates soil microbial activity and nitrogen cycling, thereby enhancing plant production. For example, Frost and Hunter (2004) found that the addition of insect frass to soils planted with red oak (*Quercus rubra*) saplings increased soil carbon and nitrogen content, and caused a pulse in inorganic nitrogen availability and nitrogen loss from soil in leachates. In a subsequent experiment, they added ^{15}N-enriched insect frass to mesocosms containing red oak saplings to follow the fate of frass-nitrogen over 2 years (Frost and Hunter 2007). They found that while some of the added frass-nitrogen was lost from soil in leachates immediately after addition, the majority was retained in soil, which is consistent with the finding that the addition of frass from larvae of the gypsy moth (*Lymantria dispar*) to soil stimulated microbial growth and nitrogen immineralization (Lovett and Rueseink 1995), and with deciduous forest soils being strong sinks for nitrogen inputs via microbial immobilization followed by incorporation into organic matter (Zak et al. 1990; Zogg et al. 2000). However, some of the added frass-nitrogen was also rapidly mineralized by soil

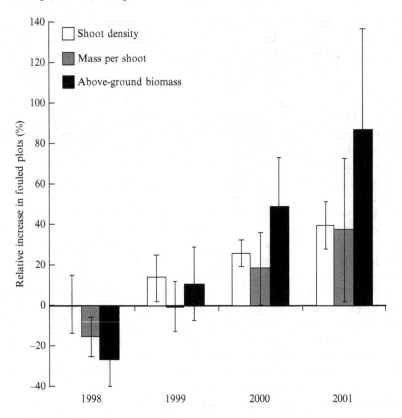

Fig. 4.4 Development of the response by grasses over 4 years to the addition of reindeer (*Rangifer tarandus*) faeces in 1997 to moss-dominated tundra in high-arctic Spitsbergen, measured as shoot density (no. m^{-2}), shoot mass (mg shoot^{-1}), and total aboveground live mass of grasses (g m^{-2}). Data are expressed as the mean ($\pm$ SE) increase in each of these measures in fouled plots relative to the untreated control plots. Soil microbial biomass, measured 4 years after the addition of reindeer feaces, was 85% greater in fouled areas than the control. From Van der Wal et al. (2004), with permission from Wiley-Blackwell.

microbes, acquired by the oaks and recycled into foliage, and then assimilated by late-season insect herbivores. Also, some of the frass-nitrogen taken up by the oaks re-entered the soil system as leaf litter at the end of the first season, and was then subject to decomposition processes in the following year. This latter finding indicates that the effects of frass deposition can carry over and affect ecosystem processes during decomposition. However, as shown by Frost and Hunter (2008a) in another related study, these effects are short lived relative to abiotic forces which also regulate decomposition process in forest soils. Collectively, these studies illustrate that, as with ungulates, invertebrate herbivores can positively influence nutrient dynamics and plant production in terrestrial ecosystems.

Another mechanism by which herbivory influences belowground processes is via plant allocation. In particular, it has been shown that foliar herbivory of grassland plants leads to short-term pulses in root exudation, which stimulate the biomass and activity of microbes (Mawdsley and Bardgett 1997; Guitian and Bardgett 2000; Hamilton and Frank 2001) and the abundance of their faunal consumers (Mikola et al. 2001, 2009; Hokka et al. 2004; but see Ilmarinen et al. 2005). These positive effects of exudation on soil biota in turn have a positive feedback to the plant through a temporary increase in soil nitrogen mineralization and plant nitrogen acquisition, ultimately benefiting plant growth in the longer term (Hamilton and Frank 2001; Hamilton et al. 2008; Sørensen et al. 2008) (Fig. 4.5). Plant and soil responses to defoliation are known to vary across plant species and with the timing and frequency of defoliation events (Guitian and Bardgett 2000; Klironomos et al, 2004; Ilmarinen et al. 2005). However, the ecological significance of these short-term responses to defoliation in grasslands can be substantial, and it has been argued that they could explain, in part, the compensatory response of grassland plants to ungulate grazing in natural ecosystems (Hamilton and Frank 2001; Bardgett and Wardle 2003; Hamilton et al. 2008). Much less is known about the significance of defoliation-induced pulses in root exudation in forest ecosystems, although pot experiments suggest that exudation of carbon from tree roots is also important in mediating the response of tree seedlings to herbivory (Ayres et al. 2004; Frost and Hunter 2008b). Therefore, it is likely that carbon release from roots and its stimulatory impact on soil microbes and nutrient availability is a widespread mechanism that enables plants to compensate for loss of tissues through herbivory.

In the longer term, effects of herbivory on plant carbon and nitrogen allocation become apparent through alterations in root biomass and nutrient content. There is

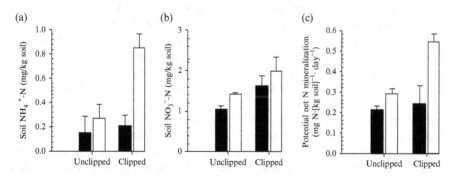

Fig. 4.5 Positive influence of defoliation of the grass *Poa pratensis* on soil nitrogen availability, measured as: (a) soil NH_4^+-N; (b) soil NO_3^--N; and (c) potential net nitrogen mineralization. Values are means ± 1 SE in unclipped and clipped bulk soils (filled bars) and rhizospheric soils (open bars) 24 h after treatment. This stimulation of soil nitrogen availability was in turn associated with elevated plant nitrogen uptake, shoot nitrogen content, and rates of photosynthesis. From Hamilton and Frank (2001), with permission from the Ecological Society of America.

mixed evidence about how herbivory affects root productivity, with positive (Milchunas and Laurenroth 1993), negative (Guitian and Bardgett 2000; Mikola et al. 2001; Ruess et al. 1998; Sankaran and Augustine 2004), and neutral (McNaughton et al. 1998) effects being reported. Moreover, there appears to be no consistent relationships between root responses to herbivory and rates of soil nutrient cycling. For example, in a study on prairie dog colonies in Wind Cave National Park, South Dakota, grazing was found to decrease the biomass and nitrogen content of roots, causing a reduction in nitrogen immobilization due to reduced microbial growth, and an increase in net nitrogen mineralization and plant nitrogen uptake (Holland and Detling 1990). In contrast, in a study on temperate sub-humid grassland in Argentina, Chaneton et al. (1996) found that grazing increased nitrogen allocation to roots, which increased root biomass and the rate of underground nutrient circulation. In the Serengeti, increases in nutrient cycling in response to herbivory have been found (McNaughton et al. 1997a), despite there being no difference in root biomass between fenced and unfenced plots (McNaughton et al. 1998).

Little is known about the relative importance of the main mechanisms described above, namely animal waste deposition and defoliation, in explaining the effects of herbivory on belowground organisms and nutrient cycling. However, a recent study by Mikola et al. (2009) sheds some light on this issue. These authors examined the relative importance of three grazing mechanisms, namely plant defoliation, dung and urine return, and physical presence of herbivores (i.e. trampling and concentrated dung and urine return), for explaining aboveground and belowground responses to cattle in dairy pasture. They found that grazer effects on plant attributes, including various measures of plant growth and nutrient allocation, were almost entirely explained by plant responses to defoliation. However, for belowground attributes, including the abundance of different faunal groups and soil nitrogen availability, the picture was more mixed in that all mechanisms played a role, although dung and urine return had major effects on soil animals and nitrogen availability when it was concentrated in patches. This finding is consistent with that of Hamilton and Frank (2001), who argued that animal wastes alone cannot explain the widely documented positive effects of herbivores on grassland nutrient cycling at large spatial scales. This is because, typically, waste patches remain localized within an ecosystem where they greatly influence only a relatively small proportion of the total surface area (Augustine and Frank 2001).

One final mechanism through which herbivores may positively affecting below-ground processes is through modifying soil temperature. This mechanism appears to be especially important in cold climes, such as the high Arctic where low soil temperatures limit soil microbial activity and nutrient cycling (Brooker and Van der Wal 2003). For example, grazing by barnacle geese (*Branta leucopsis*) and reindeer in Ny-Ålesund, Spitsbergen, has been shown to reduce the thickness of the moss layer, leading to an increase in soil temperature (Van der Wal et al. 2001). Similarly, studies in the high Arctic of Spitsbergen, have shown that reindeer grazing promotes soil nitrogen availability and plant productivity (Van der Wal et al. 2004), partly as a result of reduced the depth of the moss layer in grazed areas (Brooker and

Van der Wal 2003). Studies on upland steppe in Yellowstone National Park (Coughenour 1991), seasonally dry high country in New Zealand (McIntosh et al. 1997), and oak savanna in Minnesota (Ritchie et al. 1998), have also reported positive effects of herbivory on soil temperature. This may potentially contribute to the acceleration of decomposition and nitrogen mineralization in these situations, at least if increased soil temperature is not accompanied by a loss in soil moisture.

4.2.2 Negative effects of herbivores on belowground properties and ecosystem functioning

One of the main routes by which herbivores negatively affect nutrient cycling and plant production is through selective foraging on nutrient-rich plants. This is especially prevalent in relatively unproductive and infertile ecosystems, where low rates of consumption and selective foraging on the most nutritious plant species with the highest litter quality lead to the dominance of well-defended plants that produce recalcitrant litter (Ritchie et al. 1998). Because most nutrients are returned to the soil as plant litter in these situations, the net effect of herbivory is to reduce soil biotic activity, nutrient mineralization, and supply rates of nutrients to plants, and this overrides any positive effects of dung and urine (Fig. 4.2b) (Bardgett and Wardle 2003). A classic example of the importance of selective feeding by herbivores for soil processes is the study of Pastor et al. (1993) on moose (*Alces alces*) browsing in the boreal forests of the Isle Royale National Park, Michigan, USA. These authors showed that selective foraging by moose on deciduous hardwoods, with nutrient-rich foliage, led with time to dominance by less nutritious species such as spruce which produces litter of lower quality and hence decomposability. This slow-decomposing litter in turn accumulates on the soil surface, leading to a reduction in soil nitrogen mineralization and hence a decline in the productivity of the ecosystem (Fig. 4.6).

Several other studies across a range of ecosystems also report that selective foraging increases the dominance of less nutritious species that ultimately slow down the rate of nutrient cycling. For example, Ritchie et al. (1998) found that selective foraging by deer, rabbits, and a variety of insects in oak savanna in Minnesota decreased rates of nutrient cycling rates due to the removal of nutrient-rich woody plants and legumes. This in turn led to dominance by prairie grasses (e.g. *Andropogon geradi* and *Sorghastrum nutans*) that produce nutrient-poor litter of low decomposability, thereby slowing down rates of nutrient cycling. Likewise, Van Wijnen and Van der Wal (1999) found that grazing by geese, hares, and rabbits in salt marsh encouraged domination by nutrient-poor and tannin-rich plants such as common sea lavender (*Limonium vulgare*). The litter produced by this species decomposes slowly, impairing rates of nutrient mineralization. Similarly, Kielland and Bryant (1998) found that selective browsing by moose in Alaskan taiga led to the replacement of nutritious deciduous tree species by less nutritious evergreens with recalcitrant litter, thereby slowing down rates of nutrient turnover. Finally, Harrison and Bardgett (2004) found that selective

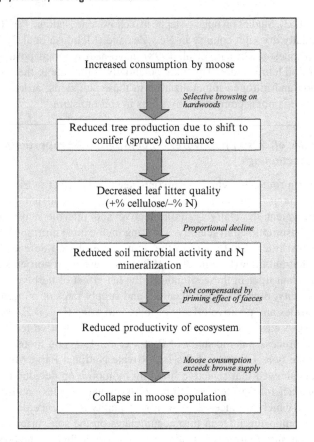

Fig. 4.6 Schematic diagram of the cascading effects of moose (*Alces alces*) browsing on ecosystem properties in the Isle Royale National Park, Michigan. Adapted from Pastor et al. (1993) by Bardgett (2005), and reproduced with permission from Oxford University Press.

browsing and the consequent reduction in growth of the dominant tree species *Betula pubescens* by European red deer (*Cervus elaphus*) in regenerating forest in the Scottish Highlands reduced rates of soil nitrogen mineralization (Fig. 4.7). This arises through the well-known ability of *B. pubescens* to positively affect soil nutrient cycling through improvements in the quality of litter inputs to soil and by the action of its roots. However, this increase in nitrogen mineralization in unbrowsed plots was not matched with an increase in soil phosphorus availability, and therefore led to a shift from nitrogen to phosphorus limitation of tree growth (Carline et al. 2005).

In situations where shifts in plant community composition resulting from selective grazing slow down rates of nutrient cycling, a feedback system may subsequently emerge. Here, reduced rates of nutrient cycling and availability will further favour the growth of nutrient-poor species that have high nutrient-use efficiency,

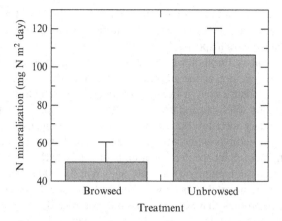

Fig. 4.7 Negative effect of red deer (*Cervus elaphus*) browsing on rates of soil nitrogen mineralization in regenerating birch forest at Creag Meagaidh National Nature Reserve in the Scottish Highlands. From Harrison and Bardgett (2004), with permission from Elsevier.

thereby enhancing their competitive dominance in nutrient-poor habitats. Such a feedback strengthens the effects of the herbivores on the ecosystem and could lead to additional reductions in plant production and rates of nutrient cycling (Ritchie et al. 1998). However, it should be noted that herbivore-induced shifts in vegetation composition that favour of nutrient-poor plants do not always result in reduced rates of nutrient cycling. For example, Wardle et al. (2001) sampled soils from inside and outside fenced enclosures across 30 locations in New Zealand forests, and found that while browsing by introduced mammals (feral goats and deer) altered vegetation in a consistent way (reducing nutritious broad-leaved species and promoting other less nutritious types), subsequent effects on measures of soil nutrient status were highly idiosyncratic, with an equal number of positive and negative effects being detected. Moreover, effects of herbivory on vegetation composition, and hence feedbacks to soil nutrient cycling, also appear to depend on soil fertility itself. For example, Buckland and Grime (2000) found by using synthesized plant communities that under low-fertility conditions preferential feeding by invertebrates on nutrient-rich early successional species accelerated vegetation succession and the growth of late successional plant species known to produce poorer-quality litter. In contrast, under fertile conditions herbivory retarded vegetation succession and led to the dominance of early-successional nutrient-rich species (presumably because their main defence against herbivory is rapid compensatory growth that is made possible by high resource availability) (Buckland and Grime 2000). Although they did not assess the consequences of these plant community changes for the belowground subsystem, it is likely that acceleration of vegetation succession under infertile conditions will have further reduced soil nutrient availability and therefore strengthened the dominance of nutrient-poor, late-successional species that thrive under these conditions. In contrast, the retardation by herbivores of succession under more fertile conditions and the continued dominance of fast-growing nutrient-rich plant species would be expected to further promote soil fertility and the dominance of nutrient-rich species.

Another mechanism through which herbivores could contribute to the slowing down of nutrient cycling is the promotion of plants that accumulate secondary metabolites within their leaves, which in turn reduces plant litter quality and thereby adversely affects the soil biota and its activity. Some plant species can produce a vast cocktail of secondary metabolites, and many of these chemicals are known to have adverse effects on herbivores through a variety of mechanisms (Hartley and Jones 1997). As such, these plant species are often avoided by herbivores, which can then dominate the plant community (Bardgett and Wardle 2003). Further, herbivore attack, especially by invertebrates, is well known to induce the production by plants of secondary metabolites that serve as defence compounds (Schultz and Baldwin 1982; Rhoades 1985; Agrawal et al. 1999; Nykanen and Koricheva 2004), with subsequent impacts on the herbivores themselves (Wold and Marquis 1997; Boege 2004). Few studies have studied the after-life effects on decomposition and nutrient cycling of herbivore-induced changes in plant secondary metabolite production. However, Findlay et al. (1996) found that cellular damage caused by spider mites to seedlings of eastern cottonwood (*Populus deltoids*) increased the concentration of polyphenols in foliage, resulting in a 50% reduction in the rate of decomposition of subsequently produced leaf litter. Also, it is well established that plant secondary metabolites such as polyphenolics can play an important role in terrestrial-ecosystem nutrient cycling (Hättenschwiler and Vitousek 2000), for example through reducing rates of nitrogen fixation (Schimel et al. 1998), nitrification (Baldwin et al. 1983), and nitrogen mineralization (Northup et al. 1995). Consequently, it is reasonable to assume that increased production of plant secondary metabolites following herbivore attack will reduce the quality of litter inputs to soil, their decomposability, and hence rates of nutrient cycling.

So far, we have considered how herbivores can reduce of rates of nutrient cycling through altering the quality of litter inputs to soil. However, reductions in NPP and hence the quantity of organic matter entering soil also has the potential to negatively influence soil biota and the functioning of the belowground subsystem. For example, Sankaran and Augustine (2004) showed that grazing by large mammals in a semiarid grassland ecosystem in Kenya reduced soil microbial biomass in both high- and low-fertility sites. This was likely due to reductions in plant carbon input to soil resulting from a depression in root production, as well as the diversion of plant carbon to herbivore respiration and growth. Likewise, the previously mentioned negative effects of browsing by red deer (*C. elaphus*) on soil nitrogen cycling in regenerating forest reported by Harrison and Bardgett (2004) were in part attributed to a reduction in the growth of the dominant tree species *B. pubescens* (Fig. 4.7), and hence the amount of carbon entering soil from its roots and foliage. These examples contrast with studies of productive grasslands, where herbivore-mediated increases in primary productivity have been shown to enhance carbon addition to soil, thereby stimulating soil biological activity because it alleviates carbon limitation of microbes

(Bardgett and Wardle 2003). However, it is important to note that knock-on effects of herbivore-induced changes in NPP to the belowground subsystem and nutrient cycling are not at all clear. For example, increasing NPP has been shown to have both positive and negative effects on both microbial biomass and higher trophic levels of the soil food web (Chapter 3; see also Wardle 2002). There are two reasons for such contrasting responses: first, the relative importance of top-down and bottom-up forces in regulating soil food web components may be context-dependent; and second, plants not only provide carbon resources for microbes but also compete with them for nutrients. Therefore, the direction of effects of herbivore-induced changes in NPP on the belowground subsystem may be governed by which of two opposing effects (stimulation of microbes by carbon addition, inhibition of microbes by resource depletion) dominates.

It should be noted that not all soil organisms are stimulated by enhanced carbon input to soil, such as occurs following defoliation. In particular, the growth of mycorrhizal fungi is often reduced following defoliation because carbon is preferentially allocated to other plant and soil pools (Gehring and Whitham 1994, 2002). Evidence for this arises from a literature synthesis by Gehring and Whitham (1994) which revealed that mycorrhizal colonization declined following defoliation or grazing by ungulates for 23 of the 37 plant species that had been examined at that time, whereas it was unaffected for ten species, positively affected for two species, and showed variable responses for two species. Moreover, the widely reported negative relationship between mycorrhizae and herbivory was found to occur across a wide range of plant types, including grasses colonized by arbuscular mycorrhizal fungi and conifers colonized by ectomycorrhizal fungi (Gehring and Whitham 1994). As emphasized by Gehring and Whitman (1994), however, it is difficult to draw firm conclusions on this topic because studies examining the effects of herbivory on mycorrhizal fungi are limited in number and scope, and cover only a small number of plants in a limited range of environments. Also, more recent studies indicate that it is difficult to generalize about the response of fungal symbionts to defoliation, in that responses vary with the frequency (Klironomos et al. 2004) and timing (Saravesi et al. 2008) of defoliation. Further, effects of defoliation on mycorrhizal fungi are not restricted to changes in root colonization, because other fungal structures such as vesicles, spores, and extra-radical hyphae of arbuscular mycorrhizal fungi (Klironomos et al. 2004), and sporocarps of ectomycorrhizal fungi (Kuikka et al. 2003), are also affected. Clearly further studies are needed to determine the ecological significance of such herbivore-induced changes in mycorrhizal communities and to test for generalities in responses across a range of plant species and environmental conditions.

One further mechanism by which herbivores can negatively impact soil biota and nutrient cycling is via physical disturbance, which is commonly associated with intensive grazing. This has been most widely documented in livestock farming systems, where high grazing pressures can lead to soil compaction, erosion, and poaching of the soil surface, with negative impacts on soil biota and ecosystem productivity. For example, increased livestock stocking densities on pasture have been found to severely reduced collembolan numbers in the surface soil due to soil compaction and associated reductions in soil pore space (King and Hutchinson 1976; King et al. 1976; Walsingham 1976). Further, simulated cattle trampling was found to strongly reduce the abundance, diversity, and species richness of oribatid mite and Collembola communities in grassland (Fig. 4.8) (Cole et al. 2008). Over longer time scales, there are numerous reports in the literature of overgrazing causing catastrophic soil degradation, especially in the world's drylands which are especially prone to desertification (Diamond 2005; Avni et al. 2006; Lal 2009). For instance, in the African Sahel, overgrazing by cattle has caused a catastrophic shift to a desertified ecosystem state that can no longer sustain livestock production (Rietkerk et al. 2004), and in the Negev Highlands, Israel, the overgrazing caused by local Bedouin population has accelerated the soil-degradation process (Avni et al. 2006).

Reports of herbivore-driven physical disturbance are also common in natural ecosystems. For example, increased grazing by lesser snow geese (*Anser caerulescens caerulescens*) along the coast of the Hudson Bay, Canada, has promoted hypersaline conditions due to enhanced soil evapotranspiration, which has caused the destruction of salt marshes and subsequent eviction of its grazers (Jefferies 1998; Srivastava and Jefferies 1996). Likewise, in the high Arctic at Svalbard, spring foraging by pink-footed geese (*Anser brachyrhynchus*) for belowground plant parts, such as rhizomes, roots, and tubers (termed grubbing), is associated with significant disturbance to vegetation and soil, causing localized declines in plant cover, carbon uptake, and soil carbon storage (Fig. 4.9) (Van der Wal et al. 2007). The main concern here, however, is that the pink-footed goose population has doubled over the last 30 years (Fox et al. 2005), and is expected to increase with climate warming (Jensen et al. 2008), potentially causing widespread vegetation loss and soil disturbance in the European Arctic (Speed et al. 2009). Further, high levels of grazing pressure by reindeer in dry boreal forest in Fennoscandinavia have been associated with the removal of lichen cover, which exposes soil biota to a less favourable microclimate (Stark et al. 2000), and with soil compaction and the crushing of plant roots, which reduces carbon supply to soil (Stark et al. 2003). While these responses to grazing are likely to have negative consequences for nutrient cycling, the net effect of reindeer grazing will depend on the relative dominance of negative and positive effects of grazing, which could also vary depending on season (Stark et al. 2000, 2003). However, a study by Sørensen et al. (2009) found that the effects of simulated reindeer trampling on decomposer communities in sub-Arctic grasslands outweighed the effects of defoliation and fertilization by reindeer. These authors also found that simulated trampling negatively affected the coverage of mosses and sedges, as well as the abundance of functionally important faunal groups such as collembolans and enchytraeids. Finally, as discussed in Chapter 5, physical disturbances propagated by invasive herbivores such as pigs (*Sus scrofa*) and beavers (*Castor canadensis*) can greatly transform ecosystems, leading to potentially substantial changes in soil organisms and processes.

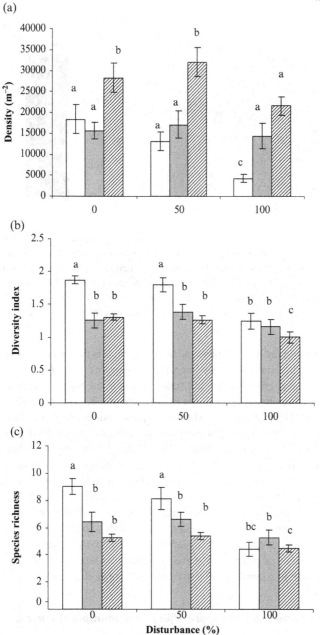

Fig. 4.8 Response of oribatid mites (open bars), mesostigmatid mites (shaded bars), and Collembola (hatched bars) to disturbance caused by simulated cattle grazing in grassland. The disturbance levels refer to 0, 50, and 100% ground-cover disturbance and data are for (a) abundance, (b) diversity (Shannon–Wiener diversity index), and (c) richness of species to these disturbance treatments. Error bars represent SE and bars within the same panels bearing the same letter do not differ at $P < 0.05$. From Cole at al. (2008), with permission from Elsevier.

Fig. 4.9 (a) Spring foraging by pink-footed geese (*Anser brachyrhynchus*) for belowground plant parts, such as rhizomes, roots, and tubers (termed grubbing), is associated with (b) significant disturbance to vegetation and soil, causing localized declines in soil carbon storage in the high Arctic. The data show that the amount of carbon stored in the surface organic horizon of the soil was reduced by goose grubbing. From Van der Wal et al. (2007) and image by Christiaane Gruss, with permission from Wiley-Blackwell.

Before leaving the topic of physical disturbance, it is important to note that, in most cases, the detrimental effects of intensive grazing are localized in both space and time. For instance, grubbing by pink-footed geese in the high Arctic is most intense in early spring when they migrate from their southern wintering grounds, and in wetter habitats in low-lying landscapes where their preferred forage is abundant (Speed et al. 2009). However, it is likely that increased disturbance of preferred habitats by these geese will force them to forage less preferred habitats, thereby extending their impact to the landscape scale with far-reaching implications for ecosystem functioning (Speed et al. 2009). Likewise, in grazed pastures, excessive trampling and associated loss of vegetation cover tends to occur in localized areas, for instance around supplementary feeding areas, gates and walkways, and at times of excessive soil wetness. However, as will be considered later in this chapter, predicted increases in grazing pressures due to climate or land use change will increase both the temporal and spatial extent of grazing, potentially leading to large-scale vegetation loss and soil disturbance in natural ecosystems (van de Koppel et al. 2005; Speed et al. 2009).

4.2.3 Landscape-scale herbivore effects and multiple stable states

Although herbivores have a great ability to modify ecosystems via the mechanisms described above, their effects on ecosystem processes are in some cases outweighed by landscape controls, such as spatial variation in topography, soil texture, and soil moisture conditions. For example, in the aforementioned study of regenerating forest in Scotland, Harrison and Bardgett (2004) found, using four fenced exclosure plots, that although browsing by red deer had significant and negative effects on measures of soil biological properties and nitrogen cycling, the topographic position of the exclosures was the dominant factor affecting these measures at the landscape scale. Likewise, studies of grazer effects in native grassland at Yellowstone National Park revealed that topographic position, and associated variation in soil conditions, was often the dominant factor affecting soil biological properties and rates of nutrient cycling (Tracy and Frank 1998; Verchot et al. 2002). Also, in the previously mentioned study by Sankaran and Augustine (2004), landscape-scale constraints on soil organic matter content were found to overarch the negative effects of grazers on microbial abundance. Further, in sub-Arctic grassland in Abisko, northern Sweden, local-scale variation in soil abiotic properties was shown to have a stronger impact on soil biota than did simulated reindeer grazing (Sørensen et al. 2009).

At the landscape scale, herbivore impacts are also likely to vary substantially across environmental gradients of soil fertility, topographic position, and climate (Olff and Ritchie 1998; Olff et al. 2002; Anser et al. 2009). For example, using an airborne remote-sensing system, Anser et al. (2009) showed that the impact of large herbivores (including elephants, buffalo, giraffe, and zebra) on vegetation structure in the Kruger National Park, South Africa, varied by topographic position, geologic substrate, and time since grazing was excluded. In particular, they found that effects of herbivore exclusion on woody vegetation structure were greater on nutrient-rich clay soils formed on basalt substrate than on sandy soils formed on granite, and in lowland (where water, nutrients, and hence forage are locally abundant) than upland areas. Similarly, Augustine and McNaughton (2006) found in Kenyan rangeland that the effects of grazers (including impala, zebra, and buffalo) on plant production and nitrogen cycling varied with both soil fertility and with annual rainfall patterns. They found that in low-rainfall years grazers reduced aboveground plant production in both high- and low-fertility sites. However, in a high-rainfall year grazers enhanced aboveground plant production and soil nitrogen availability in high-fertility sites, but suppressed these properties on low-fertility sites. These findings not only indicate that grazer effects on aboveground plant production and nitrogen cycling are tightly coupled at the landscape scale, but also that climatic variability is an important determinant of grazer effects in grassland systems.

Effects of herbivores can also vary greatly in space and time depending on population fluctuations and herbivore body size (Bakker et al. 2004). For example, the scale of redistribution of nutrients (discussed later in this chapter) and the feeding behaviour of herbivores, as well as the extent to which they influence soil physical properties, is body-size-dependent (Olff and Ritchie 1998; Bakker et al. 2004). As a consequence, different-sized herbivores have the potential to have different effects on soil nutrient cycling in the same habitat, but at different temporal and spatial scales. Very few studies have explored these issues, although Bakker et al. (2004) used an exclosure set-up in grassland grazed by cattle, rabbits, and meadow voles, in which fences with different hole sizes were used to selectively exclude different components of the grazing herbivore community based on body size. They found that that exclusion of cattle grazing led to a 1.5-fold increase in net nitrogen mineralization, which was attributed to a build-up of litter on the soil surface. In contrast, the combined exclusion of cattle and rabbit grazing (i.e. vole grazing only) altered the timing of N cycling, causing an autumn maxima in nitrogen mineralization which coincided with a peak in vole density and high levels of N input through vole faeces. In general, very little is known about how herbivore population dynamics and life histories impact on nutrient cycling at the landscape-scale or their importance relative to abiotic controls on ecosystem processes, such as topography, soil fertility, and climate. Future studies on these topics are therefore required to fully understand how the types of herbivore effects that have been well studied at local spatial scales on ecosystem processes are manifested at the landscape scale.

One area that has received recent attention is the idea that gradual changes in ecosystems resulting from herbivory can lead to dramatic stepwise transitions, or alternative stable states (Rietkerk and van de Koppel 1997; van de Koppel et al. 1997; Rietkerk et al. 2004; Van der Wal 2006). Consequently, ecosystems can contain mosaics of communities in several alternative states which are each relatively resistant to change, but can exhibit rapid shifts to another state if subject to altered grazing pressure. The concept of alternative stable states has been applied to situations where overgrazing has led to soil degradation and an associated complete and irreversible collapse in existing vegetation, and ultimately the formation of a two-phase mosaic consisting of patches of dense vegetation alternating with almost bare areas (McNaughton 1983; Belsky 1986; Rietkerk and van de Koppel 1997; van de Koppel et al. 1997). For instance, overgrazing in semi-arid ecosystems in the Sahel region of Africa, and resultant soil erosion, compaction, reduced infiltration, and increased water runoff, has been shown to trigger catastrophic shifts from a highly productive vegetated state to a severely degraded state (Fig. 4.10) (Rietkerk et al. 1997; Rietkerk and van de Koppel 1997). Similar mosaics have also been reported in Arctic coastal plant communities in Hudson Bay, where summer grazing by geese creates grass-dominated grazing lawns, while spring grubbing for roots and rhizomes causes destruction of existing plant communities and the creation of patches of bare ground with reduced nutrient availability that are slow to recover (Jefferies 1998; Srivastava and Jefferies 1996). It has been argued that improved understanding of such herbivore-driven alternative stable states could help to establish whether an ecosystem might collapse under certain conditions, or whether there are warning signs of imminent collapse (Rietkerk and van de Koppel 1997).

Although typically associated with overgrazing, the alternative stable states framework has also been applied to evaluate non-degenerative herbivore-driven vegetation change, especially in tundra ecosystems. As argued by Van der Wal (2006), the Arctic tundra can occur in essentially three different vegetation states, and relatively sudden transitions between them are driven by reindeer grazing: one is characterized by lichen dominance, one by moss dominance, and one by an abundance of graminoids. Importantly, herbivore-driven transitions from lichen-dominated to

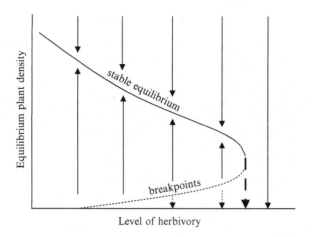

Fig. 4.10 A catastrophic fold showing the relationship between equilibrium plant density and the level of herbivory. Sudden jumps in plant density occur at distinct levels of herbivory, leading to catastrophic shifts from a highly productive vegetated state to a severely degraded state. From Rietkerk et al. (1997), with permission from Wiley-Blackwell.

moss- and graminoid-dominated states are associated with increases in plant productivity and hence the carrying capacity for reindeer populations (Fig. 4.11). This idea is consistent with the classic work of Zimov et al. (1995), described in Chapter 5, which proposes that the extinction of mega-herbivores at the end of the Pleistocene was responsible for a large-scale biome shift from productive grass-dominated steppe to unproductive moss-dominated tundra across Alaska and Russia. However, as documented by Van der Wal (2006), the more recent expansion of reindeer across the Arctic provides yet more evidence that herbivores are responsible for sudden shifts between the three vegetation states that enhance their carrying capacity. For example, grazer-driven switches from lichen to moss-dominated vegetation (caused primarily by selective grazing and trampling) have been reported in Greenland (Thing 1984), Svalbard (Cooper and Wookey 2001; Van der Wal et al. 2001), Russia (Vilcheck 1997), and North America (Manseau et al. 1996), allowing tundra to sustain a greater density of reindeer (Van der Wal 2006). Also, there are many reports across the Arctic of intensive reindeer grazing causing a transition from moss- to graminoid-dominated tundra (Olofsson et al. 2004; Van der Wal and Brooker 2004; Van der Wal et al. 2004), which as already discussed in section 4.2.1 is associated with plant–soil feedbacks that enhance soil nitrogen cycling and plant productivity (Van der Wal et al. 2004), enhancing the carrying capacity of the ecosystem. What is clear from these examples is that herbivores have the capacity to force sudden and predictable shifts in vegetation state within terrestrial ecosystems, with both catastrophic negative (Rietkerk and van de Koppel 1997; van de Koppel et al. 1997) and positive (Van der Wal 2006) consequences for their functioning and carrying capacity.

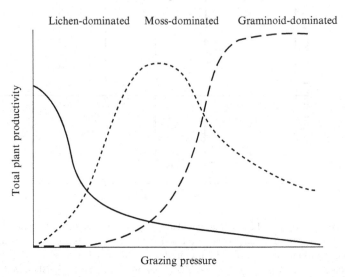

Fig. 4.11 Schematic representation of the occurrence of tundra vegetation states in relation to grazing pressure exerted by reindeer or caribou (*Rangifer tarandus*). Herbivore-driven state transitions from lichen- to moss- and graminoid-dominated state are associated with an increase in plant productivity and carrying capacity for reindeer populations. From Van der Val (2006), with permission from Wiley-Blackwell.

4.3 The role of plant traits in regulating herbivore impacts

As discussed in Chapter 3, plant traits provide a framework for understanding how changes in vegetation composition and diversity influence belowground communities and the ecosystem processes that they drive. Given that grazing is both dependent on, and influences, plant functional traits (Huntly 1991; Díaz et al. 2006), an understanding of plant trait responses to grazing helps provide a framework for investigating feedbacks between grazers, plants, soil organisms, and nutrient cycles, as well as how they vary across environmental gradients such as soil fertility and climate.

Although most studies on herbivore effects of vegetation do not explicitly consider plant traits, conceptual models of plant trait responses to grazing do exist (e.g. Diaz et al. 2006). For example, the range-succession model predicts that grazing increases the cover of annual plants over perennials, leads to the replacement of palatable by unpalatable plants, and causes replacement of tall grasses by short grasses, shrubs, and prostrate herbs (Dyksterhuis 1949; Arnold 1955). Likewise, when adapted to consider responses to grazing, Westoby's (1998) LHS (leaf-height-seed) model predicts that moderate, selective grazing promotes plants with a particular set of traits linked to unpalatability, whereas heavy, non-selective grazing promotes plants with traits associated with palatability. Other models distinguish plant trait responses to grazing in productive and less-productive ecosystems, where the proximate cause of variation in productivity is water and nutrient availability. For instance, the generalized model of Milchunas et al. (1988) predicts that plant trait responses to grazing depend on rainfall (as a proxy for productivity) and grazing history, and are the most marked in humid (productive) systems with a long grazing history where short annual plants with prostrate growth are encouraged. Likewise, models such as those proposed by Grime (1977) and Coley et al. (1985) predict that plant trait responses to grazing depend on productivity, in that grazing promotes palatable plants (i.e. those with high grazing tolerance, involving fast re-growth of high-quality tissue and low investment in structural defence) in humid situations, and unpalatable plants (i.e. those with grazing avoidance, involving slow-growing tissue and high investment structural defence) in dry systems. Moreover, as suggested by Herms and Mattson (1992), grazing history also plays an important role: the stronger the historical impact of grazing, the stronger the selection for plants that have traits associated with investment in grazing avoidance.

In order to determine whether some widely recognized plant trait responses to grazing are consistent at the global level, Díaz et al. (2006) performed a meta-analysis of 197 studies selected from around the globe. In this synthesis, they found that grazing favoured annual over perennial plants, short over tall plants, prostrate over erect plants, and plants with stoloniferous and rosette architecture over tussocks. Also, and consistent with the models mentioned above, they found that grazing consistently favoured plants with traits that confer low palatability, and that this response was strongest in dry systems with a long grazing history. An important finding of Díaz et al. (2006) was that in almost all cases that they studied the response of plant traits to grazing varied depending on both climate and grazing history, suggesting that vegetation responses to grazing are modulated by these two factors. This finding has important consequences for understanding grazer effects on belowground properties and nutrient cycles at the global scale, given the strong link between plant traits and the belowground subsystem discussed in Chapter 3. However, as noted by Díaz et al. (2006) our knowledge of the effect of grazing is restricted to a surprisingly small number of plant traits, and data availability varies markedly across regions, making it difficult to generalize too much about plant responses to herbivory across gradients of climate, grazing history, and soil fertility.

Although the above models and literature synthesis of Díaz et al. (2006) do not explicitly consider cascading effects of grazing on the belowground subsystem and nutrient cycling, they provide a very general framework for understanding variation in grazer impacts belowground across ecosystems of differing fertility and precipitation regimes. While the magnitude and direction of herbivore effects on the belowground biota, nutrient dynamics and plant production differs greatly across ecosystems (Bardgett and Wardle 2003), this variation appears to depend largely on ecosystem productivity in a way that is broadly consistent with the models of plant trait responses discussed above. For instance, positive effects of grazing on soil biota and nutrient cycling covered earlier in this chapter appear to dominate in productive ecosystems, where models predict that grazing selects for palatable, nutrient-rich plants (i.e. tolerance strategy) (Grime 1977; Coley et al. 1985) which readily supply labile substrates to soil as nutrient-rich litter and root exudates (Bardgett and Wardle 2003). In contrast, negative effects of grazing on soil biota and nutrient cycling dominate in less-productive situations, where models predict that grazing selects for unpalatable, slow-growing plants (i.e. avoidance strategy) (Grime 1977; Grime 2001; Coley et al. 1985). As highlighted earlier in this chapter, these unpalatable plants have traits that result in poor-quality litter which decomposes slowly and therefore reduces nutrient availability and plant growth (Bardgett and Wardle 2003). Moreover, selection for unpalatable plants, and hence negative effects on soil nutrient cycling, are likely to be stronger with grazing history (Bardgett et al. 2001b), which is consistent with the emphasis that is placed on grazing history in modulating plant trait responses to grazing (Herms and Mattson 1992; Díaz et al. 2006). Also, the above feedbacks resulting from herbivory will further enforce the dominance of particular plant traits, thereby strengthening the effects of herbivores on the ecosystem. For example, the selection of plant traits linked to grazer tolerance in productive situations and consequent promotion of nutrient cycling should further encourage plants with this growth strategy. In contrast, the selection of plant traits linked to grazer avoidance in less-productive systems and consequent slowing-down of nutrient cycling, would further enhance the success of plants that exhibit an avoidance strategy (Bardgett and Wardle 2003).

A handful of studies have experimentally investigated the linkages between herbivory or foliage palatability, plant traits, and belowground processes. For example, comparative studies across multiple plant species in temperate (Grime at al. 1996; Cornelissen et al. 1999) and sub-Arctic (Cornelissen et al. 2004) communities have found the palatability of foliage to herbivores and decomposability of plant litter to be significantly correlated because both processes are governed by similar suites of functional traits. Likewise, Wardle et al. (2002) used a broad range of plant species collected from New Zealand forests to show that plant species that produced high-quality litter that decomposed rapidly were also those that were disadvantaged by selective browsing (i.e. more palatable), whereas those that produced poor-quality litter were those that were advantaged by browsing (i.e. less palatable). Consequently, it was concluded that the decomposability of litter and vegetation responses to herbivory are governed by the same, or similar, suites of traits, notably the concentration of secondary metabolites and structural carbohydrates in plant tissue. However, the relationship between foliage palatability and litter decomposability is not universal, and Kurokawa and Nakashizuka (2008) found no evidence of such a relationship among a large number of tropical rainforest tree species in Sarawak, Borneo. With regard to the decomposer biota, little is known about the role of plant traits in regulating the response of soil organisms to defoliation. However, Guitian and Bardgett (2000) found that grass species varied in their allocation response to defoliation, and that the greatest soil microbial response was detected in the species that allocated most resources to roots when subjected to clipping. Further studies are clearly needed in this area to determine how these responses relate to inter-specific differences in plant traits, especially in the context of variation among grasses in their relative investment of resources into growth (i.e. tolerance) and anti-herbivore defence (avoidance). In this light, Massey et al. (2007) showed that 18 grass species varied considerably in their investment in defence and growth rate, and that the two attributes were negatively correlated. This suggests that there is likewise a spectrum of belowground responses to herbivory at the individual species level, which relate to the balance of allocation of resources to growth (i.e. tolerance) and defence (i.e. avoidance).

There is clearly much to be learned about plant trait responses to grazing and how they link to belowground biota and the processes of decomposition and nutrient cycling that they drive. However, the evidence presented here, albeit based on a limited number of traits and studies, suggests that a conceptual framework for investigating plant trait responses to herbivory could be extended to incorporate belowground responses, and hence provide a basis for improved understanding the consequences of herbivory for ecosystem function in different contexts. However, as highlighted by Díaz et al. (2006), this requires more investigations of plant trait responses to grazing that both follow a common framework (e.g. by using a common list of traits) and are conducted across several contrasting locations. Moreover, while there is much understood about how plant traits may drive aboveground–belowground linkages and feedbacks (as discussed in Chapter 3), our knowledge of how plant traits select for traits of soil organisms and vice versa is lacking. Such information is required to develop an improved mechanistic understanding of herbivore effects on aboveground–belowground linkages and how these linkages drive ecosystem processes.

4.4 Aboveground trophic cascades and consequences for belowground properties

Aboveground herbivores are consumed by their predators, which are well known to influence their densities and population dynamics. This can in turn lead to cascading effects, or trophic cascades that affect the biomass, productivity, and composition of the plant community. While there has been some historical debate about the importance of trophic cascades in terrestrial ecosystems (e.g. Strong 1992; Polis 1994), and there are many ecosystems for which they are probably unimportant, there is a cumulating number of examples for which they have been shown to be of great importance (Pace et al. 1999; Halaj and Wise 2001; Terborgh and Estes 2010). As described so far in this chapter, herbivores can exert both positive and negative effects on the functioning of the belowground subsystem through their influence on individual plants and plant communities. It therefore follows that predators of herbivores should reverse these effects, at least if they substantially reduce herbivore densities and therefore the extent of herbivory in the ecosystem (Fig. 4.12). However, despite widespread awareness of the importance of belowground trophic cascades for

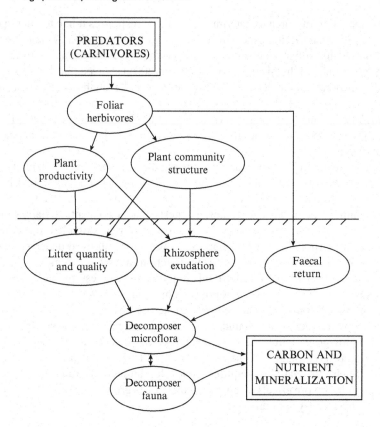

Fig. 4.12 Schematic diagram of the mechanistic basis through which aboveground carnivores can indirectly influence decomposer organisms and the ecosystem-level processes that they perform. Adapted from Wardle et al. (2005).

soil processes and ecosystem functioning (see Chapter 2), and recent interest in how predator diversity might influence ecosystem functioning (Duffy et al. 2007; Bruno and Cardinale 2008), the issue of how aboveground trophic cascades affect the functioning of the belowground subsystem has only recently attracted attention (e.g. Feeley and Terborgh 2005; Wardle et al. 2005; Maron et al. 2006; Schmitz 2008a; Wardle 2010).

Predators of herbivores have been shown in some studies to potentially alter the quality and quantity of resources entering the belowground subsystem; this in turn has the potential to alter the soil food web and processes driven by the decomposer subsystem. First, top predators can greatly alter the functional composition of the vegetation by altering the levels of herbivory that the vegetation is subjected to. For example, as will be discussed in Chapter 5, cougars (*Puma concolor*) and wolves (*Canis lupus*) in North America are well known to reverse the negative effects of their prey (cervids such as deer, elk, and moose (*A. alces*) on the forest understorey and the growth of palatable deciduous plant species (Ripple and Betscha 2008; Creel and Christianson 2009). This will in turn reverse the effects of cervids on the quality of litter entering the decomposer subsystem (e.g. Pastor et al. 1993). Second, predators of herbivores may reverse the effects that herbivores have on secondary leaf chemistry, and therefore the quality of leaf litter that the plant subsequently returns belowground. Evidence for this emerges from the study of Stamp and Bowers (1996) who found that the induction of defence compounds (iridoid glycosides) in *Plantago lanceolata* by invertebrate herbivores was reversed by the predators of these herbivores. Third, predators can reverse the effects of herbivores on the quantity of resources entering the belowground subsystem. For example, Post et al. (1999) showed on Isle Royale, Michigan, USA, that predation of moose by wolves reverses the adverse effect of moose on the radial growth of *Abies balsamea* trees, and that, over time, radial growth tracks the population dynamics of wolves. This would in turn reduce the quantity of resources that *A. balsamea* inputs to the belowground subsystem. As an example of cascading effects of predators at a whole ecosystem scale, Terborgh et al. (2001, 2006) used islands created by a hydroelectric impoundment in Venezuela to highlight the indirect impact of predator-induced trophic cascades on vegetation characteristics. Here, small islands lack predators such as large snakes, raptors, jaguar (*Panthera onca*), and puma (*Puma concolor*), and as such have hyper-abundant animals of lower trophic status such as rodents, howler monkeys (*Alouatta* spp.), iguanas, and leaf-cutter ants. The net result of this lack of predators is a substantial reduction in seedlings and saplings of canopy tree species (Terborgh et al. 2006), and an apparent shift to plant species that are less preferred by the herbivores and which probably have poorer-quality litter (Feeley and Terborgh 2005).

Although top predators indirectly control the quantity and quality of plant-derived materials entering soil, the consequences of this for the belowground subsystem have been explicitly studied only recently. For example, Frank (2008) found that rates of soil nitrogen mineralization were negatively influenced by the consumption of ungulates by wolves (*Canis lupus*), and Feeley and Terborgh (2005) showed that mammalian predators of other mammals and leaf-cutting ants reduced carbon-to-nitrogen ratios in the soil. Further, Wardle et al. (2005) showed that predators of aphids indirectly enhanced soil carbon concentrations, and Dunham (2008) showed that insectivorous birds and mammals had important indirect influences on soil phosphorus turnover. These predator-induced changes in soil chemical properties are likely to be linked to changes in the composition of the soil biota, but few studies have investigated whether aboveground trophic cascades instigated by aboveground predators trickle through to the soil food web. However, in the study of Dunham (2008), the exclusion of insectivorous birds and mammals was found to have indirect positive effects on soil-associated invertebrate taxa known to be important for driving soil processes (Fig. 4.13). Further, in a manipulative glasshouse experiment, Wardle et al. (2005) showed that aboveground predators of aphids (lacewings and a ladybird beetles) induced a trophic cascade that altered plant species composition and thereby

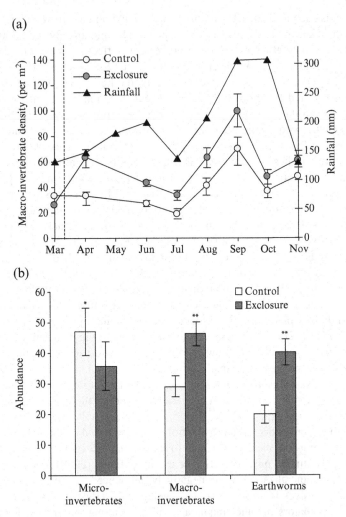

Fig. 4.13 Abundance of soil invertebrates in plots with and without exclusion of top aboveground predators (mammalian and avian insectivores) in tropical rainforest in Coté d'Ivorie. (a) Total macroinvertebrate density in exclosure and control plots over an 8 month period; (b) total abundance of invertebrate groups in exclosure and control plots summed over an 8 month period; *, ** indicates statistical significance at $P = 0.05$ and 0.01 respectively. Error bars are SE. From Dunham (2008), with permission from Wiley-Blackwell.

promoted the primary consumers (microflora) and tertiary consumers (top predatory nematodes) of the soil food web. The aboveground predators also sometimes indirectly altered microbial community structure, promoting bacteria at the expense of fungi, and reduced the diversity of herbivorous nematodes. In contrast, in an experimental study involving an understorey forest shrub, Dyer and Letourneau (2003) showed that species diversity within each of three consumer trophic levels in a decomposer food web was unaffected by manipulation of a top predatory beetle that feeds upon herbivorous and possibly detritivorous insects; diversity within each trophic level was instead driven mainly by the manipulation of basal resources. Collectively, these studies highlight the multitrophic linkages that can exist between aboveground and belowground food webs, and which may involve multiple trophic levels both above and belowground.

As discussed for foliar herbivores, aboveground predators of herbivores can vary greatly in their effects on the belowground subsystem. This arises not just through predators reversing the effects of herbivores (which themselves can have positive or negative effects), but also through the influence of predator identity in its own right. First, different predators may differentially affect the population dynamics of their herbivorous prey, and thus the temporal pattern of plant material addition to the belowground subsystem. For example, Wardle et al. (2005) found that two different predators of aphids differed greatly in their effects on the timing of peaks in aphid populations, and therefore their effects on both the plant and soil community. Second, different predator species differ greatly in prey specificity. It is expected that indirect effects of aboveground predators on the belowground subsystem are greatest when they are relatively specific, and when they feed on herbivore taxa that are themselves major drivers of the ecosystem. These effects may be less strong for generalist predators in reticulate food webs (Polis 1994) or for aboveground predators that feed on both herbivores and decomposers (Scheu 2001). Third, differences in behaviour or hunting mode between different predator species may influence herbivore feeding pattern, and therefore the plant community and belowground properties (Schmitz et al. 2004). This idea has been little explored, although Schmitz (2008b) found that two grasshopper-feeding spider species with different feeding modes, one being an active hunter and the other being a sit-and-wait ambush predator, exerted opposing effects on plant diversity, net aboveground primary productivity and rates of soil nitrogen mineralization in grassland. These effects were attributed to different responses to the two different predators by their grasshopper prey, the dominant herbivore species in this grassland system, although whether these predators had opposing effects relative to predator-free conditions is not known. Finally, some of the strongest belowground effects of aboveground trophic cascades have involved aboveground consumers being subjected to predation pressures to which they have not evolved, for example through invasions by novel predators, or through hunting by humans colonizing new land masses; trophic cascades of this type will be explored in Chapter 5.

4.5 Spatial movement of resources by consumer organisms

The preceding sections illustrate how aboveground consumers modulate the functioning of terrestrial ecosystems through a variety of mechanisms. Through their movements and foraging behaviour, consumer organisms can also transfer resources both within and between terrestrial ecosystems, largely through their faeces and carcasses, with potentially major consequences for spatial patterns of nutrient cycling and vegetation. Moreover, terrestrial ecosystems are also affected by both terrestrial and aquatic consumers that are capable of transferring resources from aquatic to terrestrial habitats, with potentially major consequences for terrestrial ecosystems both aboveground and belowground. Here, we discuss the consequences of these types of consumer-mediated resource subsidies for terrestrial ecosystems. First we explore the consequences for both the aboveground and belowground subsystem of consumer-mediated resource distribution within and between terrestrial ecosystems, and second we examine the consequences of resource transfers from aquatic to terrestrial ecosystems by aquatic and terrestrial consumers.

4.5.1 Resource transfers across land

Movement of mammalian herbivores during foraging is driven both by their behavioural repertoire and by livestock management, and operates at a range of temporal and spatial scales (Bailey and Provenza 2008; Boone et al. 2008). For example, wild herbivores may move up and down elevation zones to take advantage of temporal variation in plant phenology and forage quality, or they might migrate from one region to another due to forage or water shortages, or changes in precipitation patterns (Bailey and Provenza 2008). At finer scales, wild herbivores may move relatively short distances (metres to kilometres) to select feeding sites where forage is more abundant or nutritious, or to carry out non-foraging activities such as resting and ruminating (Bailey and Provenza 2008). Likewise, in livestock systems, herbivores are moved from field to field to maximize forage use, and herders move their animals to varying degrees in response to extreme climatic conditions or to allow them to cope with seasonal changes in forage supply (Boone et al. 2008). For example, in the Mongolian steppe, herders can move livestock over distances ranging from 10 km several times a year, to over 300 km between the mountains and lowlands, to make the best use of forage supply across a range of habitats. Meanwhile, in mountainous regions, herders move their livestock from sheltered valleys to higher elevations during the summer months to make use of snow-free mountain meadows (Boone et al. 2008). Given the potential for herbivores to modulate soil nutrient cycling and vegetation dynamics (as discussed so far in this chapter) and also to transfer resources from one site to another in their faeces and bodies, such herbivore movements will inevitably have important consequences for spatial patterns of nutrient cycling and vegetation.

Despite the tremendous research effort that has gone into understanding the behavioural mechanisms that result in grazing distribution patterns (reviewed by Bailey and Provenza 2008), comparatively little is known about the consequences of herbivore movement for nutrient transfers within and across ecosystems. Moreover, most of what is known is from agricultural situations, where several studies have shown that livestock promotes the patchy return of nutrients through using different parts of their habitat for feeding and for excretion. For example, Edwards and Hollis (1982) found that grazing by horses on pastures in the New Forest in southern England created a pattern of latrine and grazing areas, with generally higher concentrations of soil nutrients and greater plant biomass in the latrine areas. This is similar to the previously mentioned example of Willot et al. (2000), who found that although rabbits deposit dung throughout their range, dung deposition is greatest in places where they congregate, and a large amount of dung is deposited in a small number of latrines. Likewise, although cattle tend to deposit their dung haphazardly in level pastures (White et al. 2001), a large proportion of this dung is often input to areas that they use for resting and ruminating. Consequently, soil nutrients become concentrated in relatively small areas where the animals spend a large proportion of their time (Syers et al. 1980; Afzal and Adams 1992; Kohler et al. 2006).

Several studies have also explored the consequences of redistribution of nutrients through faeces for nutrient budgets and pasture production. For example, in a study of *Brachiaria* pastures in Brazil, Boddey et al. (2004) found that increasing cattle stocking rates caused a decline in pasture quality because more nitrogen was deposited in the form of urine and faeces, leading to substantial gaseous and leaching losses of nitrogen. Moreover, much of this N was deposited in rest areas and around drinking troughs where the vegetation was too trampled to derive much benefit from it, while the availability of nitrogen in other areas was correspondingly reduced (Boddey et al. 2004). Similarly, in a study of mountain pastures in the Swiss Alps, Jewell et al. (2007) found that cattle redistribute substantial amounts of phosphorus by feeding and defecating in different places. They found that cattle dropped most of their dung in relatively small areas where they gathered for ruminating and resting, and, as a result, a few small patches had accumulated high amounts of phosphorus (in excess of 50 kg P ha^{-1} year^{-1}), while much larger areas of pasture had become depleted. They concluded that by concentrating nutrients in small areas, cattle promote nutrient loss from mountain pastures and thus cause a gradual decline in their productivity. However, they also note that the significance of this type of nutrient redistribution for long-term pasture production will depend upon factors such as the stability of the spatial pattern caused by cattle and the magnitude of the phosphorus depletion relative to other ecosystem inputs and outputs.

Relatively few studies have explored the influence of herbivores on nutrient translocation and spatial distribution of soil nutrients in natural ecosystems, although similar mechanisms as described above are likely to be at play, albeit over larger spatial scales. For example, Schütz et al. (2006) examined the role of red deer in the redistribution of phosphorus (the primary limiting nutrient) in a subalpine grassland ecosystem in the Swiss Alps. They compared phosphorus removal by grazing of aboveground plant biomass with phosphorus input due to the deposition of faeces, across an extensive sampling gird that covered the entire ecosystem. Using this approach, they found that the proportion of heavily grazed short-grass vegetation (dominated by *Festuca rubra* and *Briza media*) in the landscape increased with increasing soil phosphorus content, indicating that red deer preferably grazed patches that were more phosphorus-rich. They also found that rates of phosphorus removal by grazing in the phosphorus-rich areas was greater than from phosphorus-poor areas, and that this enhanced loss was in excess of the increase of phosphorus input due to greater faecal deposition, leading to an average annual net phosphorus loss. This net loss, was, however, very low (i.e. 0.083 kg ha^{-1}) and it was estimated that it would take 1660 years for the soil phosphorus pool in the most phosphorus-rich parts of the ecosystem to be depleted to the levels observed in the phosphorus-poor parts. They also noted that the yearly net phosphorus loss caused by grazing was comparable to that reported for vegetation communities grazed by elk in the Rocky Mountain National Park, USA (Schoenecker et al. 2002), indicating that phosphorus translocation by herbivores in these types of grazing systems may impact on the spatial dynamics of vegetation only in the long-term.

While the above example concerns soil phosphorus, there is also evidence that large herbivores in natural grassland influence the spatial distribution of soil nitrogen and rates of nitrogen mineralization across a wide range of spatial scales. For example, McNaughton et al. (1997b) found that free-ranging resident grazers, principally Thomson's (*Gazella thomsoni*) and Grant's (*Gazella granti*) gazelles, topi (*Damaliscus korrigum*), and kongoni (*Alcelaphus buselaphus*), enhance rates of nitrogen mineralization and sodium availability in areas where they preferentially graze, and thus contribute to the spatial heterogeneity of soil fertility at the landscape scale. Also, Augustine and Frank (2001) found that native migratory ungulates, namely elk, bison, and pronghorn antelope, influenced the spatial distribution of soil nitrogen and rates of nitrogen mineralization at a range of spatial scales in grasslands of Yellowstone National Park, USA. Through geostatistical analyses of soils collected inside and outside long-term fenced exclosures, they found that grazers altered the spatial distribution of soil nitrogen properties at all scales ranging from individual plants (<10 cm), to whole plant communities (30 m), to topographically variable landscapes. Finer-scale heterogeneity was attributed to the promotion of local plant species diversity, variation in plant turnover by grazers, and possibly the patchy return of urine and dung. In contrast, landscape-scale variability, which was significantly enhanced by grazers, was attributed to variation in grazing intensity, bedding-site selection, and dunging across the topographically variable site. Indeed, it is well known that elk respond to heterogeneity in forage production by concentrating in areas of high productivity at any one time (Frank and McNaughton 1992) and by landscape-scale selection of winter grazing locations based on elevation and grassland type (Pearson et al. 1995). Overall, these results were taken to demonstrate that, in addition to the effects of topography and vegetation on spatial patterning of soil nutrients, biotic components of ecosystems such as large herbivores not only respond to resource heterogeneity, but also play a significant role in contributing to it.

A further mechanism by which herbivores can enhance spatial heterogeneity in nutrient cycling and vegetation at the landscape scale is through their carcasses, which cause concentrated pulses of nutrients into the soil. This mechanism is poorly explored, but some studies clearly point to significant and lasting effects of carcasses in terrestrial ecosystems. For example, Towne (2000) studied the effects of ungulate carcasses in tallgrass prairie in Kansas, USA, and found that they created hotspots of soil fertility and plant production, thereby increasing community heterogeneity. However, as emphasized by Towne (2000), the significance of ungulate carcasses as contributors to grassland vegetation dynamics and nutrient cycling becomes especially significant when considered from a historical perspective: episodic disasters such as severe droughts, harsh winters, or endemic disease can produce high rates of herbivore mortality, and hence such historical events have enriched grasslands for eons (Towne 2000). In another study, Bump et al. (2009c) showed that carcasses of white-tailed deer (*Odocoileus virginianus*) created biogeochemical hotspots that contributed to resource heterogeneity and shifting tree competitive interactions in northern hardwood forest in the Upper Peninsula of Michigan, USA. Moreover, Bump et al. (2009a) showed that heterogeneity in soil nutrients and plant nutrient content caused by carcasses, in this case of moose (*A. alces*) in boreal forest of Isle Royale National Park, Michigan, USA, was modulated by wolves (*C. lupus*). Through their predatory behaviour, wolves influenced the spatial distribution of moose carcasses, which in turn determined spatial patterns of soil nutrients, microbial communities, and plant tissue quality (Fig. 4.14). Meanwhile, Parmenter and MacMahon (2009) studied the effects of carcass decomposition of a wide range of vertebrates (including mammals, two birds, a snake species, and one frog species) on soil nutrients in semiarid shrub-steppe in Wyoming, USA. Regardless of source, they found that carrion-decomposition processes enriched soils in nitrogen in the vicinity of the carcass, causing localized changes in the plant community. Although this contributed minimally to the total nitrogen budget of the ecosystem, it was argued that vertebrate carcasses have an impact on soil nutrient status and turnover at the local patch-scale level, thereby contributing to community heterogeneity.

Finally, it is important to note that carcasses of invertebrates can also significantly impact on terrestrial ecosystems. This is best illustrated by the study of Yang (2004), who examined the ecological effects of resource pulses resulting from periodic deposition of carcasses of cicadas of the genus *Magicicada*. These insects, which can be the most abundant herbivore in North American deciduous forests, spend most of their lives belowground feeding on xylem in tree roots, but every 17 years they emerge in vast numbers, mate, lay eggs, and die. Yang (2004) simulated a cicada emergence event by applying cicada carcasses to the forest floor, and found this to stimulate the biomass of soil bacteria and fungi, soil inorganic nitrogen availability, and hence the nitrogen content and seed mass of an understorey plant, the American bellflower (*Campanulastrum americanum*) (Fig. 4.15). Given that the spatial distribution of cicadas is highly variable and dynamic at the landscape scale, Yang (2004) argued that the patchiness of cicada distributions may contribute to spatial and temporal heterogeneity in resource pulse effects in forest ecosystems. Also, as highlighted by Yang (2004), while these findings result from the unusual life history of a single insect genus, they illustrate a more general potential consequence of resource pulses for aboveground and belowground subsystems.

(a)

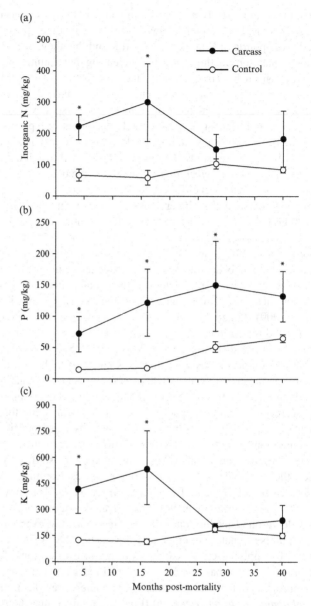

(b)

(c)

Fig. 4.14 Impact of carcasses of moose (*Alces alces*) killed by wolves (*Canis lupus*) on soil nutrients in the boreal forest of Isle Royale National Park, Michigan, USA. Data show soil concentrations of (a) inorganic nitrogen (nitrate + ammonium), (b) phosphorus, and (c) potassium in soils from wolf-killed moose carcass sites (solid circles) and paired control sites (open circles) at 4, 16, 28, and 40 months post-mortem. Asterisks indicate significant ($P < 0.05$) differences between carcass and control sites at each postmortem sampling time. Error bars show means $\pm$ SE (some are too small to be seen). Note the different *y*-axis scales.

From Bump et al. (2009a), with permission from the Ecological Society of America.

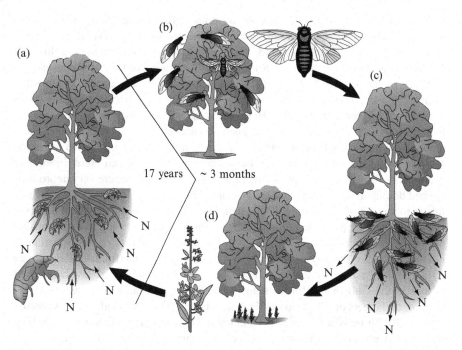

Fig. 4.15 Schematic of the belowground and aboveground effects of a 17-year pulse of nitrogen from cicada carcasses: (a) for 17 years, cicada nymphs feed on tree xylem, slowly incorporating belowground nitrogen (N) absorbed by the tree's roots; (b) upon emergence, adult cicadas mate and lay eggs within a several-week period, and then die and drop to the forest floor; (c) the accumulated nitrogen in their carcasses is released after a burst of activity by microbial decomposers; and (d) this spike of available soil nitrogen leads to increased nitrogen content and seed size of the American bellflower (*Campanulastrum americanum*), an under-storey plant. From Ostfeld and Keesing (2004) and derived from Yang (2004), with permission from the American Association for the Advancement of Science.

4.5.2 Resource transfers from aquatic to terrestrial ecosystems

Most terrestrial ecosystems occur in the proximity of aquatic ecosystems, such as oceans, lakes, rivers, and streams. Aquatic and terrestrial communities do not function independently of one another, and there are numerous examples of interactions between these two communities, especially at their interface. As such, terrestrial ecosystems are affected not just by terrestrial consumers that derive their energy from terrestrial plants, as discussed so far in this chapter, but also by both terrestrial and aquatic consumers that derive at least part of their resources from the aquatic environment. It is well known, for example, that many riparian terrestrial predators can obtain a large proportion of their nutrition from aquatic prey (Sabo and Power 2002; Ballinger and Lake 2006), and that emerging aquatic insects which spend their larval phase in the water can have major effects on terrestrial communities during their adult phase (Knight et al. 2005; Gratton and Vander Zanden 2009). Both aquatic and terrestrial consumers are capable of transferring resources from aquatic to terrestrial habitats, with potentially major consequences for terrestrial ecosystems both aboveground and belowground. We now discuss the consequences of this type of consumer-mediated resource subsidy for the functioning of terrestrial ecosystems.

Many of the strongest and best documented examples of aquatic to terrestrial resource transfer involves seabird activity on coastal and island ecosystems (Polis et al. 1997; Anderson et al. 2008). Here, seabirds feed on aquatic organisms (mainly fish) in the ocean, and deposit them as guano on nearby terrestrial ecosystems. These inputs can represent the main source of nutrient inputs for unproductive terrestrial ecosystems adjacent to productive marine ecosystems (Polis and Hurd 1996). For example, in coastal ecosystems in Antarctica, plant communities are very unproductive and dominated by lichens and mosses, and they therefore input very little organic matter to the soil. Here, nesting Adélie penguins transfer large quantities of nutrients and organic matter from the ocean to their nesting colonies, leading to the formation of ornithogenic soils that can contain up to 100 times the density of soil bacteria and microbial biomass that occurs in nearby non-ornithogenic soils (Ramsay 1983; Roser et al. 1993). Another system which has been extensively studied involves desert islands in the Gulf of California, some of which support large colonies of nesting seabirds and others of which do not (Polis and Hurd 1996; Wait et al. 2005; Anderson et al. 2008). Those islands which contain seabird colonies, and therefore receive high inputs of ocean-derived resources, have concentrations of soil nutrients that are several times greater than those that lack seabirds (Anderson and Polis 1999). This enhanced nutrient input has important bottom-up effects on a range of arthropod taxa and causes a multiple-fold increase in tenebrionid beetles that serve as the main macrofaunal detritivore on the islands (Sanchez-Piñero and Polis 2000). Subsequent studies have also pointed to important positive effects of seabirds on soil nutrient availability on more productive island and coastal systems, for example in New Zealand (Mulder and Keall 2001; Fukami et al. 2006), the Aleutian Islands in Alaska (Maron et al. 2006), the north-eastern USA (Ellis et al. 2006), and the Great Barrier Reef in Australia (Schmidt et al. 2004). Also, on offshore islands in northern New Zealand, high seabird densities have been shown to greatly enhance densities of a range of belowground groups of soil invertebrates (Towns et al. 2009) and indirectly alter rates of plant litter decomposition (Fukami et al. 2006). Likewise, nitrogen enrichment from seabird colonies on the Isle of May, in the Firth of Forth, east Scotland, was found to stimulate microbial biomass and abundance of nematodes in soil, and increase the abundance of bacteria relative to fungi, although no effects on soil functioning (e.g. rates of nitrogen mineralization and decomposition) were detected (Wright et al. 2010).

Nutrient transfer by seabirds from ocean to land in turn has important consequences aboveground for plant nutrient acquisition and growth (Ellis 2005). For example, plants in soils associated with seabird colonies have been shown to grow much more rapidly (Fukami et al. 2006), and contain greater levels of foliar nutrients (Maron et al. 2006; Mulder et al. 2009) than plants growing in soils without seabird colonies. Plants in seabird colonies may also produce litter that decomposes and

(a)   (b)

Fig. 4.16 Photographs of (a) arid vegetated island with seabird nutrient inputs and (b) vegetation-poor island without seabird nutrient inputs derived from the ocean in the Gulf of California (Wait et al. 2005; Anderson et al. 2008). Images by A. Wait.

releases nutrients more rapidly (Wardle et al. 2009b), and foliage that is more palatable to foliar herbivores (Mulder and Keall 2001). These positive effects of seabird colonies on plant growth and nutrition are particularly pronounced for ruderal or annual plant species (Anderson and Polis 1999; Ellis 2005) (Fig. 4.16). Seabird colonies can also exert important negative effects on plant seedling recruitment, growth, and biomass production, largely through increased physical disturbances (Maesako 1999). For example, while high seabird densities on forested oceanic islands in north-east New Zealand lead to greater nutrient uptake by plants, these effects are offset by the adverse effects of soil perturbation associated with burrowing during nesting, leading to both lower densities of tree seedlings (Fukami et al. 2006) and a large net reduction in aboveground standing biomass (Wardle et al. 2007). As such, while high seabird densities alleviate nutrient limitation of plants, they also increase limitation of plants by physical disturbance and tissue damage, although the relative importance of these two opposing forces has been seldom explored. The combined effects of less nutrient stress and greater disturbance has important consequences for plant community composition, and there is a common trend for seabird activity to favour domination by short-lived and fast-growing ruderal species at the expense of longer-lived and slow-growing species (Ellis 2005). As discussed in Chapter 3, these types of plant community-level shifts can potentially feedback to the belowground community through greater quality of plant-derived resources, although this has yet to be investigated for seabird-dominated ecosystems.

Resources transported by consumers from fresh water to land can also serve as important ecosystem drivers. For example, moose in Isle Royale, USA, feed on aquatic macrophytes and deposit them as faecal material in nearby riparian forests, thereby increasing soil nitrogen availability (Bump et al. 2009b). This stimulation may help counter the negative effects of moose on nitrogen availability through other mechanisms described in this chapter (Pastor et al. 1993). Another example involves fertilization of riparian forests by dead salmon (*Oncorhynchus* spp.) in Alaska and

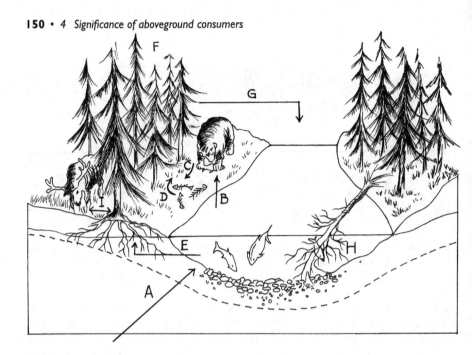

Fig. 4.17 Cycling of salmon-derived nitrogen and effects on river and riparian ecosystems: (a) spawning salmon transport nitrogen upstream; (b) bears (*Ursus* spp.) and other piscivores consume salmon; (c) bears disseminate salmon-enriched wastes and partially eaten salmon carcasses in the riparian forest; (d) terrestrial and aquatic insects colonize salmon carcasses, enhancing decomposition and nitrogen diffusion; (e) dissolved nitrogen downwells beneath the riparian forest and is taken up by tree roots; (f) salmon-derived nitrogen enhances foliar nitrogen and growth rates of riparian trees; (g) riparian trees provide shade, bank stabilization, and coarse woody debris, enhancing the quality of instream habitat for salmonid fishes; (h) coarse woody debris retains post-spawn salmon carcasses in streams, further enhancing nitrogen availability; and (i) increased foliar nitrogen enhances palatability of riparian plants, potentially altering patterns of browsing, which in turn affects patterns of riparian productivity and species composition. Redrawn from Helfield and Naimann (2006) by D.W. Colquhoun.

British Colombia (Helfield and Naiman 2006) (Fig. 4.17). Here, bears (*Ursus* spp.) capture salmon from rivers during spawning, and can carry over 70% of captured salmon into the forest to avoid other bear (Quinn et al. 2009). A large proportion of the captured salmon biomass is never consumed (Gende et al. 2001; Holtgrieve et al. 2009). Therefore, dead salmon serve as an important nitrogen input into these forests, accounting for up to 24% of ecosystem nitrogen input, which is greater than that derived from nitrogen fixation by alders (Helfield and Naiman 2006). Further, this nitrogen is highly labile, and fertilization by dead salmon can therefore cause substantial local increases in soil mineral nitrogen concentrations (Gende et al. 2007), leading to increased nitrogen uptake by soil microbes and plants (Helfield and Naiman 2002, 2006; Wilkinson et al. 2005) and enrichment of $\delta^{15}N$ in litter-dwelling invertebrates (Hocking and Reimchen 2002). It has been proposed that these responses in turn cause an increase in the nitrogen content and hence palatability of plant foliage, thereby potentially altering patterns of browsing by herbivores (Helfield and Naiman 2006) (Fig. 4.17).

Aquatic-derived invertebrates also have the potential to serve as terrestrial ecosystem drivers. Indeed, a handful of studies have shown that insects that spend their developmental stages in rivers (Sabo and Power 2002; Ballinger and Lake 2006) and lakes (Hyodo and Wardle 2009; Jonsson and Wardle 2009) may serve as a major source of prey for terrestrial predators when they emerge on land as adults. Little is known about the consequences of aquatic-derived invertebrates for the functioning of terrestrial ecosystems either above or below ground, although a recent study performed in lakes (including Lake Myvatn) in Iceland offers some insights (Gratton et al. 2008). Here, very high densities of midges (chironomid flies) can emerge from the lakes, with numbers varying greatly among years (Ives et al. 2008). Midges often die on land, and their input of nitrogen to the surrounding terrestrial ecosystem is probably in the order of 200 kg N ha^{-1} $year^{-1}$ (data in Gratton et al. 2008), which could represent a major fertilization effect, with important consequences for both aboveground and belowground properties (Gratton and Vander Zanden 2009). Stable isotope analyses have shown that this input of midges probably serves as an important resource for key decomposer groups such as collembolans and soil-associated predators, including opilionids and lycosid spiders (Gratton et al. 2008). It is not known how widespread this type of effect is, although recent evidence from stable isotope studies suggests that dead chironomids could also fertilize island ecosystems in lakes in northern Sweden (Hyodo and Wardle 2009).

Studies of the impact of aquatic organisms on the structure and function of adjacent terrestrial ecosystems are still in their infancy, with most studies to date being focused on the transfer of dead marine organisms to land by either seabirds or bears. As such, it is difficult to draw generalizations about how and when aquatic to terrestrial inputs are important. However, two factors may influence the extent to which terrestrial ecosystems are affected by inputs from aquatic systems. First, it is expected that terrestrial ecosystems may benefit most from aquatic inputs when the terrestrial system is much less productive than the aquatic one (Polis and Hurd 1996). However, there is evidence that aquatic resource subsidies can exert important effects on terrestrial communities even when the terrestrial and adjacent communities are of comparable productivity (Paetzold et al. 2008). Second, it is unclear as to whether these effects are restricted only to terrestrial ecosystems very close to the shore or whether they can extend some distance inland. For example, in the abovementioned case of midges from Icelandic lakes, while the highest density of midges occurs at the lake edge, a high density nevertheless still occurs at 150 m from the shore (Gratton et al. 2008). Also, the transport of marine-derived nutrients by seabirds can occur much further from the coast, given that seabirds in New Zealand may nest 50 km or more inland (Worthy and Holdaway 2002). Much remains to be learned about whether large effects of aquatic subsidies on terrestrial ecosystems represent a small collection of special cases for specific types of ecosystems, or are more widely spread.

4.6 Aboveground consumers, carbon dynamics, and global change

As discussed in Chapters 2 and 3, the effects of global change phenomena on belowground communities and the processes that they drive operate via both direct and indirect mechanisms, with often far-reaching consequences for ecosystem-level properties and carbon-cycle feedbacks at regional and global scales. Indeed, as we argue in Chapter 3, an understanding of the effects of global change phenomena on terrestrial ecosystems and feedbacks, and also their potential mitigation, requires explicit consideration of linkages between aboveground and belowground subsystems. Given that herbivores can strongly modify these linkages, and that herbivore behaviour and performance may be affected by global change both directly and indirectly (i.e. via shifts in vegetation composition), it follows that herbivores should act as key modulators of global change impacts on the functioning of terrestrial ecosystems. Here, we illustrate this using soil carbon dynamics and climate change as examples. We first examine how herbivores can modulate terrestrial ecosystem carbon cycling, with potential consequences for carbon-cycle feedbacks, and then consider how global change-induced shifts in vegetation composition could affect herbivore populations with consequences for ecosystem functioning. We emphasize that very little is known about these two issues; indeed, as noted recently by Gough et al. (2007), most studies that have examined responses of vegetation and soil properties to climate change have not explicitly considered interactions with herbivores.

A considerable amount of research effort has gone into understanding how herbivores influence soil biotic interactions and the functioning of terrestrial ecosystems. As discussed so far in this chapter, most of this of this work has focused on the cycling of plant growth-limiting nutrients (i.e. nitrogen and phosphorus) and plant productivity. However, the same pathways by which herbivores impact on soil biotic interactions and nutrient cycling will also influence soil carbon dynamics, and potentially the amount of carbon that is stored in and lost from soil. Although a growing number of studies have evaluated the effects of domestic livestock grazing on grassland soil carbon stores (e.g. Schuman et al. 1999; Han et al. 2008; Gollusccio et al. 2009), our understanding of how herbivores influence on ecosystem-scale carbon stores and fluxes in natural ecosystems is limited. Further, only a handful of studies have considered the role of herbivores in regulating soil biotic interactions and carbon fluxes relative to climate change. We consider here some of what is known about the role of herbivores in regulating soil carbon stores and fluxes to the atmosphere in managed and natural ecosystems, and how such activities might contribute to climate change via carbon-cycle feedbacks.

As outlined in Chapter 2, the amount of carbon stored in soil, and hence the potential for soils to act as carbon sinks, is determined by the balance between the input of carbon to soil via primary productivity (dead leaves, roots, and exudates) and its loss via heterotrophic and autotrophic respiration (De Deyn et al. 2008). Although abiotic factors (especially temperature and moisture) act as primary determinants of these components of soil carbon flux, they are also strongly regulated by aboveground consumers, as already outlined in this Chapter (Fig. 4.18). In particular, herbivores can regulate the amount and quality of organic matter entering soil as plant litter (shoot and roots) and as root exudates, and the rate of decomposition of plant-derived organic matter by heterotrophic soil organisms. Given the wide range of mechanisms by which herbivores can influence soil carbon dynamics, it is unsurprising that the effects of grazing on soil carbon stores and fluxes are highly variable. For instance, Milchunas and Lauenroth (1993) carried out a review of 34 studies involving grazed and ungrazed sites around the world and found that 40% of them reported a decrease and 60% an increase in soil carbon stocks as a result of grazing exclusion. Consistent with this variability, more recent studies of grasslands likewise report variable responses of soil carbon content to livestock grazing, with increases (e.g. Schuman et al. 1999; Reeder and Schuman et al. 2002), decreases (e.g. Frank et al. 1995; Bardgett et al. 2001b; Han et al. 2008; Golluscio et al. 2009; He et al. 2009), and neutral effects (Shrestha and Stahl 2008) being reported, even within the same study (Piñeiro et al. 2009).

Few studies have attempted to explain such differences in the response of soil carbon storage to grazing. However, these differences have been attributed in part to

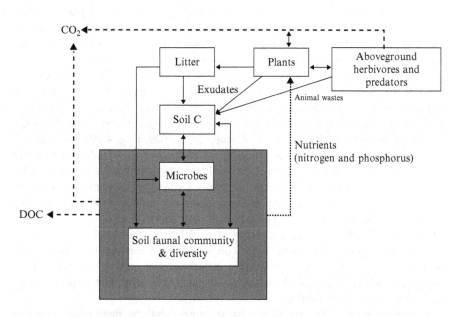

Fig. 4.18 Schematic diagram showing the various routes by which aboveground consumers influence carbon fluxes in terrestrial ecosystems. Solid lines represent inputs of carbon the soil system, for instance from plant litter, root exudates, and animal wastes. Thick dashed lines represent outputs of carbon in the form of dissolved organic carbon (DOC) and carbon dioxide (CO_2). Fine dashed lines represent a feedback to plant production and carbon input via altered nutrient availability.

variation in soil physical properties (e.g. soil texture and depth), depth of soil sampling, and responsiveness of the plant community to grazing (which in turn determines the amount and quality of carbon allocated belowground) (Schuman et al. 1999; Welker et al. 2004; Piñeiro et al. 2009). For instance, it has been proposed that grazing can enhance soil carbon in situations where root biomass, and hence the allocation of carbon belowground, is enhanced (Schuman et al. 1999; Piñeiro et al. 2009), or when declines in aboveground plant inputs to the soil are offset by manure inputs (Conant et al. 2001). In contrast, declines in soil carbon are likely when heavy grazing reduces plant biomass and the input of carbon to soil (Han et al. 2008; Pei et al. 2008), or when grazing selects for palatable, nutrient-rich plants which readily supply labile substrates to soil as nutrient-rich litter and root exudates. As discussed earlier in this chapter, such labile substrates, as well as dung and urine, promote the activity of decomposer organisms and bacterial-based energy channels, and hence carbon loss from soil (Bardgett and Wardle 2003).

This latter idea, which is broadly consistent with the trait-based framework discussed in section 4.3 and the acceleration/deceleration hypothesis presented in Figure 4.2, was recently tested by Klumpp et al. (2007, 2009). These authors set up a series of grassland monoliths which were subjected to simulated high and low grazing pressure for 14 years, and found that monoliths exposed to low grazing pressure were characterized by slow-growing, high-stature plant species, fungal-dominated soil communities, and greater soil carbon content. In contrast, monoliths exposed to high grazing were dominated by small-stature, fast-growing plant species, bacterial-dominated soil communities, higher aboveground plant productivity, and lower soil carbon storage (Klumpp et al. 2007). Further exploration of the mechanistic basis for these trends revealed that high grazing pressure caused a decline in root biomass and proliferation of soil bacteria, and an associated acceleration in the decomposition of old particulate organic carbon in soil, leading to a reduction in soil carbon storage (Klumpp et al. 2009) (Fig. 4.19). Collectively, these findings indicate that heavy grazing of productive grasslands can promote 'fast'-cycling microbial communities in soil, thereby accelerating decomposition and carbon loss from soil. In contrast, situations where grazing selects for unpalatable, slow-growing plants that produce low-quality litter which decomposes slowly, or where it promotes belowground allocation to roots, should lead to enhanced soil carbon storage (Schuman et al. 1999; Welker et al. 2004; Piñeiro et al. 2009). However, in making these conclusions we stress that there is much to be learned about the ways that grazing influences soil carbon storage, and, as highlighted above, many mechanisms are at play, suggesting that responses may vary greatly among ecosystems.

The above studies all refer to soil carbon stocks, but the main challenges that ecologists face in the context of global warming include quantifying the rates at which ecosystems exchange carbon dioxide with the atmosphere, and determining how these rates are controlled by abiotic, biotic, and management factors (Wohlfahrt et al. 2008). A number of studies have examined how grazing influences soil respiration (i.e. evolution of carbon dioxide from soil) in the field, and have reported

Grazing pressure

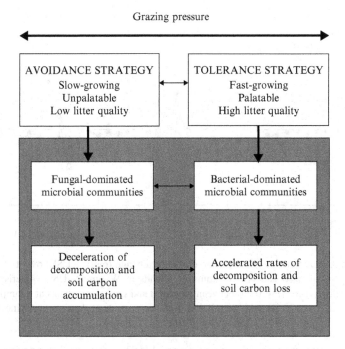

Fig. 4.19 Schematic diagram showing how shifts in vegetation and soil microbial community composition influence soil organic matter decomposition and carbon storage in temperate grassland. In this model, heavy grazing promotes the dominance of small-stature, fast-growing plant species, bacterial-dominated soil communities, higher aboveground plant productivity, faster rates of organic matter decomposition (especially of old particulate organic matter), and lower soil carbon storage. In contrast, low grazing pressure promotes the dominance of slow-growing, high-stature plant species, fungal-dominated soil communities, reduced rates of organic matter decomposition, and greater soil carbon content.

negative (Bremer et al. 1998; Knapp et al. 1998; Johnson and Matchett 2001; Van der Wal et al. 2007), positive (Ward et al. 2007), and neutral (Risch and Frank 2006; Ward et al. 2007; Susiluoto et al. 2008) responses. However, only a handful of studies have considered how grazing influences net ecosystem exchange (NEE), namely the balance between net uptake of carbon dioxide by photosynthesis and its loss by respiration, which determines whether an ecosystem is a source or sink for carbon. One such study is that of Welker et al. (2004), who measured NEE over two consecutive years in alpine grassland in Wyoming. They found that over this period, rates of NEE were generally lower (i.e. lower carbon sink activity) in grazed than ungrazed grassland, and that the grazed area was a net carbon source of 170 g C m^{-2}, whereas the ungrazed area was a net carbon sink of 83 g C m^{-2} (Fig. 4.20). Despite this, soil carbon stocks were greater in grazed than ungrazed areas, indicating that integrative measures of ecosystem carbon budgets (i.e. soil

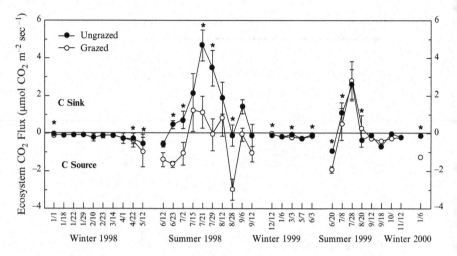

Fig. 4.20 Annual net ecosystem carbon dioxide exchange (mean $\pm$ 1 SE) over two consecutive years in ungrazed and grazed alpine grasslands in Wyoming, USA. Asterisks signify significant differences ($P < 0.05$) between ungrazed and grazed treatments at a sampling date. From Welker et al. (2004), with permission from the Institute of Arctic and Alpine Research, University of Boulder, Colorado.

carbon contents) may not always correspond to shorter-term, instantaneous measures of carbon dioxide exchange. As noted by these authors, soil carbon sequestration is a dynamic process, and while the long-term trajectory might be positive in grazed areas, periods of net carbon source activity are intermixed with periods of when they are net sinks (Fig. 4.21). This argument is consistent with finding that ecosystems can oscillate from being carbon sources to sinks for extended periods in different ecosystems (Oechel et al. 1993, 1995; Flanagan et al. 2002; Frank 2002), and that grazing can trigger short-term shifts in NEE in grassland systems (Wohlfahrt et al. 2008).

Other studies that have compared NEE in grazed versus ungrazed areas also point to grazer effects on NEE being driven by changes in vegetation biomass and composition, albeit in contrasting ways. For example, Polley et al. (2008) found that in four out of five growing seasons studied, cattle grazing of mixed prairie in North Dakota, USA, increased mean NEE (leading to a greater carbon sink) by reducing night-time respiration more than daytime carbon dioxide exchange. Moreover, they found that grazing reduced inter-annual variability in mean carbon dioxide fluxes, which was attributed to a reduction in the response of respiration to variation in soil temperature. The mechanism for these responses was not studied, but given that grassland respiration is greatly affected by the availability of recently fixed carbon, and that grazing has been shown to reduce root biomass in mixed prairie, it was suggested that grazing reduced respiration and its responsiveness to soil temperature in part through reducing the input of labile carbon to soil (Polley et al. 2008). Long-term grazing by sheep was also

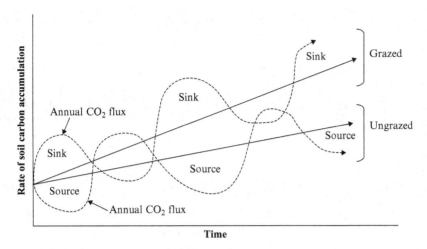

Fig. 4.21 Conceptual model illustrating inter-annual variation in net ecosystem carbon dioxide exchange as it relates to long-term carbon sequestration in grazed and ungrazed alpine grasslands in Wyoming, USA. While both grazed and ungrazed areas have periods of carbon gain and carbon loss, the net response is that grazed areas accumulate more soil carbon than ungrazed areas, as indicated in the long-term trajectory. Derived from Welker et al. (2004).

found to increase NEE and hence ecosystem carbon sink strength in a study of British peatlands by Ward et al. (2007) (Fig. 4.22). This effect, which was caused by a greater increase in photosynthesis relative to respiration, was attributed to an increase in the cover of fast-growing grasses relative to slower-growing dwarf-shrubs and bryophytes, and hence a greater assimilation of carbon by the plant community (Ward et al. 2007). In contrast, Van der Wal et al. (2007) found that belowground foraging (i.e. grubbing) by pink-footed geese (*Anser brachyrhynchus*) reduced NEE and soil carbon stocks in high Arctic tundra in Svalbard (Fig. 4.23). However, consistent with the above studies, this effect was attributed to a shift in the plant community, and especially a reduction in the cover of bryophytes and vascular plants, leading to a decline in ecosystem photosynthesis and thus carbon gain. The decline in soil carbon stocks due to grubbing, however, was largely attributed to the exposure of the humus layer to wind and water erosion rather than to reduced NEE (Van der Wal et al. 2007). Finally, and in contrast to the above studies, Susiluoto et al. (2008) found no detectable effects of grazing by reindeer on NEE in Finnish alpine tundra, despite significant shifts in vegetation composition. They found that lichen cover was significantly reduced by grazing; however, lichens are relatively unproductive and contribute little to NEE. Here, NEE was largely dependent on dwarf-shrub cover which was unaffected by grazing, and so the overall effect of grazing on NEE was neutral.

It is important to note that year-to-year variation in ecosystem carbon budgets are mostly attributed to climatic variability (e.g. Barford et al. 2001; Flanagan et al. 2002).

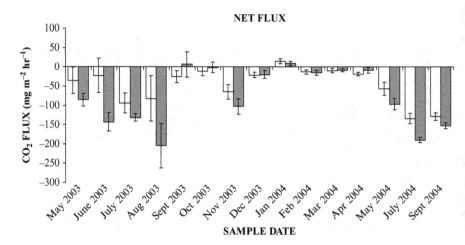

Fig. 4.22 Influence of grazing on net ecosystem carbon dioxide exchange over two growing seasons in peatland the north of England. Values are means ± SE. Positive values indicate a net loss of carbon dioxide and negative numbers imply carbon uptake. Filled bars are grazed plots whereas empty bars are ungrazed plots. Data from Ward et al. (2007), with permission from Springer Science+Business Media.

However, it is becoming apparent that grazing (and presumably other biotic factors) not only affects ecosystem carbon exchange, but also alters the influence of climate on land–atmosphere carbon fluxes. For instance, in the abovementioned study of Polley et al. (2008), grazing was found to change the identity of the primary climatic driver of year-to-year variability in carbon dioxide flux, and to alter the responsiveness of respiration to soil temperature. Likewise, while climatic factors were the main factor explaining seasonal variation in gross and net fluxes of carbon dioxide in the above-mentioned study of Ward et al. (2007), significant grazer effects were detected and, moreover, they altered the seasonal dynamics of net and gross carbon fluxes. As argued by Polley et al. (2008), these findings indicate that predictive models need to accommodate biotic factors such as grazing in order to accurately simulate the dynamics of carbon dioxide fluxes in terrestrial ecosystems. However, this is hampered by a lack of understanding of the relative roles of climatic versus biotic drivers of carbon dynamics, and the potential for grazing (and other biotic factors) to modulate the response of carbon flux to climatic variability and hence climate change.

Very few studies have experimentally tested the relative role of herbivory and climate change in driving carbon fluxes in terrestrial ecosystems. One such study, by Sjögersten et al. (2008), examined the combined roles of herbivory (by captive Barnacle geese, *Branta leucopsis*) and climate warming (through open-top chambers) as drivers of carbon dioxide fluxes in two Arctic habitats, namely mesic heath and wet tundra. They found that in wet tundra, goose grazing significantly reduced NEE of carbon dioxide (and hence carbon sink strength) while warming had little impact during the growing season. However, warming did increase carbon dioxide efflux

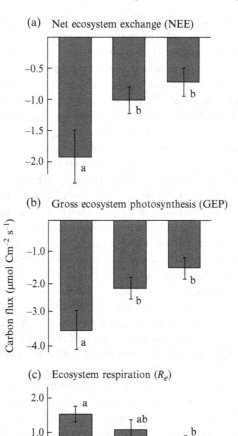

Fig. 4.23 Impact of goose grubbing on (a) carbon dioxide NEE, (b) gross ecosystem photosynthesis (GEP), and (c) ecosystem respiration (R_e) in Arctic tundra. Data are means $\pm$ SE for an intact control (C), goose-grubbed (G), and experimentally grubbed (E) wet tundra. Negative numbers imply carbon uptake and positive numbers imply carbon losses from the system. Values with different letters are significantly different at the $P < 0.05$ level. Data from Van der Wal et al. (2006), with permission from Wiley-Blackwell.

from wet tundra during the winter, further contributing to carbon loss from the system. In the mesic heath, warming reduced carbon dioxide efflux, and the combination of warming and low-level grazing altered the carbon balance during the growing season from a source to a weak net sink of carbon dioxide. Collectively, these results indicate that grazing by geese can result in either increased or decreased carbon sink strength in the high Arctic, depending on the habitat and climatic conditions; however, in agreement with Van der Wal et al. (2007), grazing exerted the strongest negative effect on carbon dioxide sink strength in wet tundra. The negative effect of grazing on NEE at the wet site (which is the type of feeding ground preferred by geese), was attributed to declines in aboveground plant biomass, and hence carbon assimilation by photosynthesis. In contrast, the interactive effect of warming and grazing at the mesic site was likely due to increased plant biomass and hence assimilation of carbon dioxide caused by both warming and grazing (Sjögersten et al. 2008). Overall, this study indicates that grazing has the potential to modulate the effects of climate warming on the ecosystem carbon balance, but that this effect is site specific and strongly driven by vegetation responses.

Two other studies on the combined effects of grazing and warming also indicate the potential for grazers to modulate the effects of climate change on plant communities and belowground processes, albeit not in the context of carbon dioxide fluxes. First, Rinnan et al. (2009) performed a factorial warming and herbivory-simulation (clipping) experiment in Finnish sub-Arctic tundra, and showed that long-term warming enhanced the growth of the dominant dwarf-shrub *Vaccinium myrtillus* and reduced soil ammonium-nitrogen availability and microbial immobilization of nitrogen only in the absence of simulated herbivory. These results point to the role of defoliation in modulating the responses of both plants and soil microbes to warming. Second, Post and Pedersen (2008) incorporated natural herbivory by muskoxen and caribou into a 5-year experimental investigation of Arctic plant community response to warming in West Greenland. In accordance with other studies, they found that warming increased total community biomass by promoting growth of birch (*Betula nana*) and willow (*Salix glauca*). However, muskoxen and caribou grazing reduced the total community biomass response to warming by 19%, and the response of birch and willow by 46 and 11% respectively. Further, under warming alone, the plant community shifted after 5 years from graminoid-dominated towards dwarf birch-dominated. In contrast, when herbivory was present, plant community composition on warmed plots after 5 years did not differ from that on ambient plots. These results, along with those of Rinnan et al. (2009) and Sjögersten et al. (2008), highlight the potentially important and often overlooked influences of herbivores on plant community and belowground responses to warming. As noted by Post and Pedersen (2009), conservation and management of herbivores may therefore be an important, though little understood, component of mitigating ecosystem responses to climate change.

As highlighted recently by Wookey et al. (2009), climate change-induced shifts in the functional composition of vegetation also have the potential to affect herbivore populations through a variety of mechanisms. First, large-scale shifts in vegetation productivity, phenology and composition will substantially alter the quantity, quality, and temporal availability of herbivore diet, thereby affecting herbivore performance and behaviour. Second, climate-induced alteration of growth and allocation patterns in individual plant functional types, for example in terms of secondary metabolite production and allocation to foliage, will alter the quality and quantity of forage (Coley et al. 1985). Third, the potential direct effects of climate change, together with changes mediated by shifts in vegetation composition, will influence soil biological properties that regulate the availability of nutrients to plants, thereby creating a potential feedback on plant productivity and forage supply (Bardgett and Wardle 2003). Finally, changes in vegetation composition will modify habitat availability and architecture, and spatial arrangement, with potential effects for herbivores and fauna that use plants for habitat, such as rodents, insects, and birds (Wookey et al. 2009).

Such indirect effects of climate change on herbivore performance and behaviour could in turn feedback to the plant community and the belowground subsystem, and serve to inhibit or facilitate vegetation change. For example, evidence from long-term studies in northern-temperate elk/aspen (*Cervis canadensis/Populus tremuloides*) and moose/balsam fir (*A. alces/Abies abies*) systems suggests that woody-plant expansion in response to warming might be constrained by large herbivores (Post et al. 1999; Ripple and Betscha 2004), and recent studies in the Arctic likewise suggest that shrub expansion due to warming might be limited by reindeer grazing (Olofsson et al. 2009). Moreover, climate-induced changes in vegetation productivity and composition could exert multitrophic effects aboveground, influencing densities of not just herbivores, but also their predators. Changes in predator numbers or behaviour in response to climate change are in turn likely to cause cascading effects on ecosystem properties by further modulating herbivore effects on vegetation change and resulting feedback mechanisms (Terborgh et al. 2001; Creel et al. 2005; Gunn et al. 2006). As discussed earlier in this chapter, these cascading effects have the potential to indirectly influence belowground food webs and the ecosystem processes that they drive. However, the impact of climate change on multitrophic feedbacks between the aboveground and belowground subsystems, and the consequences of this for ecosystem functioning, remains essentially unexplored.

Clearly there is much to be learned about the response of herbivores and other trophic groups to climate change and the consequences of this for ecosystem processes, including ecosystem carbon exchange and carbon-cycle feedbacks. Such responses are likely to be complicated and site-specific, but as illustrated in this and the preceding section, they are likely to play a significant role in modulating the response of ecosystems to climate and other global changes. Given the dearth of studies in this area, there is a clear need for more research aimed at an improved understanding of how herbivores, and other biotic factors, modulate ecosystem responses to climate change, and ultimately how herbivore management might be altered to mitigate climate change.

4.7 Conclusions

In this chapter we have examined how aboveground consumers, including mammalian and invertebrate herbivores and their predators, affect the belowground subsystem through a variety of mechanisms, with often far-reaching consequences for the structure and functioning of terrestrial ecosystems. Moreover, we have examined these responses in a global change context, exploring how herbivores can modulate terrestrial ecosystem carbon cycling with potential consequences for carbon-cycle feedbacks and climate change. Although it has been recognized for some time that aboveground herbivores can strongly modulate ecosystem processes through their influence on the belowground subsystem, there have been several important recent advances in understanding in this area. Not only has our knowledge of the various mechanisms by which herbivores influence the belowground subsystem and aboveground–belowground feedbacks improved, but there is also a growing understanding of the relative importance of these different mechanisms, both biotic and abiotic, in explaining belowground and ecosystem-level responses to herbivory (Mikola et al. 2009; Sørensen et al. 2009; Veen et al. 2010). Our understanding of the influence of aboveground consumers on ecosystem properties has also extended to consider the multitrophic linkages. In particular, there is growing recognition that aboveground predators of herbivores can induce trophic cascades that affect the functioning of the belowground subsystem with consequences for ecosystem functioning (e.g. Feeley and Terborgh 2005; Wardle et al. 2005; Maron et al. 2006; Schmitz 2008a; Wardle 2010).

Another advance has been an improved mechanistic and predictive understanding about why different ecosystems differ in their response to herbivory. This has been helped through an improved understanding of plant trait responses to grazing (e.g. Díaz et al. 2006), which we argue could provide a framework for investigating feedbacks between grazers, plants, soil organisms, and nutrient cycles, and how they vary across environmental gradients such as soil fertility and climate. Further evidence that we present in this chapter supports the notion that a conceptual framework for investigating plant trait responses to herbivory could be extended to incorporate belowground responses, and hence provide a basis for improved understanding the consequences of herbivory for ecosystem function in different contexts. Another important and related development has been an improved understanding of how herbivore impacts vary across environmental gradients at the landscape scale, for instance of soil fertility, topographic position, and climate (Augustine and McNaughton 2006; Anser et al. 2009). Moreover, there is growing recognition that herbivores can force predictable shifts in vegetation state at the landscape scale, with both catastrophic negative (Rietkerk and van de Koppel 1997) and positive (Van der Wal 2006) consequences for their functioning and carrying capacity. Studies in these topics are still in their infancy, and further work is needed to better understand how herbivore effects on ecosystem processes that are becoming increasingly well understood at local spatial scales may be manifested at the landscape scale.

A related development is the recognition that herbivore movements during foraging and associated transfers of resources across sites as animal wastes can have significant consequences for landscape-scale spatial patterns of nutrient cycling and vegetation. Most work on this issue has historically been done in agricultural situations, where livestock can promote patchy return of nutrients through using different parts of their habitat for feeding and for excretion. As a result, nutrients become concentrated in relatively small areas where animals spend a large amount of their time (Afzal and Adams 1992; Kohler et al. 2006; Jewell et al. 2007). However, there is growing evidence that wild herbivores can also influence nutrient transloca-tion and the spatial distribution of soil nutrients in natural ecosystems, with long-term consequences for the spatial dynamics of vegetation (e.g. Augustine and Frank 2001; Yang 2004; Schütz et al. 2006). Moreover, there is emerging evidence that carcasses of vertebrate and invertebrate herbivores have significant and lasting effects on the functioning terrestrial ecosystems, by creating pulses of nutrients into the soil which create spatial heterogeneity in nutrient cycling and vegetation at the landscape scale (Towne 2000; Yang 2004; Bump et al. 2009a, 2009c; Parmenter and MacMahon 2009). Also, as shown by Bump et al. (2009a), such spatial heterogeneity in terrestrial ecosystems can be modulated by predators, whose behaviour influences the spatial distribution of carcasses and their impact on ecosystem processes; such findings further emphasize the important roles that predators play in modulating the function-ing of terrestrial ecosystems. There is also growing recognition that aboveground and belowground components of terrestrial ecosystems are affected by both aquatic and terrestrial consumers that transfer resources from water to land. For example, the input of seabird guano to terrestrial ecosystems can have strong positive effects on soil nutrient availability (Polis and Hurd 1996; Mulder and Keall 2001; Fukami et al. 2006) and the abundance of soil organisms (Towns et al. 2009), resulting in enhanced plant nutrient acquisition and growth (Anderson and Polis 1999; Fukami et al. 2006; Maron et al. 2006; Mulder et al. 2009), and increased plant litter quality (Wardle et al. 2009b). Similarly, the capture of salmon from rivers by bears, and their subsequent transport into adjacent riparian forests, can serve as a significant nitrogen input in these forests, with consequences for local patterns of soil nitrogen availability, plant nutrient uptake, and ultimately browsing patterns of herbivores (e.g. Helfield and Naiman 2006; Gende et al. 2007). As we stress in this chapter, however, studies of the impact of aquatic organisms on the structure and function of adjacent terrestrial ecosystems are still at their infancy, so generalizations about the extent to which aquatic to terrestrial inputs are important are still poorly understood.

As highlighted in both this and the preceding two chapters, a topic that has attracted much recent attention is the role that biotic interactions play in regulating carbon dynamics in terrestrial ecosystems, especially in the context of land–atmosphere carbon exchanges and carbon-cycle feedbacks that could impact on climate change. It is clear from this chapter, for example, that climate change-induced shift in vegetation composition can affect herbivore populations, which could in turn feedback to the plant community and the belowground subsystem, thereby inhibiting or facilitating vegetation change (Wookey et al. 2009). Moreover, recent studies indicate that herbivores themselves can modulate the effects of climate change on plant commu-nities and belowground processes (Rinnan et al. 2009; Olofsson et al. 2009; Post and Pedersen 2009), suggesting that conservation and management of herbivores may be an important, though little understood, component of mitigating ecosystem responses to climate change (Post and Pedersen 2009). Another important development in this general area is the recognition that grazing (and presumably other biotic factors) not only affects ecosystem carbon exchange, but also alters the influence of climate on land-atmosphere carbon fluxes (Polley et al. 2008). Although more studies are needed in this area, this finding is especially important because it indicates that predictive models need to accommodate biotic factors such as grazing in order to accurately simulate the dynamics of carbon dioxide fluxes in terrestrial ecosystems. However, this is hampered by a lack of understanding of the relative roles of climatic versus biotic drivers of carbon dynamics, and the potential for grazing (and other biotic factors) to modulate the response of carbon flux to climatic variability and hence climate change. Given this, there is a clear need for more research aimed at an improved understanding of how herbivores, and other biotic factors, modulate ecosystem responses to climate change, and ultimately how herbivore management might be altered to mitigate climate change.

5

Aboveground and belowground consequences of species losses and gains

5.1 Introduction

In Chapters 2–4 we discussed three major groups of biotic drivers of terrestrial ecosystem functioning, namely the belowground biota, the plant community, and the aboveground consumers. We also explored the ecological significance of changes in community composition within each of these groups of biotic drivers resulting from human-induced atmospheric global change phenomena, such as climate change and nitrogen deposition. However, global change also involves biotic interchange, resulting from both the loss and gain of species in the community (Vitousek et al. 1997b). This occurs on two fronts. First, human activity is contributing to significant extinctions of species on both local and global scales, and it is widely recognized that the Earth is undergoing its sixth major extinction event. Indeed, for many groups of biota, the current rate of extinction of species is probably around 100–1000 times their pre-human levels (Pimm et al. 1995; Millennium Ecosystem Assessment 2005). These species losses are a consequence of several human-induced factors, the most important of which is arguably habitat destruction and land-use change (Sala et al. 2000). However, other important causes of species loss include climate change (Thomas et al. 2004; Thuiller et al. 2005), nitrogen deposition (Stevens et al. 2004; Phoenix et al. 2006), invasive predators (Beggs and Rees 1999; McKinney and Lockwood 1999), and harvesting of desirable plant and animal species (Bodmer et al. 1997; Wardle et al. 2008c). Second, human migration throughout the world has greatly facilitated the introduction of plant and animal species to new habitats; these introductions have been both accidental and deliberate, and have frequently led to biological invasions of natural ecosystems. Ecological changes caused by biological invasions have occurred whenever humans have colonized new land, and are currently occurring most rapidly on the most recently settled land masses, such as New Zealand (Allen and Lee 2006) and Hawai'i (Vitousek et al. 1997b). Further, many ecosystems are losing some species through extinction and gaining others through invasion, and therefore the net result can be either a net gain or net reduction in species depending on context (Sax and Gaines 2003; Sax et al. 2005; Phoenix et al. 2006; Van Calster et al. 2008).

When species are gained by or lost from a community, there can be important consequences for both the aboveground and belowground subsystems, especially when these species differ in key functional attributes from the rest of the community. Some of the more extreme effects of species loss or gain are widely known. For example, extinction of megaherbivore populations can lead to major shifts in vegetation type and soil fertility (Zimov et al. 1995; Wardle and Bardgett 2004). Similarly, invasions of ecosystems by novel types of organisms such as top predators (O'Dowd et al. 2003; Fukami et al. 2006) or nitrogen-fixing plants (Vitousek and Walker 1989) can induce fundamental shifts in both aboveground and belowground ecosystem processes. However, although these represent extreme cases driven by losses or gains of species with key attributes, more subtle ecological effects of species extinctions and invasions might be widespread. In this light, there is growing evidence for communities of both animals (e.g. Bodmer et al. 1997; Cardillo et al. 2005) and plants (e.g. Duncan and Young 2000; Van Calster et al. 2008) that those species lost through extinction may collectively differ in particular traits or attributes from those that are not lost. Similarly, for plant communities at least, comparative studies often point to significant differences in key traits between the native and alien components of the resident flora (e.g. Funk and Vitousek 2007; Leishman et al. 2007). When the traits that differ between invasive and non-invasive species, or between species that vary in their susceptibility to extinction, are the same as those traits which influence key ecological processes, there is the potential for changes in ecosystem functioning to occur.

The goal of this chapter is to outline the changes in aboveground and belowground communities, and the ecosystem processes that they drive, that occur as a consequence of biotic interchange. To do this, we first discuss how species losses from communities may affect aboveground and belowground biota and ecosystem properties, and hence how biological extinctions may impact upon ecosystems. We will then outline the consequences of ingress of new species into communities for the aboveground and belowground components of ecosystems. Finally, we discuss how global change phenomena affect aboveground and belowground properties and organisms through causing gains or losses of key species. In doing this, our overall goal is to draw upon the concepts developed in Chapters 2–4 to better understand how biotic interchange affects ecosystems by influencing aboveground–belowground linkages.

5.2 Species losses through extinction and aboveground–belowground linkages

5.2.1 The diversity-function issue from an aboveground–belowground perspective

The issue of how variation in species richness of organisms may affect the rates and stability of ecosystem processes (e.g. productivity, decomposition and nutrient cycling) has long been a research focus of both agronomists (e.g. Trenbath 1974; Vandermeer 1990) and ecologists (e.g. Odum 1969; McNaughton 1977). However, since the mid-1990s, components of this topic have attracted substantial attention from ecologists and have also generated a considerable amount of debate (Huston 1997; Kaiser 2000; Hooper et al. 2005). Over that time, many experimental studies have been performed to study this topic, mostly involving the setting-up of artificial communities in which organism diversity (usually at the species or functional-group level) is experimentally varied (often by drawing random assemblages of species from a pool of species), with ecosystem response variables then being monitored over time. Various points of view have been expressed as to how relevant such studies are for understanding the effects of human-driven species losses in real ecosystems (Huston 1997; Tilman 1999; Wardle 1999; Leps 2004; Ridder 2008; Duffy 2009). Because studies investigating biodiversity effects on ecosystem functioning frequently interpret their results in terms of the ecological consequences of species extinctions, we provide a brief overview of the current state of knowledge of this topic, especially in the context of aboveground–belowground linkages. We emphasize that it is not possible to provide an exhaustive review of the topic given the volume of recent literature on it; as such, our choices of examples are representative, not exhaustive. For more detailed treatment we refer the reader to reviews by Hooper et al. (2005), and for the belowground subsystem Wardle and Van der Putten (2002) and Hättenschwiler et al. (2005).

The effects of plant species or functional-group richness (or other measures of diversity) on NPP have been extensively investigated. Many experimental studies have found positive effects of diversity on NPP (see Balvanera et al. 2006; Cardinale et al. 2006), although the interpretation and mechanistic basis underlying the results of these studies continues to be debated (Hooper et al. 2005). Further, the importance of richness as a driver of productivity varies greatly among and even within studies (Fridley 2002; Hooper and Dukes 2004), and may be only minor in natural ecosystems when compared to other biotic and abiotic factors (e.g. Wardle et al. 1997b; Grace et al. 2007). For those studies that do point to positive effects of live plant diversity on NPP (see Balvanera et al. 2006; Cardinale et al. 2006), an increased quantity and diversity of resources should in turn enter the belowground subsystem, potentially stimulating decomposer organisms and processes through various mechanisms (Fig. 5.1). However, of the 30 or more studies to date that have explicitly investigated the effects of plant diversity on the performance of the decomposer subsystem, the vast majority have found diversity to have weak or non-existent effects, and most of these have instead shown plant species identity to be the main belowground driver (e.g. Wardle and Van der Putten 2002; Porazinska et al. 2003; Hedlund et al. 2003; De Deyn et al. 2004). A small number of experimental studies have detected a strong effect of plant diversity on the decomposer subsystem (notably Stephan et al. 2000 and Zak et al. 2003), but in those cases plant diversity also had strong effects on NPP and therefore presumably the input of resources belowground. The fact that most studies involving experimental manipulation of plant diversity find some positive effects on NPP, but little or no consistent effect on belowground organisms and their activities, is reflective of relatively weak or inconsistent coupling between the producer and decomposer subsystems.

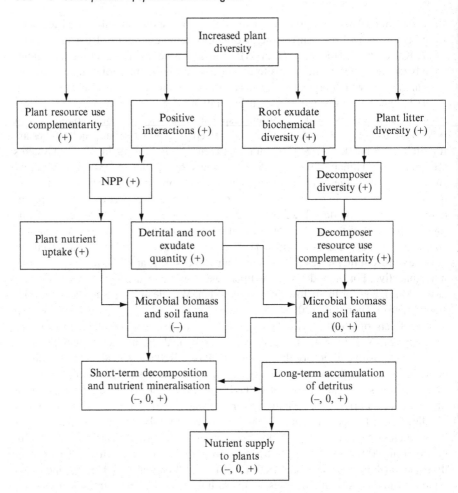

Fig. 5.1 Hypothesized mechanisms by which increasing plant species diversity may affect decomposer-mediated processes. Symbols: +, 0, and − indicate positive, neutral, and negative effects, respectively. From Wardle and van der Putten (2002), with permission from Oxford University Press.

As discussed in Chapter 3, plants affect the belowground subsystem not only when they are alive, but also through 'afterlife effects' manifested through the litter that they produce. Therefore, the question emerges as to what effects plant litter diversity might exert on the decomposer subsystem. This has been extensively studied through the so-called litter mixture experiments in which the decomposition of plant litter in multiple species mixtures is compared with that of the component species decomposed singly. Such an approach has now been used in well over 50 studies (see Wardle and Van der Putten 2002; Gartner and Cardon 2004), and collectively these studies show highly variable effects of litter mixing on litter-mass loss, ranging from strongly positive to strongly negative, although positive (i.e. synergistic) effects of litter mixing are more numerous than negative effects (Gartner and Cardon 2004). The mechanistic basis of litter-mixing effects remains elusive, and evidence is mixed as to whether or not non-additive mixing effects become greater when litter types are more dissimilar in terms of their quality (Wardle et al. 1997a; Hoorens et al. 2003; Quested et al. 2005). Further, litter mixing can influence the density of decomposer organisms, though not in an easily predictable way (Hansen 2000; Wardle 2006). For example, Blair et al. (1990) found that litter mixing often promoted the density of fungivorous nematodes, but reduced the density of mesofaunal groups, living within the litter (Fig. 5.2). Finally it is important to note that decomposer invertebrates can in turn greatly influence the magnitude of non-additive litter-mixing effects on mass loss (Hättenschwiler and Gasser 2005; Schädler et al. 2005), pointing to the possibility of complex interactions between multiple plant species and the decomposer community in regulating decomposer processes.

As highlighted in Chapter 2, soil organisms also have a major role in driving belowground processes that influence nutrient availability and therefore plant growth. The issue therefore emerges as to whether diversity of detritivorous organisms affects processes in the soil that may impact on aboveground organisms and processes. There is a modest but growing number of studies that have explicitly addressed the effects of decomposer diversity on soil processes (Wardle 2002; Hättenschwiler et al. 2005). For example, and as also emphasized in Chapter 2, there is some evidence that decomposition of organic materials and other related soil processes can be positively influenced by the species diversity of saprophytic fungi (Robinson et al. 1993; Setälä and McLean 2004; Tiunov and Scheu 2005) and arthropods (Liiri et al. 2001), probably through different taxa preferentially utilizing different organic substrates and leading to greater resource use complementarity. However, these effects are mostly apparent at the low-diversity end of the spectrum (see Chapter 2), and at levels of diversity that are likely to be well below that which is found in real ecosystems (Wardle 2002). There has to date been little exploration of how decomposer diversity effects on soil processes might in turn indirectly influence aboveground processes, such as plant nutrient acquisition and growth. However, as discussed in Chapter 2, Laakso and Setälä (1999a) found little effect of soil mesofaunal species diversity on growth of *Betula pendula* seedlings, and Cole et al. (2004) showed that nitrogen acquisition by the grass *Agrostis capillaris* was unresponsive to soil microarthropod species richness.

A handful of studies have also focused on the aboveground effects of the diversity of soil organisms that interact with plants through the direct pathway. As discussed in Chapter 2, some studies have investigated the effects of diversity of ectomycorrhizal fungi (Jonsson et al. 2001) or arbuscular mycorrhizal fungi (Van der Heijden et al. 1998b; Vogelsang et al. 2006; Maherali and Klironomos 2007) on plant growth, and have collectively documented responses ranging from strongly positive to neutral, even within the same study (Jonsson et al. 2001). Despite the likely importance of root pathogens and root herbivores as regulators of plant growth (Chapters 2 and 3), little is known about how the diversity of these organisms impact upon plants. However, in a

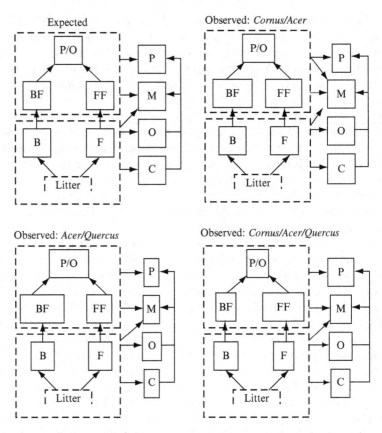

Fig. 5.2 Effects of plant litter diversity on microbial and faunal components of the detritus food web inhabiting the litter, in a forest ecosystem in Georgia, USA. For each component, the area of each square or rectangle represents its observed mass or population in two- or three-species litter mixes, relative to expected values derived from litter monocultures, assuming effects of all litter types are purely additive. B, bacteria; F, fungi; BF, bacterial-feeding nematodes; FF, fungal-feeding nematodes; P/O, predatory/omnivorous nematodes; C, Collembola; O, oribatid mites; M, mesostigmatid mites; P, protostigmatid mites. Calculated from data of Blair et al. (1990), as presented by Wardle and Lavelle (1997), with permission from CAB International.

study in which species richness of root-feeding nematodes was experimentally varied, Brinkman et al. (2005) found that growth of the dune plant *Ammophila arenaria* was influenced primarily by the identity of the nematodes, rather than the number of species present. This is consistent with other studies showing plant growth to be responsive to species identity of root-feeding nematodes (Wurst and Van der Putten 2007) and root pathogenic fungi (De Rooij-van der Goes 1995). Little is also known about the ecological effects of diversity of soil bacteria that perform specialized belowground functions, such as nitrogen fixation and nitrification. However, these types of function are often performed by a physiologically and/or phylogentically narrow range of taxa, when compared to 'broad' processes such as nitrogen mineralization (Schimel et al. 2005). Hence, it is generally thought that processes such as nitrogen fixation and nitrification will be more sensitive to changes in microbial community composition and diversity than is nitrogen mineralization. For example, symbiotic nitrogen fixation in legumes is characteristically performed by a small number of strains of *Rhizobium*, and losses of a subset of these strains can greatly impair symbiotic nitrogen fixation and legume growth (Giller et al. 1998).

Despite the importance of aboveground consumers (foliar herbivores and their predators) in driving the aboveground and belowground subsystem (see Chapter 4), we are not aware of any studies that have investigated how the diversity of soil organisms may affect the performance of aboveground consumers. Further, despite growing interest in how consumer diversity may drive ecosystem functioning (Johnson 2000; Duffy et al. 2007), the belowground consequences of aboveground consumer diversity has received scant attention. It is, however, theoretically possible for herbivore diversity to either increase or decrease the densities of belowground organisms and rates of soil processes, depending on the mechanisms involved (Bardgett and Wardle 2003). In the one study to date that has investigated this issue, Wardle et al. (2004c) performed a microcosm experiment in which the diversity of foliar-feeding aphids was experimentally varied from one to eight species. Although aphid species identity emerged as an important driver of both the plant community and multiple trophic levels of the soil food web, aphid species richness had few effects, and those that did occur were at the low-diversity end of the spectrum (two versus one species).

There is a long-standing interest in how biodiversity affects not just ecosystem processes, but also their stability. As such, even when species richness may be unimportant as a driver of ecosystem processes at a given point of time, the possibility remains that richness may influence the temporal dynamics of the ecosystem through altering its resistance and resilience to external perturbation (Hooper et al. 2005). There is a long history of theoretical studies (May 1973; McNaughton 1977) and many empirical studies (e.g. Sankaran and McNaughton 1999; Tilman et al. 2006; Bezemer and Van der Putten 2007) that have explicitly addressed this issue. A detailed discussion of this topic is beyond the scope of this book, and we refer the reader to reviews of this topic by Cottingham et al. (2001), Hooper et al. (2005) and Ives and Carpenter (2007). It is notable that the effect of organism species richness on the temporal variability of belowground organisms and processes has seldom been explored. However, controlled glasshouse studies by Wardle et al. (2000) and Orwin and Wardle (2005) provide evidence that while plant species richness has some positive effects on NPP, the stability of a range of aboveground and belowground properties, assessed by measuring ecosystem responses to an experimentally applied drying disturbance, was unaffected by plant diversity. Moreover, both these studies found that the stability of these properties was influenced by plant species identity, which is consistent with other studies that have shown plant species identity and functional attributes influence the temporal variability of aboveground and belowground properties (MacGillivray et al. 1995; Wardle et al. 1999).

In sum, when the diversity-function issue is considered from a combined above-ground–belowground perspective, there is considerable variability among studies: some studies find strong effects of diversity on ecosystem properties, whereas others find weak, neutral, or inconsistent effects. This conclusion contrasts to that drawn from recent meta-analyses of experimental studies on diversity-function relationships (e.g. Balvanera et al. 2006; Cardinale et al. 2006), and it has been claimed that those analyses show diversity effects to be 'surprisingly consistent across taxa, trophic levels and habitats' (Duffy 2009). However, given the wide variability among studies in terms of experimental design, levels of diversity used, and results obtained, it is unlikely that meta-analyses of this kind would have the statistical power to identify important differences that exist in diversity-function relationships between taxa or habitats (Wardle and Jonsson 2010). Further, diversity-function relationships are themselves driven by the balance of resource partitioning, competition, and facilitation among coexisting species, and this balance in turn varies greatly across both habitat and taxa. As such, there are good theoretical reasons for predicting that the relationship between diversity and function should not be consistent (Wardle and Jonsson 2010). The most powerful means of looking at the effect of habitat conditions on the diversity-function relationship is to experimentally vary both habitat factors and diversity within the same study. A handful of studies that are relevant to understanding aboveground–belowground linkages have taken precisely this approach, and these consistently show that the effect of diversity on ecosystem processes depends strongly on environmental context (Fig. 5.3). These include studies that have shown that the effects of litter diversity on the rate of decomposition are influenced by other trophic levels (Hättenschwiler and Gasser 2005; Schädler et al. 2005) and soil fertility (Jonsson and Wardle 2008), that the effects of live plant diversity on NPP depends on soil fertility (Fridley 2002) and other trophic levels (Mulder et al. 1999), and that the effects of mycorrhizal fungal diversity on host plant growth depends on both soil fertility and host plant species (Jonsson et al. 2001).

5.2.2 Removal experiments for studying effects of species losses

As discussed above, a large number of studies have reported experiments involving random assemblages of species that vary in species richness, and many of them have claimed to be directly relevant to understanding the consequences of human-induced extinctions in real ecosystems. However, in reality, communities do not consist of random assemblages of species, and species are not lost at random from communities during local extinction (Wardle 1999; Solan et al. 2004; Zavaleta and Hulvey 2004). As a consequence, the relevance of studies that involve random assemblages of species for understanding the effects of human-driven species losses in real ecosystems and for conservation management has been questioned (Huston 1997; Leps 2004; Ridder 2008). During both local and global extinction events, species with particular traits are often lost from a community more readily than others (Cardillo et al. 2005; Van Calster et al. 2008). Importantly, losses of species from a community are more likely to influence the functioning of that community when those traits that

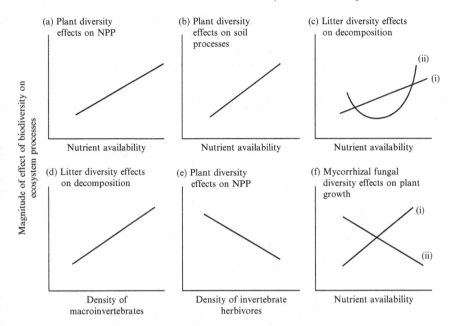

Fig. 5.3 Studies showing that the magnitude or occurrence of effects of biodiversity on ecosystem level processes (vertical axes) depends upon environmental context such as soil fertility and organisms in other trophic levels (horizontal axes). Curves depict only the directions of trends and do not reflect relative magnitude or precise shape of the curve. Data are sourced from: (a) Fridley (2002); (b) Wardle and Zackrisson (2005); (c) Jonsson and Wardle (2008), where (i) = plots without shrubs and (ii) = plots with shrubs; (d) Schädler and Brandl (2005); (e) Mulder et al. (1999); and (f) Jonsson et al. (2001), where (i) = *Pinus* as host plant and (ii) = *Betula* as host plant.

predispose particular species to extinction are also those that are important in driving ecosystem processes. For example, atmospheric nitrogen deposition results in the disproportionate loss of species with functional attributes that also influence nutrient cycling and promote conservation of nutrients within the ecosystem (Berendse 1998; Nilsson et al. 2002). Similarly, those species that are most likely to be lost through exploitation, for example through tree harvesting (Wardle et al. 2008c) or over-hunting (Zimov et al. 1995), are likely to disproportionately influence ecosystem processes because they are often the biomass dominants within their trophic level. The ecosystem-level effects of these sorts of non-random losses are not captured in experiments in which communities are assembled at random. However, a growing number of studies are employing alternative approaches, such as the non-random experimental removals of species or functional groups, to better understand what happens when species are lost from natural ecosystems in a non-random manner (reviewed in Díaz et al. 2003). Such removal experiments have employed various approaches to remove most main groups of terrestrial biota from communities, including plants, microbes, soil fauna, and aboveground animals (Fig. 5.4).

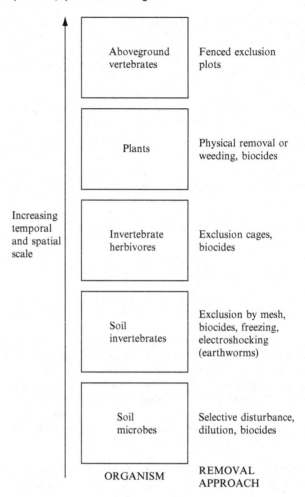

Fig. 5.4 Approaches that have been used for studying the ecosystem effects of removals of organisms at contrasting spatial and temporal scales.

Removal experimental approaches applied to the plant community usually involve physical removal of key species or functional groups. Removal approaches have long been used for studying plant–plant interactions such as competition and facilitation (e.g. Abdul-Fatih and Bazzaz 1979; Armesto and Pickett 1985), but are increasingly being used for determining the consequences for ecosystem processes of losses of subsets of the resident flora (Díaz et al. 2003) (Fig. 5.5). For example, Wardle et al. (1999) performed a removal experiment in a grazed grassland ecosystem in New Zealand to show that losses of the main plant functional groups in the community (all C_3 grasses, annual C_3 grasses, C_4 grasses, forbs) sometimes exerted important effects on the composition of the soil biotic community, with C_3 grasses usually having the

(a)

(b)

(c)

(d)

Fig. 5.5 Removal experiments for studying the effects of plant species losses on ecosystem properties. (a, b) Ongoing experimental plots on lake islands in northern Sweden, in which mosses, ericaceous dwarf-shrubs, and tree roots have been continually excluded from plots in various combinations since 1996 (Wardle and Zackrisson 2005). (c, d) Experimental plots on peatland in northern England, where mosses, ericaceous dwarf-shrubs (c), and graminoids (d) have been removed from plots in various combinations (Ward et al. 2009). Images: (a, b) D.A. Wardle; (c, d) R.D. Bardgett.

strongest effects. However, functional group removals frequently had very weak effects on soil processes, because the remaining functional groups often compensated for the effects of those functional groups that were removed. As discussed in Chapter 3, Suding et al. (2008) also used species removals in combination with nutrient additions to show that the slow-growing forb *Geum rossii* promoted soil microbial feedbacks that slowed rates of nitrogen cycling, while the fast-growing grass *Deschampsia caespitosa* promoted microbial feedbacks that enhanced nitrogen cycling. Further, in a peatland in northern England, Ward et al. (2009) used a removal experiment coupled with $[^{13}C]CO_2$ pulse labelling to demonstrate that losses of plant functional groups, notably ericaceous shrubs, can greatly increase total ecosystem CO_2 fluxes. Meanwhile, an ongoing removal experiment initiated in 1996 on forested lake islands in northern Sweden (Wardle and Zackrisson 2005; Wardle et al. 2008a; Fig. 5.6) revealed that losses of particular functional groups (notably ericaceous dwarf shrubs) and species (notably *Vaccinium myrtillus* and *Vaccinium vitis-idaea*) significantly impaired several belowground properties and promoted growth of planted seedlings, but only on productive islands with relatively fertile soil. In contrast, vascular plant removal treatments impaired nitrogen fixation by cyanobacteria living in feather mosses (the main source of nitrogen input to these ecosystems) only on small and unproductive islands (Gundale et al. 2010). These types of removal studies point to potentially important ecosystem-level consequences of plant species and functional groups (as well as the loss of these plants from the ecosystem), and emphasize the key role of plant functional traits in driving aboveground and belowground processes, as also highlighted in Chapter 3.

Removal experiments are especially useful for understanding the effects of loss of dominant species in forested ecosystems. The lengthy life span of most tree species combined with their large size creates particular logistic difficulties in using them for conventional species-richness/ecosystem-functioning experiments. Several approaches have been applied to study the ecosystem effects of forest tree diversity (Díaz et al. 2009), but most of these are able to address the issue only indirectly. However, there are many instances worldwide in which forests have been selectively logged and from which particular tree species have been removed, and comparisons of selectively logged stands with those that have not been logged serve as unintended removal experiments experiments for investigating the ecosystem consequences of losses of particular tree species (Díaz et al. 2003). Using such an approach, Wardle et al. (2008c) studied the aboveground and belowground consequences of loss of a highly desirable timber tree species, namely the podocarp *Dacrydium cupressinum*, through selective logging in New Zealand rainforest, approximately 40 years after the logging had been performed. This tree species has contrasting functional traits (e.g. litter quality) to those of the other dominant tree species present in these forests, and loss of this species was found to cause important effects on understorey vegetation, soil carbon sequestration, soil nutrients, and the structure of the soil microbial community. The ecosystem-level effects of selective loss of a tree species from a forest will, however, be case-specific and depend upon the differences in traits between those species that are lost and not lost (Díaz et al. 2009), and possibly the sequence of species loss (Bunker et al. 2005). In any case, measurements performed on selectively logged forests may have considerable potential for understanding the consequences of losses of long-lived species from an ecosystem arising from direct human intervention.

With regard to belowground biota, it is not possible to determine the extent to which many groups of soil organisms are being subjected to extinction, even on local

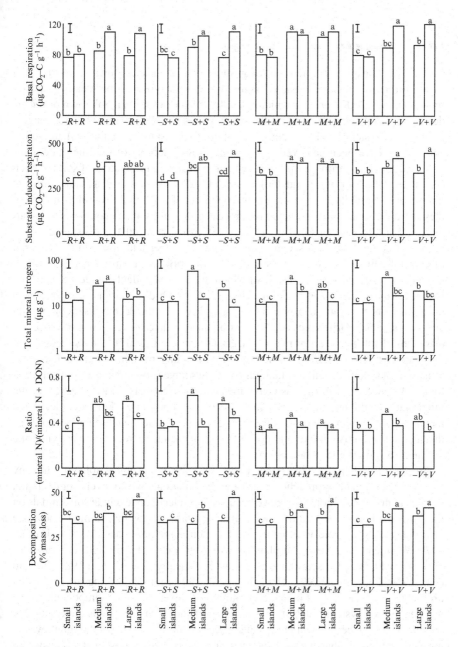

Fig. 5.6 Interactive effects of island size and removals of functional groups or species on selected belowground properties after 7 years of an ongoing plant-removal experiment. Removal treatments shown are: tree roots removed (−R) or not removed (+R); all shrubs removed (−S) or not removed (+S); *Vaccinium myrtillus* removed (−M) or not removed (+M); and *Vaccinium vitis-idaea* removed (−V) or not removed (+V). DON, dissolved organic nitrogen. Within each panel, bars topped by the same letter are not significantly different at $P = 0.05$ (least significant difference (LSD) test), and vertical bars represent LSD values at $P = 0.05$. From Wardle and Zackrisson (2005), with permission from Macmillan Publishers Ltd.

scales. This is especially the case for groups such as bacteria, fungi, protozoa, and some components of the nematode fauna, which may often not be greatly dispersal limited (Finlay 2002), and for which over 90% of species remain unknown or under-described (Klopatek et al. 1992; Coleman and Crossley 1995). Nevertheless, insights into the ecosystem-level consequences of losses of subsets of the soil microbial community have been achieved through removal of subsets of the microbial community through experimental perturbations. For example, fumigation of soil has been shown by Degens (1998) and Griffiths et al. (2000) to cause detectable shifts in microbial community composition and reductions of microbial diversity. In both of these studies, such shifts impacted on processes driven by the decomposer community, and in the case of Griffiths et al. (2000) they reduced the resistance of the microbial community to a second stress, namely copper addition. Similarly, serial dilution of soil results in progressive loss of microbial diversity, and on statistical grounds it is expected that those taxa which are least abundant should be lost first. However, the reduction of microbial diversity through dilution has been shown to have little effect on rates of key soil processes driven by the decomposer community and their stability (Griffiths et al. 2001), or on the resistance or resilience of processes driven by denitrifiers and nitrate oxidizers (Wertz et al. 2007). This suggests that those species removed from the microbial community by dilution are either functionally replaceable or unimportant. Other studies have shown that the application of toxic substances to soil, such as heavy metals, can selectively remove subsets of the microbial community, and thereby impair ecosystem processes. For example, heavy metal application can reduce the diversity of rhizobial strains, including those strains that are most effective at fixing atmospheric nitrogen, and therefore impair symbiotic nitrogen fixation (Giller et al. 1998).

Groups of soil fauna with larger body sizes are more amendable to study by removal or exclusion approaches, at least at the functional group level. For example, meshes with different hole sizes have been used to selectively exclude soil animals on the basis of body size; studies using this approach show that animal body-size distribution plays a role in affecting both decomposer processes (Vossbrink et al. 1979; Wardle et al. 2003c) and plant growth (Setälä et al. 1996). Selective biocides have also widely been used as a tool, at least historically (Santos et al. 1981; Ingham et al. 1986; Beare et al. 1992; Heneghan et al. 1999), for quantifying the effect that different subsets of the soil faunal community may exert on other components of the decomposer subsystem. Such studies have significantly advanced our understanding about how loss of certain soil groups alters ecosystem functioning, although they also have limitations in that the biocide may also exert side effects on the ecosystem response variable of interest. Removal of larger-bodied soil organisms is also possible using other approaches. For example, removal of earthworms has been performed by 'electroshocking' the soil, which drives most of the earthworms to the soil surface (from where they can then be collected) while keeping other soil properties and organisms intact (Bohlen et al. 1995; Staddon et al. 2003). Larger surface-dwelling predators can also be kept out of study plots by appropriately designed exclusion fences; as discussed in Chapter 2, these have been used to quantify the indirect effects of predatory spiders (Kajak et al. 1993; Lensing and Wise 2006) and salamanders (Wyman 1998) on plant litter decomposition through their influence on detritivores. In sum, removal approaches applied to the soil faunal community offer important insights about the effects of losses of specific functional groups of soil fauna (or subsets of the faunal community with particular functional attributes) on the decomposer subsystem, and consistently point to impairment of ecosystem processes when these organisms are lost.

As highlighted in Chapter 4, aboveground herbivores operate as important ecosystem drivers both aboveground and belowground, and this has been shown through a large number of studies that have taken a removal or exclusion approach. For example, several studies have used insecticides to illustrate the importance of invertebrate herbivory in influencing plant community structure (Siemann et al. 2003), and the indirect consequences of this for groups of soil organisms (e.g. Brown and Gange 1989, 1990) and decomposer-driven soil processes (e.g. Mulder et al. 1999). However, as mentioned above, one problem with such studies involves non-target effects of the insecticide on the response variables of interest (Siemann et al. 2003). The aboveground and belowground effects of larger bodied herbivores (i.e. vertebrates) have been very commonly investigated through removal experiments that involve the use of fenced exclusion plots. Studies that have used this approach for this purpose have been discussed extensively in Chapter 4 and so will not be reviewed in depth here. However, in the context of this discussion, we emphasize that fenced exclusion studies are very useful both for predicting the effects of losses of large herbivores that may occur through local extinction, and for determining the effects of re-introduction of locally extinct herbivores, such as was used to study the consequences of re-introduction of bison (*Bison bison*) into tallgrass prairies in Kansas, USA (Knapp et al. 1999). Exclosure studies also lend themselves to selective removal of vertebrate herbivores based on body size, through the use of fences with different hole size or height (e.g. Bakker et al. 2004, 2006). Such studies potentially offer insights into the ecological effects of losing some vertebrate species (e.g. large-bodied ones) while retaining others, and also enable determination of how the loss of some herbivores might influence the ecosystem impact of others that remain (Bakker et al. 2004).

It is apparent from the above discussion that removal experiments have been applied to a diverse range of taxa and ecosystems to understand what happens when species or groups of species are lost from real ecosystems. Despite this, the extent to which effects of species losses are influenced by species attributes and how they vary across ecosystems remains relatively little understood. However, as discussed above, there is some evidence from the studies of Wardle and Zackrisson (2005), Wardle et al. (2008a), and Gundale et al. (2010) that whether and how losses of forest understorey species affect belowground properties and feedbacks depends on soil fertility and ecosystem productivity (Fig. 5.6), which points to the role of environmental context in governing whether and how species losses affect ecosystem functioning. Such findings suggest that the nature of ecosystem-level responses to the loss of a particular species resulting from human activity could differ greatly among contrasting ecosystems. If this is the case, then an improved understanding of the effects of species removal on ecosystem properties may be achieved by an increased focus on how ecosystem properties and species attributes (i.e. traits) determine the ecosystem-level consequences of species loss.

5.2.3 Effects of species losses in real ecosystems

As discussed so far in this chapter, there has been a tremendous amount of activity directed to understanding how biodiversity affects ecosystem functioning, and a range of approaches have been used, from large-scale, observational studies to highly controlled experimental studies (Díaz et al. 2003). These approaches vary greatly in their applicability for understanding how species losses affect ecosystem-level properties in the real world. As such, it is important to note that the question of how species richness affects ecosystem functioning is different to that of how species losses during real extinction events affect ecosystem functioning, and different approaches may be best suited for each of the two questions (Fig. 5.7). Although random assembly experiments have been the most widely used approach for exploring how biodiversity losses affect ecosystem functioning, they are only relevant to this question if it is assumed that ecological communities are randomly assembled and if species from that community are lost at random (i.e. independent of species attributes) during extinction. As discussed earlier, whenever species traits that predispose a species to extinction are the same as those that are important for driving ecosystem functioning, then the effects of species loss are likely to be much greater than would be predicted through random assembly experiments. It is likely that this is frequently the case, and there is evidence for this from empirical (Petchey et al. 1999; Jonsson et al. 2002; Zavaleta and Hulvey 2004) and theoretical (Solan et al. 2004; Bunker et al. 2005) studies. As such, random assembly experiments may greatly

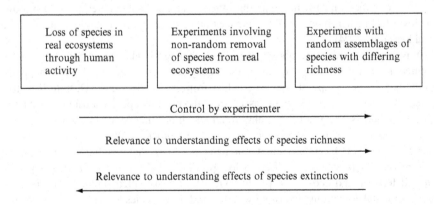

Fig. 5.7 Approaches that have been used to address the issue of how biodiversity losses (i.e. extinctions) through human activity may influence the functioning of ecosystems.

underestimate effects of species losses on ecosystem properties, and despite strong assertions to the contrary (e.g. Duffy 2009) this diminishes the relevance of such experiments for understanding the ecosystem effects of species losses (Wardle and Jonsson 2010).

Determination of how species losses through human activity may affect ecosystem processes is arguably most realistically and directly studied by measuring how the ecosystem responds to the actual species losses that have occurred. This is analogous to the large number of studies that have directly assessed how ecosystems respond to the gain of new species through biological invasion, an approach that has frequently proven its worth and advanced our understanding about invasion ecology, as discussed later in this chapter. With regard to plant communities, few studies have directly quantified the effects of actual species losses resulting from human activity. Those that have been done so far have focused largely on the selective removal of tree species that have occurred through logging as discussed above, and such studies have shown that these losses can have particularly important consequences for ecosystem functioning when the harvested species are the structural dominants of the forest (Díaz et al. 2003; Wardle et al. 2008c). Natural or unintended 'removal experiments' of this type arguably represent the most direct means available for assessing the effects of real species losses in ecosystems resulting from human-induced alterations of either abiotic or biotic environmental conditions.

Some of the most convincing studies of how actual losses of species affect ecological processes involve situations in which aboveground mammal species have become locally extinct. As outlined in Chapter 4, herbivorous mammals and their predators often have major effects on ecosystem processes, meaning that the loss of these species is also likely to have important consequences. For example, the North American bison (*Bison bison*), which was once the dominant mega-herbivore of the North American Great Plains, was reduced to a few thousand individuals by the 1880s and had become extinct from much of its natural range. Re-introduction of bison to the Konza Prairie in the mid-1980s, together with the use of fenced plots from which reintroduced bison were excluded, provides evidence that the loss of bison causes altered floristic composition and reduced plant diversity, reduced spatial variability of vegetation, reduced rates of soil nitrogen mineralization, and reduced plant-tissue nitrogen concentrations (Knapp et al. 1999; Johnson and Matchett 2001). Further, the loss of large predators, such as has occurred in many ecosystems worldwide due to human activity, can set a cascade in motion through causing population growth (and overabundance) of herbivorous prey (Wardle and Bardgett 2004). For example, the loss of the cougar (*Pumus concolor*) and wolf (*Canis lupus*) in many natural ecosystems in North America has led to large increases in cervid species (white-tailed deer (*Odocoileus virginianus*), elk (*Cervus canadensis*), and moose (*Alces alces*)). These increases have been shown to alter consumption by cervids of palatable plant species (Creel and Christianson 2009), greatly alter forest vegetation composition (Ripple and Beschta 2008), reduce understorey and riparian vegetation, promote erosion, and alter the hydrology of whole watersheds (Ripple and Betscha 2006; Betshsa and Ripple 2008). Given the impact of mammalian herbivores on nutrient cycling, as considered in Chapter 4, it is likely that losses of their predators would also exacerbate these effects. For example, studies on the reintroduction of wolves into Yellowstone National Park (Frank 2008) provide evidence that the loss of wolves and resulting increases in ungulate densities in turn indirectly promote soil nitrogen mineralization and presumably the soil organisms that drive this process.

The effect of loss of large mammals on the present-day functioning of ecosystems is also apparent through extinctions that occurred hundreds to thousands of years ago. Human colonization of new regions has often coincided with the extinction of mega-herbivore species, although there is some uncertainty about the extent to which these extinctions have been caused directly by humans or by other factors such as climate and vegetation change (Zimov et al. 1995; Guthrie 2003). Nevertheless, there is evidence that mega-herbivore extinction is likely to have caused major shifts in the types of vascular plants present and the functioning of the belowground subsystem. For example, it has been proposed that the extinction of mega-herbivores in Alaska and Russia over the past 10 000–12 000 years has led to a shift from grazed steppe grassland to wet moss-dominated tundra (Zimov et al. 1995). This has led to domination by plant species that produce poorer-quality litter, greater water-logging, and impairment of soil nitrogen mineralization, setting a feedback in motion that maintains domination by tundra vegetation (Fig. 5.8). Wherever human colonization has been associated with losses of mega-herbivore species, there are likely to have been large and irreversible consequences for aboveground–belowground linkages and therefore the functioning of the ecosystem.

Before leaving the topic of how species losses impact on ecosystem functioning, we highlight that the most widely documented and convincing examples involve ecosystem impacts of losses of large organisms, such as trees and some mammal species. As a result of their large size, they are both more likely to be utilized by humans for resources (e.g. trees and herbivorous mammals) or deliberately exterminated (e.g. large predators), and to have disproportionate effects on the functioning

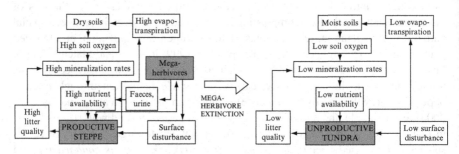

Fig. 5.8 Effects of mega-herbivore extinction in northern Russia in the late Pleistocene on feedbacks between vegetation and nutrient cycling, as hypothesized by Zimov et al. (1995). From Wardle, David A. 'Communities and Ecosystems' © (2002), Princeton University Press, reprinted by permission from Princeton University Press.

of the ecosystem. There are few examples of human-induced losses of smaller bodied organisms, including soil organisms, impacting on community and ecosystem processes. This may be because losses of smaller-bodied species often have smaller effects on ecosystem properties, or because when smaller organisms are lost their ecological impact (and even the fact that the species has been lost) is less likely to be detected.

5.3 Species gains through invasion and aboveground–belowground linkages

Ingress of alien (including invasive) organisms into a new ecosystem is to some extent the reverse of extinction, as it results in the gain of novel species by a community. As is the case for extinction, invasive species are most likely to exert important community and ecosystem-level effects when they become biomass dominants within their trophic level, or have traits that differ greatly from those of the native species. There has been much research activity, and significant recent conceptual advances, in understanding the impacts of the ingress of new species into biological communities, which we now discuss.

5.3.1 Invasions by plants

5.3.1.1 Functional Differences between Invasive and Native Species

The majority of studies that have considered invasive species from a combined aboveground–belowground perspective involve invasive plants. In plant communities, most alien species become minor and functionally unimportant components of the invaded community (Thompson et al. 1995), but a subset of species do have major effects on the functioning of ecosystems due to key trait differences. In this light, studies that have compared native and alien components of floras have often found significant differences in key functional traits between the two groups. For example, Baruch and Goldstein (1999) found that for 63 plant species in the Hawai'ian archipelago, invasive species as a group had leaf traits linked to greater resource capture (i.e. high specific leaf areas, carbon dioxide assimilation rates, and concentrations of foliar nutrients) relative to native species (Fig. 5.9). Similarly, Leishman et al. (2007) and Peltzer et al. (2009) have found, for floras near Sydney, Australia, and Kaikoura, New Zealand, respectively, that alien species on average had higher specific leaf areas and nitrogen and phosphorus concentrations than did native species (Fig. 5.9). However, differences in leaf traits between invasive and native species may depend greatly upon habitat conditions (Baruch and Goldstein 1999). For example, Funk and Vitousek (2007) found, through comparing pairs of phylogenetically related plant species in Hawai'i, that alien plant species on average had greater photosynthetic rates and resource (i.e. water, nutrient, and light) use efficiencies than did native species. However, they also found that several of these differences were mostly apparent only over short timescales and were strongly dependent upon the

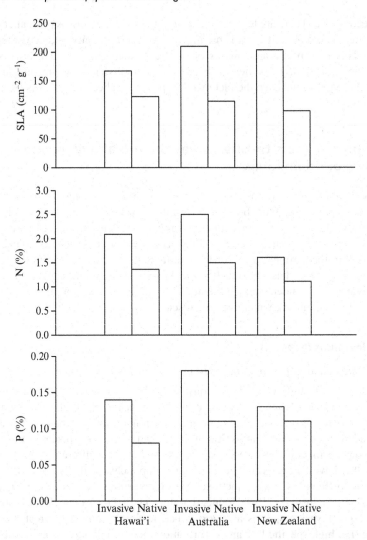

Fig. 5.9 Comparison of specific leaf area (SLA) and leaf nitrogen (N) and phosphorus (P) concentrations between native and exotic components of floras along an elevational gradient in Hawai'i (64 species; Baruch and Goldstein 1999), in bushland near Sydney, Australia (55 species; Leishman et al. 2007), and on a river floodplain near Kaikoura, New Zealand (41 species; Peltzer et al. 2009).

nature of resource limitation in the habitat in which they were growing. Nevertheless, evidence to date does generally point to native and alien species showing overall and consistent differences in key ecophysiological traits (Rejmanek et al. 2005).

The key traits that often differ between native and alien species are well known to control litter-decomposition processes, as discussed in Chapter 3. For this reason, it is expected that decomposability of litter should differ between the two types of plants, although few studies have performed comparative studies on rates of litter decomposition of large numbers of native and invasive species. A meta-analysis of 94 studies by Liao et al. (2008) showed that on average litter from invasive species decomposed 2.17 times faster than that from non-invasive species, although this effect diminishes when nitrogen-fixing plant species are not included in the analyses. Further, in that analysis it is difficult to ascertain whether or not this difference was mostly due to phylogenetic differences between the two groups. Meanwhile, Allison and Vitousek (2004) compared litter-decomposition rates for 11 understorey Hawai'ian plant species (five native and six invasive) and found that litter from the natives did decompose more slowly. However, all but one of the natives were ferns and all but one of the invaders were angiosperms, and it is recognized that fern litter usually decomposes more slowly than that from angiosperms (see Chapter 3). Meanwhile, Kurokawa et al. (2010) compared rates of litter decomposition for 41 New Zealand riverplain shrub species, and found, that as a group, litter from invasive nitrogen-fixing species decomposed faster than that from native nitrogen fixers. However, there was no difference in litter decomposability between native and exotic non-nitrogen fixers. Other studies that have compared rates of litter decomposition for smaller numbers of coexisting native and invasive species (reviewed by Ehrenfeld 2003) have often found litters of invaders to decompose more rapidly (e.g. Cameron and Spencer 1989; Standish et al. 2004), although there are many exceptions (e.g. Kourtev et al. 2002b; Güsewell et al. 2006), especially when the native and invasive species have different life forms. For example, exotic pine tree species, a major invasive problem in many ecosystems worldwide (Richardson 2006), are often likely to produce litters that decompose more slowly than those produced by the native species (Ågren and Knecht 2001; Ehrenfeld 2003) (Fig. 5.10).

Functional differences between coexisting invasive and non-invasive species also have consequences belowground. The most substantial effects of invaders often arise when the invader is able to form nodules that fix atmospheric nitrogen while the native species cannot. In a classic study, Vitousek and Walker (1989) found that the invasion of the actinorhizal shrub *Myrica faya* (native to the Azores and Canary Islands) into a montane forest site in Hawai'i, which lacked nitrogen-fixing plant species, led to an over four-fold increase in ecosystem nitrogen input. This should in turn have large ecosystem-level consequences through alleviating nitrogen limitation at the site. However, non-nitrogen-fixing invasive species also cause important changes in soil properties, and often (though not always) promote nutrient flux rates (Ehrenfeld 2003) and the availability of plant-available nitrogen (Zou et al. 2006), phosphorus (Chapuis-Lardy et al. 2006), and other minerals (Vanderhoeven et al. 2005). As a result of alteration of the quality and quantity of resources entering the soil, invasive plants also cause profound changes in soil food webs. For example, Kourtev et al. (2002a) found that invasive understorey plant species in forests in New Jersey, USA, exerted substantial effects on the composition of the soil microbial community. Further, populations of soil fauna (nematodes and microarthropods) in grasslands in south-eastern Utah have been shown to respond negatively to invasion

(a)

(b)

Fig. 5.10 Invasive *Pinus contorta* trees, native to North America, invading grassland in the Craigieburn range of New Zealand. Invasive pine trees produce poor-quality litter and acidify soils, leading to impairment of belowground processes and alteration of the belowground community. Image by D. Peltzer.

by the exotic grass *Bromus tectorum* (Belnap et al. 2005). Similarly, Yeates and Williams (2001) found invasion of native ecosystems by each of three exotic plant species in New Zealand to influence soil microfaunal communities, although the extent of these effects depended on both the plant species and location. Belowground impacts of invasive plants may be apparent even when the invaders do not dominate the plant biomass, especially if they have sufficiently distinct functional characteristics. For example, Peltzer et al. (2009) showed that for a New Zealand floodplain community, non-native species exerted disproportionate effects on the soil microbial community and on microbe-feeding and predatory nematodes, despite occupying only 3% of the standing plant biomass (Fig. 3.10).

Invasive plants can also potentially exert important effects both aboveground and belowground by altering the disturbance regime of the ecosystem (Mack and D'Antonio 1998). This is especially apparent when invasive and native species differ in key traits that affect their relative flammability, because this has important consequences for the ecosystem's fire regime. Notably, invasive grasses that invade woody ecosystems in Hawai'i, Australia, and North and South America produce highly flammable tissues and greatly increase the fuel load for fires at the ground layer, leading to an increased occurrence of fires in these ecosystem (Brooks et al. 2004; Bradley et al. 2006; Pauchard et al. 2008). For example, in Hawai'i, invasion of perennial C_4 grasses into sub-montane forest results in rapid accumulation of highly flammable material, resulting in fires that kill most of the natural vegetation. The grasses recover rapidly after fire while the native species do not, setting in motion a positive feedback between fire and the exotic grasses (D'Antonio and Vitousek 1992; Brooks et al. 2004). Given that fire exerts a multitude of effects on the belowground subsystem (Certini 2005), it is inevitable that invader-induced shifts in fire regime would have important belowground consequences, although these have seldom been addressed. However, Ley and D'Antonio (1998) found in Hawai'i that rates of nitrogen fixation were greater in fire-driven grassland dominated by alien grasses than in adjacent uninvaded woodland. This was because the main substrate upon which the nitrogen fixation occurs, namely leaf litter from native woody species, is largely absent from the grassland.

5.3.1.2 Invasive Plant Species and Plant–Soil Feedbacks

Differential effects of native and invasive plant species on soil biota can in turn have important feedback effects on plant performance. As discussed in Chapter 3 (see also Fig. 3.11), a large number of recent studies have experimentally investigated feedbacks between plants and soil organisms. One significant application of this approach has been the study of relationships between invasive plant species and their associated soil biota (see reviews by Wolfe and Klironomos 2005; Van der Putten et al. 2007). In a landmark study, Klironomos (2002) utilized plant–soil feedback experiments to show that invasive plant species in Canadian grassland and meadow ecosystems consistently showed positive plant–soil feedbacks, while rare native plant species consistently showed negative feedbacks (Fig. 5.11). This work indicates that invasive species may derive a benefit from entering positive feedbacks with their soil biota in their new habitat. As evidence for this, feedback experiments have since been used to provide evidence that invasive species respond more favourably (or less unfavourably) to their soil biota in their new habitat than in their native range

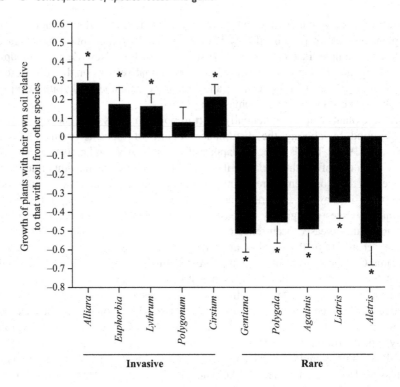

Fig. 5.11 The results of a soil-feedback experiment for five invasive and five rare native plant species from Canadian grasslands and meadows. Bars represent the mean ± 1 SE (*n* = 10). Note that invasive plants undergo positive feedbacks with their associated soil biota, while rare species undergo negative feedbacks. Modified from Klironomos (2002), with permission from Macmillan Publishers Ltd.

(e.g. Reinhart et al. 2003; Callaway et al. 2004; Knevel et al. 2004). For example, Reinhart et al. (2003) used feedback experiments to show that black cherry (*Prunus serotina*) was negatively influenced by the soil community that developed under it in its native range in North America, while it was positively influenced by its soil community in its new range in north-western Europe. Such studies provide evidence that invasive plant species may perform better in their new range through escaping soil antagonists (e.g. pathogens) that regulate their populations in their native range, supporting the enemy release hypothesis (Elton 1958; Keane and Crawley 2002). However, some caution is needed in interpreting the results of such experiments, because few studies (but see Reinhart et al. 2003) have utilized both soils and plants from both native and non-native ranges, which is an important requirement for providing convincing evidence of this hypothesis (Van der Putten et al. 2007).

The belowground effects of invasive plant species can also have important consequences for coexisting native species. One route by which this may occur is through invading plant species releasing allelochemical compounds that in turn impact negatively on native plant species (Wardle et al. 1998b, Ridenour and Callaway 2001). A development of this idea is the so-called novel weapons hypothesis which proposes that invasive plant species can release phytotoxic compounds to the soil that are novel to the native plant species and to which the native species are non-adapted and therefore vulnerable (Callaway and Ridenour 2004). There are several claims in the literature of species invading new communities in turn exerting phytotoxic effects against the resident species, although there are difficulties in providing unequivocal evidence of these effects (Harper 1977; Stowe 1979; Keeley 1988). One example that has gained much attention is the ability of the invasive Eurasian plant *Centaurea maculosa* (spotted knapweed) to produce (−)-catechin, which reportedly has a greater adverse effects on grass species in its invasive range (i.e. in North America) than on related grass species in its native range (e.g. Bais et al. 2003; Callaway and Ridenour 2004; Thorpe et al. 2009). However, subsequent investigations have suggested that (−)-catechin released from *C. maculosa* may not reach sufficient concentrations in the soil to exert negative effects on its neighbours (Blair et al. 2005; Perry et al. 2007; Duke et al. 2009). Other studies have presented evidence that invasive plant species may adversely affect coexisting species by disrupting their mycorrhizal symbionts. This has recently been shown for the invasive non-mycorrhizal plant *Alliaria petiolata* (garlic mustard), which adversely affects ectomycorrhizal fungi associated with seedlings of coexisting native species when it invades forests in North America (Stinson et al. 2006; Callaway et al. 2008; Wolfe et al. 2008). It has been proposed that mycotoxoins ('novel weapons') produced by *A. petiolata* are responsible for its effects in the habitats that it invades in each of these studies. However, as with (−)-catechin, it is unknown whether these toxins actually accumulate in soils in the invaded areas in sufficient concentrations to exert these effects.

Although many studies have focused on the success of invasive plant species in new habitats through escaping their belowground enemies, plants invading new habitats can also leave behind their belowground mutualists which could impair their success. This is likely to be important for invasive plant species that rely on ectomycorrhizal fungal mutualists, such as pine trees in much of the Southern Hemisphere (Richardson et al. 1994, 2000). Here, pine species often invade native communities that lack ectomycorrhizal hosts, and rapid invasion of pines may therefore occur only after arrival of windborne ectomycorrhizal fungal spores, for example from nearby pine plantations where the trees have been deliberately inoculated (Richardson et al. 2000; Nuñez et al. 2009). In contrast, invasive plants that form arbuscular mycorrhizal associations are usually less likely to be impaired by the availability of fungal inocula, because arbuscular mycorrhizal fungi are very widely distributed and have lower host specificity (Richardson et al. 2000). In the case of invasive plants that form mutualistic symbioses with nitrogen-fixing bacteria (i.e. *Rhizobium* and *Frankia*), the bacterial symbionts are often widespread even in soils that lack hosts, and the deliberate introduction of bacterial symbionts to many regions worldwide contributes to the success of these plants as invaders (Richardson et al. 2000; Van der Putten et al. 2007). In some cases, it is unclear as to where these symbiotic bacteria arise from. For example, it is unknown whether the highly successful actinorhizal invasive plant *Myrica faya* in Hawai'i (Vitousek and Walker 1989) arrived with its own strains of *Frankia*, or whether compatible strains of *Frankia* were already present in the soil (Richardson et al. 2000). The most widespread mutualism between plants and soil organisms is arguably between plants and their associated saprophytes that mineralize nutrients required for plant growth (see Chapter 3). Here, the saprophytic community consists of organisms that have a wide range of functional capacities, and most soils contain most functional groups of saprophytes. For this reason, it is highly unlikely that invasive plant species will 'escape' saprophytic microbes with particular functional capabilities in their old environment or encounter new types of saprophytes in their new environment (Van der Putten et al. 2007).

Before leaving the topic of invasive plants, it is important to stress that there are three factors that collectively contribute to the total ecological impact of any invasive plant species, namely, range, abundance, and per-capita or per-biomass effect of the invader (Parker et al. 1999). As such, while most studies of the effects of invasive plants belowground (and their feedback effects aboveground) have focused on the third of these factors, the invader can only have important ecological impacts if it also occupies a sufficient range and becomes abundant. The ability of a potential invader to increase its range and abundance in a new habitat will be determined by the demographic and physiological traits of the invader, and by the characteristics of the habitat and plant community being invaded. As such, the extent to which a native plant community can resist invasion will reduce the magnitude of the impact of the invader, both aboveground and belowground. As a consequence, native plant community properties and traits of the dominant plant species (Fridley et al. 2007), the nature of resource supply and availability (Huston 1994; Davis et al. 2000), and disturbance regime (Burke and Grime 1996) are all important determinants of whether, and how much, the community is likely to be invaded in the first place. An improved understanding of how invasive plant species can impact upon ecosystems will therefore gained by considering not just the effect of the invader once it has become established, but also whether the invader may increase its range and abundance sufficiently to exert important effects in the first place.

5.3.2 Belowground invaders

In general, there is relatively little known about the occurrence or ecological significance of invasive microorganisms. However, as discussed earlier in this chapter, many groups of saprophytic microbes may not be dispersal-limited at a global scale (Finlay 2002), thereby reducing the scope for microbes to invade new habitats. However, were invasions of species of saprophytic microbes to occur, they would likely remain undetected because most saprophytes have not been described at the species level (Van der Putten et al. 2007). Further, given the considerable functional diversity of the microbial saprophytic community, it is unlikely that an invasive microbe would possess sufficiently novel functional attributes for its ecological effects to become detected. However, there are some well-documented examples of pathogenic soil-borne fungi invading new habitats and causing widespread effects on vegetation in native ecosystems (reviewed by Desprez-Loustau et al. 2007; Loo 2009). For example, dieback of natural vegetation in Australia has resulted from invasion by the root pathogenic fungi such as *Phytophthora cinnamomi* (Peters and Weste 1997) and *Armillaria luteobubalina* (Shearer et al. 1998). Similarly, in California, invasion by *Phytophthora ramorum* is well known to influence a range of vascular plant species and cause sudden oak death syndrome (Venette and Cohen 2006). Such examples point to invasive pathogens having novel modes of attack that the native flora is poorly adapted to withstand. There are also examples of invasive mutualistic fungi that form associations with native tree species. For example, the distinctive European ectomycorrhizal fungus *Amanita muscaria* has invaded native forests in New Zealand and Australia (Fig. 5.12), while *Amanita phalloides* has spread from Europe to forested ecosystems elsewhere around the world (Pringle and Vellinga 2006). However, the impact of invasive ectomycorrhizal fungi on native tree growth, the native ectomycorrhizal community, or the decomposer subsystem remains little understood.

With regard to belowground invertebrates, little is known about invasions in natural ecosystems involving microfauna, such as nematodes and protozoa, largely because the vast majority of species have not been described. Further, it is possible

Fig. 5.12 The well known ectomycorrhizal fungus *Amanita muscaria* or fly agaric, native to Europe, is a distinctive invader of native forests in New Zealand and Australia, where it forms associations with native tree species. However, the impact of invasion by this fungus on native tree growth, the native ectomycorrhizal fungal community, or other components of the ecosystem, remains little understood. Image by Ian Dickie.

that, like microbes, at least a subset of the soil microfaunal community may not be dispersal-limited at global scales (Finlay 2002; but see Foissner et al. 2008). However, larger-bodied soil invertebrates that are constrained by biogeographical boundaries are certainly capable of invading new ecosystems. With regard to mesofauna, some of the best-documented examples involve invasion of sub-Antarctic islands by European species of Collembola (Frenot et al. 2005). For example, invasive collembolan species can reach very high densities on sub-Antarctic Marion Island, and it has been suggested that this could be causing displacement of native collembolan species (Convey et al. 1999). Further, invasion of ecosystems by larger-bodied saprophagous soil organisms, such as millipedes and isopods, and insects such as beetles and dipterans that spend at least part of their life cycle belowground, are well documented in many parts of the world (e.g. Gaston et al. 2003; Arndt and Perner 2008). However, the effects of these invaders on native decomposer fauna, soil processes, and plant growth and nutrition remain largely unexplored.

The one belowground faunal group for which the effects of invasive species have been comparatively well studied are the earthworms (see reviews by Bohlen et al. 2004b; Hendrix et al. 2008). Novel earthworm species have been introduced to many temperate and tropical regions worldwide (Hendrix et al. 2008), and their ecological impacts are particularly apparent when they invade ecosystems that lack functionally comparable species. As such, strong impacts of invasive earthworms have been shown in regions that lack a native earthworm fauna through Pleistocene glaciations eliminating previous earthworm species (Hendrix et al. 2008). The belowground consequences of earthworm invasions can be wide-ranging (Bohlen et al. 2004b; Fig. 5.13). For example, the geophagous South American earthworm *Pontoscolex corethurus* has invaded many forested and agricultural regions throughout the tropics, where it alters and sometimes greatly enhances soil compaction and thus reduces soil porosity (Chauvel et al. 1999). This can in turn result in a compact soil crust that potentially negatively impacts both on other soil organisms and plant growth (Lapied and Lavelle 2003; Gonzalez et al. 2006). Meanwhile, the epigeic European earthworm *Dendrobaena octaedra* has invaded natural forests in Alberta, Canada, and this has favoured domination by faster growing fungal species and reduced the abundance of many taxa of mites (McLean and Parkinson 2000a, 2000b). Further, earthworm invasion in temperate North American forests can strongly influence soil physical structure, enhance mineralization of organic matter, and lead to loss of soil organic mater and nutrients; these effects are, however, far more pronounced for burrowing than epigeic earthworm species (Bohlen et al. 2004a; Hale et al. 2005). Stimulation by invasive earthworms of organic matter mineralization can cause short-term improvements in plant nutrition and growth (Scheu and Parkinson 1994), but in the longer term it leads to removal of the surface organic layer, adversely affecting plant species that are adapted to thick forest floors (Gundale 2002; Frelich et al. 2006). As such, earthworm invasion may induce loss of native herbs while promoting some invasive plant species (Bohlen et al. 2004b), and impair seedling recruitment of some forest tree species (Frelich et al. 2006; Hale et al. 2006). However, the nature of effects of earthworm invasion on aboveground and belowground properties depends

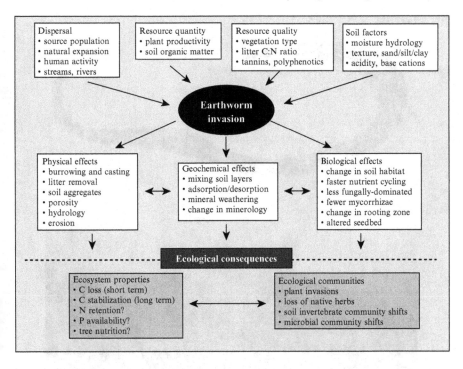

Fig. 5.13 Conceptual model demonstrating the wide-ranging effects that invasive earthworm species can exert in North American temperate forests that lack native earthworm faunas. From Bohlen et al. (2004b), with permission from the Ecological Society of America.

on the functional attributes of the earthworm species involved, as well as the properties of the ecosystem being invaded. Earthworm invasion of natural forested ecosystems in many regions is comparatively recent, and the long-term consequences of their effects for forest dynamics remains largely unknown (Bohlen et al. 2004b).

Invasive predators of belowground organisms may also exert important ecosystem-level effects, at least in cases where they consume organisms that have a functionally irreplaceable role in the ecosystem. One of the most convincing examples is the predatory New Zealand flatworm *Arthurdendyus triangulata*, which has been accidentally introduced to the British Isles and Faroe Islands where it has subsequently spread and reduced the density of lumbricid earthworms upon which it preys (Boag and Yeates 2001), potentially altering the invaded ecosystem. For example, the loss of earthworms due to flatworm invasion in Scottish grasslands has been shown to adversely affect soil porosity and drainage, leading to greater water-logging, greater dominance of vegetation by rushes, and reduced densities of burrowing moles (Boag 2000) (Fig. 5.14). Another example is the introduction of the house mouse to the sub-Antarctic islands (Smith et al. 2002; Angel et al. 2009), such as Marion Island. Here, mice consume many taxa of invertebrates, including the

(a)

(b)

Fig. 5.14 (a) The predatory New Zealand flatworm (*Arthurdendyus triangulate*), which invades grasslands in the British Isles and Faroe Islands, where it greatly reduces the density of burrowing lumbricid earthworms upon which it preys, thus reversing the ecosystem level impact of earthworms in these ecosystems. (b) Removal of earthworms by the flatworm leads to large reductions in soil porosity and water drainage, leading to waterlogging and increasing dominance by rushes (*Juncus* spp.). Images by B. Boag.

endemic flightless moth *Pringleophags marioni* which otherwise functions as the main marofaunal detritivore. In this system, predation of this moth by mice has been estimated to reduce litter processing by the moth by up to 40% (Crafford 1990), potentially leading to greatly increased rates of peat accumulation on the island (Smith and Steenkamp 1990). Ground-dwelling predatory insect groups that contain invasive members, such as ants and carabid beetles, are also likely to feed heavily on detritivorous fauna and thus affect the processes that they drive. However, while these types of predators are well known to adversely affect related native predators

(Gotelli and Arnett 2000; Niemelä et al. 1997; Snyder and Evans 2006), their impact upon belowground organisms and the processes that they drive remains unexplored (Kenis et al. 2009).

5.3.3 Invasions by aboveground consumers

As discussed in Chapter 4, aboveground primary consumers have wide-ranging effects on both the aboveground and belowground subsystems. These effects are particularly strong for invasive consumers, especially when they have escaped their natural enemies and when the new host species are not adapted to the invader. As such, there are several examples of major changes in forest ecosystems brought about by the invasion of both aboveground fungal pathogens (Desprez-Loustau et al. 2007; Loo 2009) and herbivorous insects (Liebhold et al. 1995; Kenis et al. 2009). As a well-known example of an invasive pathogen, chestnut blight (*Cryphonectria parasitica*) from Asia has resulted in widespread loss of the American chestnut (*Castanea dentata*) in forests of the eastern USA. Because chestnut produced tannin-rich wood that was of lower quality than that of co-occurring hardwood species that replaced it, it has been suggested that invasion by the blight will have greatly altered the decomposer subsystem and nutrient cycling (Ellison et al. 2005; Kenis et al. 2009). Although not experimentally tested, significant belowground effects are also likely to have resulted from the selective removal of single tree species in mixed-species forest by other invasive aboveground pathogens, such as the widespread loss of elm (*Ulmus*) trees following invasion by the Dutch elm disease (*Ophiostona* spp.). With regard to insect herbivores, severe defoliation of native oak (*Quercus*) species in North America by the invasive gypsy moth (*Lymantria dispar*) during outbreak events can have substantial belowground effects that may feedback aboveground (Lovett et al. 2006). In particular, following outbreaks there is a large pulse of nitrogen and labile carbon to the forest floor that results from insect faeces, dead caterpillars, and unconsumed fallen foliage; much of this nitrogen is in turn immobilized by soil microbes or incorporated into soil organic matter (Lovett and Ruesink 1995). Further, the hemlock woolly adelgid (*Adelges tsugae*; an aphid-like invader) is currently causing widespread death of hemlock (*Tsuga*) species in the north-eastern USA. As hemlock species produce much poorer-quality litter than do most tree species that are likely to replace it, the resulting shift in tree species composition will probably long-term impacts on the structure and functioning of the ecosystem (Lovett et al. 2006).

Mammalian herbivores, for example deer, rabbits, and goats, have been introduced to many parts of the world where they can have major impacts on natural ecosystems, both above and below ground (Wardle et al. 2001; Vázquez 2002; Spear and Chown 2009). For example, several species of deer, and domestic goats (*Capra aegagrus hircus*), were deliberately introduced to New Zealand between the 1770s and 1920s, where they have caused widespread compositional change in the forest understorey through removing broad-leaved, fast-growing dicotyledonous plant species and promoting other unpalatable species. New Zealand lacks native browsing mammals, and the native mega-herbivores (i.e. the moa birds), which were hunted to extinction a few hundred years ago, probably had much smaller effects on the forest than the introduced browsers currently do (McGlone and Clarkson 1993). Using long-term deer exclusion plots located in forests throughout New Zealand, Wardle et al. (2001, 2002) showed that invasive deer and goats caused consistent reductions of those understorey plant species which produce high-quality and readily decomposable litter, and generally promoted those species that produce poor-quality litter. Also, browsing mammals were found to have consistently negative effects on large-bodied soil organisms (Fig. 5.15), probably through adverse physical disturbance effects resulting from trampling (Wardle et al. 2001). However, most groups of small-bodied soil organisms, including microflora and microfauna, showed idiosyncratic responses to invasive browsers, with both positive and negative effects being detected depending on the location (Fig. 5.15). Ecosystem properties driven by soil biota, such as soil carbon mineralization and carbon and nitrogen sequestration also showed context-dependent responses. These varied responses emerged because browsers can exert either positive or negative effects on decomposers depending on environmental context, as discussed in Chapter 4.

Ecosystem impacts resulting from the physical effects of invasive herbivores may be especially significant when the disturbance that they introduce is novel to the ecosystem. One globally widespread example of this is the feral pig (*Sus scrofa*), which is now present on all continents except Antarctica, and on many island systems. When present, they cause significant physical disturbance through soil turnover caused by foraging for roots and invertebrates. In forested ecosystems, this activity can promote tree death, while in grasslands it can cause replacement of perennial grasses with shorter-lived grasses and forbs, and greater incidence of invasive plant species (Tierney and Cushman 2006). This disturbance regime has also been shown to adversely affect major groups of microarthropods, whose densities recover only after pigs are removed from the system (Vtorov 1993). One of the most impressive examples of the ecosystem effects of an invasive hervivore involves the beaver (*Castor canadensis*) that has been deliberately introduced to *Nothofagus*-dominated forest ecosystems in southern South America (Anderson et al. 2006, 2009) (Fig. 5.16). Tree-felling and dam formation by these beavers has been described as having the largest landscape-level impact on these ecosystems since the last ice age (Anderson et al. 2009). Here, beavers eliminate riparian forest and reduce forest canopy up to 30 m from stream edges (Anderson et al. 2006). This is accompanied by the formation of meadows with higher herbaceous species richness, but with a much greater incidence of exotic species. The belowground impacts of this transformation, while unexplored, are likely to be substantial.

Some of the strongest ecosystem effects of invasive aboveground consumers involve predators. Entry of a novel predator into an ecosystem often involves not just the invasion event, but also the reduction or loss of its prey, sometimes to the point of extinction. Invasive predators are present in most parts of the world, and their impacts are most acute when they result in the removal of a consumer organism that has an important role in the ecosystem itself. One such example involves

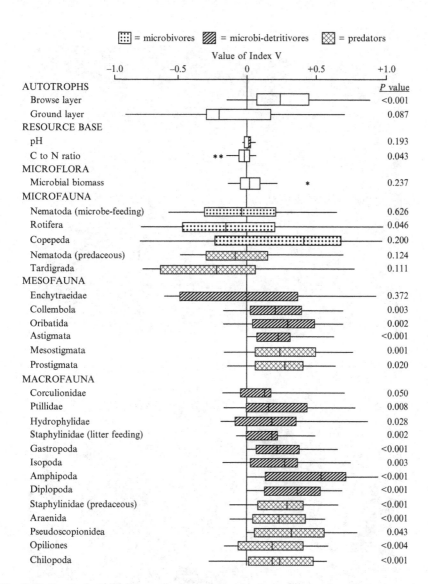

Fig. 5.15 Box-and-whisker plots summarizing data for the response to browsing mammals of resources, autotrophs, and components of the decomposer food web in the litter layer, as assessed by measurements performed inside and outside fenced exclosure plots for each of 30 locations throughout New Zealand (see Wardle et al. 2001). The index V (Wardle et al. 2001), determined for each response variable at each location, becomes increasingly negative if the value is increasingly greater outside the exclosure relative to inside it, and increasingly positive if it becomes increasingly greater inside the exclosure than outside it; the index ranges from −1 to +1, with 0 indicating no difference. For each variable, the box encompasses the middle half of the data (values of V) between the first and third quartiles (i.e. 15 of the 30 locations); the bisecting line is at the value of the median, and the horizontal line outside the box represents the typical range of data values. Asterisks indicate outlier values. P values are for paired *t*-tests comparing the significance of difference of the value of the variable inside compared with outside the exclosure across the 30 locations. From Wardle et al. (2001), with permission from the Ecological Society of America.

(a)

(b)

Fig. 5.16 Substantial ecosystem transformation resulting from tree felling in *Nothofagus* forests by invasive North American beavers (*Castor canadensis*) on Navarino Island in southern South America. This disturbance is recognized by Anderson et al. (2009) as having the largest impact in these forests since the last ice age. The images show (a) widespread death of high-elevation riparian *Nothofagus antarctica* forest caused by beavers with no apparent regeneration and (b) a low-elevation stream impounded by beavers, causing sedimentation and inundation of riparian *Nothofagus pumilio* and *Nothofagus betuloides* forest. Images by C. Anderson.

forested ecosystems on Christmas Island in the Indian Ocean. Here, the red land crab (*Geracoidea natalis*) serves as the main consumer of seeds and seedlings, and as an important processor of leaf litter (O'Dowd et al. 2003; Green et al. 2008). This island has been invaded by the yellow crazy ant (*Anoplolepis gracilipes*) which serves as a major predator of the crabs, thus eliminating the ecological role that the crabs perform. The net result is enhanced tree seedling recruitment and reduced leaf litter decomposition (O'Dowd et al. 2003). Another example involves the introduction of predatory foxes to the Aleutian Island chain, where they have caused substantial reductions in nesting seabird colonies, thereby thwarting nutrient transfer by the birds from the ocean to the land. Comparisons of islands that have been invaded by foxes with those that have not have shown that predation of seabirds causes a switch in the dominant vegetation from graminoids to low-lying forbs and dwarf shrubs, reduced levels of soil and foliar nutrients, and altered flows of nitrogen through aboveground consumer trophic levels (Croll et al. 2005; Maron et al. 2006).

Invasive aboveground predators can also exert cascading effects on the belowground food web. This is apparent through studies on forested oceanic islands off the coast of northern New Zealand. When seabirds occur on these islands in high densities, they transfer nutrients from the ocean to the land, and cultivate soil with their extensive burrowing activity during nesting periods. Several of these islands have been invaded by alien rat (*Rattus*) species while several others have not, and when present, the rats feed on seabird chicks and eggs, thereby severely reducing seabird densities. As a consequence, invasion by rats reduces nutrient inputs by seabirds to the soil, thereby adversely affecting many components of the soil food web (Fukami et al. 2006; Towns et al. 2009) (Fig. 5.17) and processes driven by the soil biota such as litter decomposition (Fukami et al. 2006). Further, reduced nutrient inputs and densities of decomposer organisms on rat-invaded islands causes a reduction in plant growth rates (Fukami et al. 2006), plant foliar and litter nutrient concentrations, and rates of nutrient release from decomposing litter (Wardle et al. 2009b). Trees on rat-invaded islands also show greater resorption of nutrients by plant leaves prior to litter fall (Wardle et al. 2009b), which is indicative of greater nutrient limitation of the plants resulting from rat invasion. However, rats also reduce soil disturbances and tree root damage caused by seabird burrowing activities, and for this reason invasion by rats also results in a greater density of establishing tree seedlings (Fukami et al. 2006; Mulder et al. 2009) and a higher tree standing biomass (Wardle et al. 2007).

Before we leave the topic of invasive organisms, it is important to note that there are many conspicuous examples of invasive organisms transforming both the aboveground and belowground components of the ecosystem. These effects are inevitably greatest when the invader has a particular capability that is lacking in the native biota, and there are many examples of this across all major trophic groupings. Invasive producers have particularly strong effects when they have an ability to fix nitrogen that is lacking in the native biota, or when they can enter novel associations with soil biota. Some invasive animals have important ecosystem effects through altering the

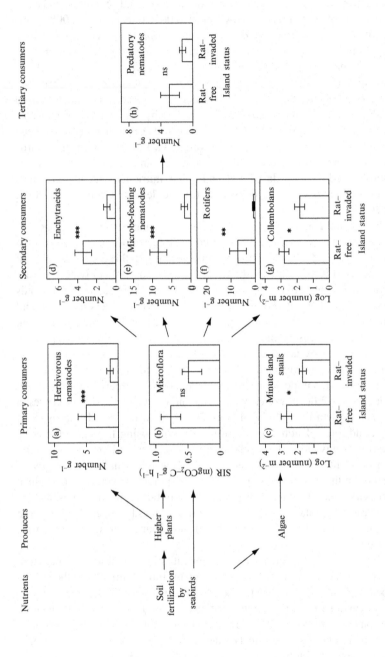

Fig. 5.17 The response of components of the soil food web to invasion of rats (*Rattus* spp.) to oceanic islands in northern New Zealand. Here, the rats eat the chicks and eggs of the seabirds and therefore thwart the transfer by seabirds from the ocean to the land. Data are means ± SE for nine rat-free (seabird-dominated) islands and nine rat-invaded islands, *, **, and *** mean $P < 0.05$, 0.01, and 0.001 respectively; SIR, substrate-induced respiration, used as a relative measure of microbial biomass. From Fukami et al. (2006), with permission from Wiley-Blackwell.

physical structure of ecosystems (e.g. invasive earthworms, beavers, and pigs), or through consuming native plant and animal species that themselves perform important functions in the ecosystem. Although most alien organisms entering a new ecosystem probably perform a relatively minor role (Thompson et al. 1995), there are now many examples, including those described above, that provide convincing evidence that a subset of the alien biota is nevertheless able to radically transform ecosystems, often to an alternative stable state.

5.4 Consequences of global change through causing species gains and losses

In the preceding sections of this chapter, we have considered the consequences of species losses and species gains (principally through invasions), for aboveground and belowground biota and the ecosystem processes that they drive. As we have discussed, the most important causes of species loss and gain are habitat destruction and land-use change, and biological invasion. However, there is also mounting evidence that other global change phenomena, including climate change, can act as important drivers of species loss with potentially far reaching consequences for aboveground and belowground properties of ecosystems, and climate feedbacks. For example, climate envelope models (which are based on species environmental preferences) predict severe loss of biodiversity due to climate warming (Thomas et al. 2004; Thuiller et al. 2005), largely because species are unable to disperse during the timescale of climate change. Moreover, climate change can cause species loss by affecting species phenology and interaction strengths, which can lead to mismatches in the life histories of consumers and their resources (Post and Forchhammer 2008) and decoupling of trophic interactions on which they rely (e.g. Fox et al. 1999; Visser and Both 2005; Memmott et al. 2007). As discussed earlier in this chapter, such losses of species and resulting changes in community composition could have important implications for the belowground subsystem and ecosystem processes, especially when the physiological traits of those species that are lost differ markedly from those that remain.

Another important route by which climate change can indirectly influence ecosystems processes is through causing range expansion of species into new territories. Although species range shifts can be caused by changes in land use, the well-documented expansion of many species towards higher latitudes and altitudes over the past few decades strongly suggests that climate warming plays a significant role (Walther et al. 2002; Parmesan and Yohe 2003). For example, climate warming has been proposed as the cause of widespread upward movement of alpine plant species (Klanderud and Birks 2003; Walther et al. 2005; Lenoir et al. 2008; but see Wilson and Nilsson 2009) and advance in the alpine treeline (Kullman 2002; Kullman and Öberg 2009), although a recent global meta-analysis indicates that treelines are not universally responding to climate warming (Harsch et al. 2009). As discussed in Chapter 3, climate warming has also been proposed as the cause of the northward expansion of boreal forest into Canadian tundra (Danby and Hik 2007) and pan-Arctic shrub encroachment in Arctic tundra (Sturm et al. 2001; Epstein et al. 2004; Tape et al. 2006; Wookey et al. 2009). Also, the northward and upward migration of many vertebrate and invertebrate species in Britain (Hickling et al. 2006; Menendez et al. 2007) (Fig. 5.18), and small mammals (Moritz et al. 2008) and invertebrates, such as the mountain pine beetle (Logan and Powell 2001; Williams and Liebhold 2002), in North America has been attributed to climate warming. As will be discussed below, some of these range shifts and consequent changes in the diversity and composition of communities have the potential to strongly modify ecosystem process of the territories that they invade, with, in some cases, far-reaching consequences for carbon-cycle feedbacks.

Many biotic factors determine the ability of different species to shift their ranges under climate warming, such as dispersal ability, habitat specificity (Warren et al. 2001; Menendez et al. 2006, 2007) and availability (Hill et al. 2001), and escape from natural enemies (Menendez et al. 2008). Most work on this has been done from an aboveground perspective, but there is now emerging evidence that belowground biota

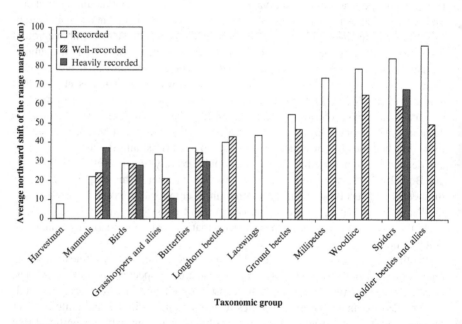

Fig. 5.18 Latitudinal shifts since the 1960s in the northern range margins of 12 taxonomic groups of terrestrial environments in Great Britain, at a 10-km-grid-square resolution. The groups analysed were grasshoppers and allies (Orthoptera), lacewings (Neuroptera), butterflies (Rhopalocera), spiders (Araneae), mammals (Mammalia), woodlice (Isopoda), ground beetles (Carabidae), harvestmen (Opiliones), millipedes (Diplopoda), longhorn beetles (Cerambycidae), soldier beetles and allies (Cantharoidea and Buprestoidea), and birds (Aves). Results are given for three levels of sub-sampling of data, namely recorded, well-recorded, and heavily recorded (see key). From Hickling et al. (2006), with permission from Wiley-Blackwell.

might also play a role in affecting range expansion of species under change. For example, it was recently shown by Engelkes et al. (2008) that range-expanding plant species are better defended against shoot and root enemies than are related native plant species growing in the same habitat, suggesting that successful range expanders may experience less control by aboveground or belowground enemies than the natives. Another mechanism that might influence the ability of certain species to expand into new territories under climate change concerns species-specific coupling of plants with soil biotic communities. As discussed in Chapter 3, there is often a high degree of specificity between individual plants species and the soil biotic community; for example, there is some evidence that certain plant species can preferentially select for decomposer taxa that enhance the decomposition of their own litter (e.g. Hansen 1999; Vivanco and Austin 2008; Ayres et al. 2009), indicative of feedback mechanisms. As a consequence, it is possible that the expansion of such species under climate change will disrupt such specialism in aboveground–belowground interactions, with possible consequences for their growth and competitive interactions in new territories. This idea has, however, only started to be tested very recently (e.g. Van Grunsven et al. 2007, 2010; Engelkes et al. 2008).

Relatively little is known about the consequences of range-expanding species for soil biota and ecosystem processes. As discussed in Chapter 3, the main route by which climate-driven shifts in plant species influence soil organisms is likely to be via changes in the quality and quantity of organic matter entering the soil, and through modifying the soil physical environment, for example by changes in root architecture and rooting depth (Jackson et al. 1996). One of the best documented examples of the consequences of climate-driven range expansion for ecosystem processes and carbon-cycle feedbacks is the expansion of dwarf-shrubs northward in the Arctic upwards in alpine regions (see Wookey et al. 2009). As discussed in Chapter 3, dwarf-shrubs produce woody litter of much poorer quality, and hence lower decomposability, than the graminoids and forbs that they replace (Cornelissen 1996; Quested et al. 2003; Dorrepaal et al. 2005). This may in turn slow down rates of decomposition and, potentially, counteract the direct enhancement of decomposition by warming and consequent loss of carbon from Arctic soils (Cornelissen et al. 2007). However, as highlighted in the discussion of this topic in Chapter 3, shrub expansion in the Arctic will also affect several other factors that influence rates of carbon cycling in Arctic soils, such as rooting depth, the incidence of fire, and snow cover. Moreover, as recently shown by Olofsson et al. (2009), shrub expansion in a warmer Arctic might be limited by biotic factors such as reindeer grazing. Therefore, as emphasized in Chapter 3, the role of biotic factors and interactions should be considered in order to understand how a changing climate will influence shrub expansion and carbon-cycle feedbacks in tundra ecosystems (Olofsson et al. 2009; Wookey et al. 2009).

Another case of shrub expansion concerns that which is occurring in arid and semi-arid grassland ecosystems throughout the world (Schlesinger et al. 1990; Peters et al. 2006; Throop and Archer 2008; Maestre et al. 2009). Shrub expansion in these ecosystems has been attributed to a variety of factors, including overgrazing, the suppression of fire and climate change (Archer et al. 1995; Van Auken 2000), and can lead to increased heterogeneity in soil resources due to the formation of 'islands' of fertility beneath shrub patches and impoverished inter-canopy areas (Schlesinger and Pilmanis 1998). In some situations, such as in the Chihuahuan Desert, USA, this process has been shown to create a feedback that ultimately leads to the desertification of the ecosystem (Schlesinger et al. 1990), a process that can be further exacerbated by factors such as climate change and over-grazing (Verstraete et al. 2009). However, this process does not appear to be universal. For instance, in the Mediterranean region, recent shrub encroachment into *Stipa tenacissima* dominated grasslands has been shown to be an important step in the reversal of desertification processes (Maestre et al. 2009). This encroachment, albeit caused by land abandonment rather than climate change, was found to enhance amounts of carbon and nitrogen in soil, and rates of nitrogen mineralization in soil beneath both shrub and *Stipa* canopies, and also in bare ground areas.

In general, there is much uncertainty about how shrub encroachment affects the belowground subsystem and soil carbon and nutrient pools, with both negative (Schlesinger et al. 1990; Jackson et al. 2002) and positive (Throop and Archer 2008; Maestre et al. 2009) responses being detected. Moreover, differences in shrub impacts on ecosystem carbon storage appear to vary under different climatic conditions. For example, in a study of eight North American graminoid-dominated ecosystems invaded by shrubs that covered a four-fold range in mean annual precipitation, from Arctic tundra to Atlantic coastal dunes, Knapp et al. (2008) found that shrub invasion decreased aboveground NPP, and hence carbon input, in drier sites, but dramatically increased aboveground NPP in regions with high mean annual precipitation (Fig. 5.19). In contrast, however, Jackson et al. (2002) studied woody-plant invasion along a precipitation gradient in south-western USA, and found a strong negative relationship between precipitation and changes in soil organic carbon content: woody-plant invasion into drier sites resulted in soil carbon gain, whereas in wetter sites it caused soil carbon loss. As suggested recently by Maestre et al. (2009), such variation in the impact of shrub encroachment on carbon dynamics is likely due to differences in the traits of the woody vegetation in different climatic regions, which as discussed in Chapter 3 act as a key determinant of plant effects on the belowground subsystem. It is widely acknowledged that much remains unknown about the influence of shrub encroachment on the belowground subsystem and carbon-cycle feedbacks (Jackson et al. 2002; Knapp et al. 2008; Maestre et al. 2009), and further work is needed on this issue, especially given the global extent of shrub encroachment and the potential for climate change to exacerbate this phenomenon.

Recent evidence also points to significant effects of range expansion of herbivores on ecosystem carbon dynamics. With regard to herbivorous insects, Kurz et al. (2008) used simulation models to estimate that the recent outbreak of the mountain pine beetle (*Dendroctonus ponderosae*) in British Columbia, Canada, which is an order of magnitude larger in area and severity than all previous outbreaks (Taylor et al. 2006; Williams and Liebhold 2002), converted the forest from a small net carbon sink to a large net carbon source, both during and immediately after the

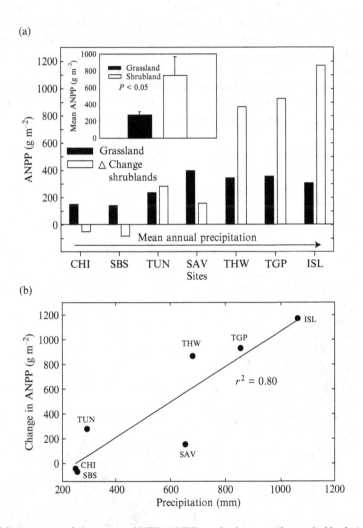

Fig. 5.19 Response of aboveground NPP (ANPP) to shrub encroachment in North American grasslands. (a) Patterns of ANPP in grasslands (black bars) and former grassland sites now dominated by shrubs (open bars) along a gradient of mean annual precipitation, showing that the largest responses in ANPP to shrub conversion occur in the most mesic sites (the inset shows the mean increase in ANPP over all sites with shrub encroachment into grasslands). (b) The relationship between mean annual precipitation and the change in ANPP for seven sites across North America where grassland has been converted to shrubland. Solid line indicates a significant positive relationship between mean annual precipitation and the change in ANPP. Sites codes, along the gradient of increasing mean annual precipitation, are: Chihuahuan Desert (CHI); sagebush steppe (SBS); tussock tundra (TUN); subtropical savanna (SAV); subtropical thorn woodland (THW); tallgrass prairie (TGP); and Barrier Island (ISL). From Knapp et al. (2008), with permission from Wiley-Blackwell.

outbreak (Fig. 5.20). This was attributed to a reduction in photosynthetic capacity (and hence carbon uptake) caused by widespread tree mortality, combined with an increase in heterotrophic respiration derived from the decomposition of dead trees. Importantly, the impacts of the outbreak on carbon emissions over 21 years was calculated to exceed that due to other factors, such as forest fire, and to be comparable to 5 years of greenhouse gas emissions from Canada's transportation sector (i.e. 200 Mt carbon dioxide equivalents in 2005). Although the increased extent and severity of the outbreak has been attributed in part to an increased area in

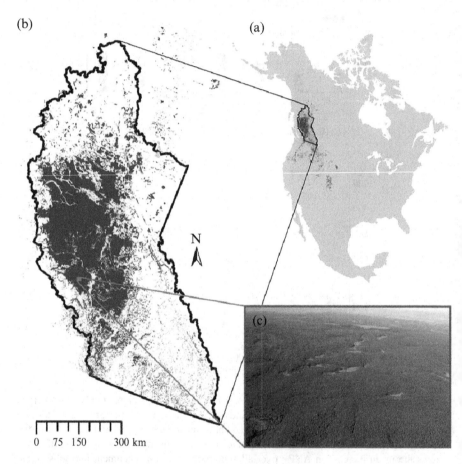

Fig. 5.20 Geographic extent of mountain pine beetle (*Dendroctonus ponderosae*) outbreak in North America. (a) Extent (dark grey) of mountain pine beetle. (b) The study area includes 98% of the current outbreak area. (c) An area of pine forest subject to tree mortality caused by pine beetle outbreak. Although not visible in the image, pine trees turn red in the first year after beetle kill and grey in subsequent years. From Kurz et al. (2008), with permission from Macmillan Publishers Ltd. © 2008 Her Majesty the Queen in right of Canada, Natural Resources Canada, Canadian Forest Service as originally published in *Nature*.

the host tree (i.e. mature pine stands), it is also widely attributed to climate change, which has led to an expansion of the outbreak northward and into higher elevation forests (Taylor et al. 2006; Williams and Liebhold 2002). As noted by Kurz et al. (2008), evidence of climate change related increases in the extent or severity of forest insect disturbance is mounting, and future climate change could facilitate further range expansion; failure to account for their impacts could therefore result in overestimation of the potential for forests to offset anthropogenic carbon dioxide emissions.

Range expansion and subsequent overabundance of mammalian herbivores has also frequently been observed, particularly for various species of deer in forest, grassland and tundra (Côte et al. 2004). For example, sika deer (*Cervus nippon*) have expanded their range through Japan by nearly 70% over the past two decades (Takatsuki 2009). The relative importance of different agents of global change in influencing range expansion of deer is poorly understood, but climate change is predicted to contribute to both range expansion and contraction of the major herds of caribou (*Rangifer tarandus*) on the Québec-Labrador peninsula through affecting snowfall patterns, food (notably lichen) availability, local-scale temperature, fire regime, and abundance of blood-sucking insects (Sharma et al. 2009). Enhanced range expansion of deer species in both forest and tundra may lead to large shifts in vegetation composition (Côte et al. 2004; Takatsuki 2009), which as described in Chapter 3 can have important knock-on effects on ecosystem processes, including those relevant to carbon sequestration. While the impacts of deer on ecosystem processes in newly colonized areas following range expansion and subsequent population growth has been little studied, we would expect these effects to have fundamental similarities to those frequently observed as a consequence of deer invasion and overabundance (Wardle and Bardgett 2004), as described earlier in this chapter.

Much is still to be learned about the different ecological traits and mechanisms responsible for range shifts, which are taking place in the context of not just climate warming, but also land use and other environmental changes (Hickling et al. 2006). Moreover, little is known about the consequences of range shifts for community and ecosystem properties, although, as discussed above, evidence is mounting that species-range shifts are widespread across taxonomic groups and geographic regions. It is also increasingly recognized that, in some cases, these shifts can have significant consequences for both the composition and functioning of the aboveground and belowground subsystems, and ultimately climate carbon-cycle feedbacks. However, as also highlighted above, the scale of such impacts will vary greatly depending on a variety of factors, such as the traits of those species that are shifting in their range and the environmental conditions and heterogeneity of the habitats into which they invade, which may also be affected by climate change, as discussed in Chapters 2 and 3. Therefore, a major challenge for the future will be to understand the mechanisms by which climate change simultaneously impacts on terrestrial ecosystems through range shifts and through its direct and indirect effects on resident aboveground and belowground communities.

5.5 Conclusions

In Chapters 2–4 we emphasized how the functioning of terrestrial ecosystems is driven by the community composition of plants, the soil biota, herbivores, and their predators, as well as the interactions among these organisms. However, as emphasized throughout this chapter, the species composition of communities does not remain constant over time, and species can be both gained and lost. While these losses and gains do occur naturally, the rates of both processes are being enhanced by several orders of magnitude through human activities. Consequently, communities can be simultaneously losing some species and gaining others, leading to substantial shifts in both structural and functional attributes of the community. In this chapter, we have provided multiple examples across a range of taxa and ecosystems to highlight both how species losses and gains can impact upon aboveground and belowground biota and their linkages, as well as the consequences of this for the functioning of the ecosystem.

Most studies that have aimed to understand how species losses in real ecosystems affect processes have utilized experiments in which species richness is experimentally varied as a treatment, with species at each richness level often chosen at random from a species pool. Despite the enormous research effort devoted to this approach over the past decade, our understanding of the effects of biodiversity loss that occurs in real ecosystems through human-induced changes in both biotic and abiotic factors remains poor. Moreover, as we highlight in this chapter, evidence for strong and consistent causative effects of species diversity on ecosystem processes remains limited, especially when the belowground subsystem is considered. Instead, there is growing evidence for diversity effects being 'context-dependent' in that biodiversity effects on ecosystem processes vary greatly according to environmental context (e.g. Fridley 2002; Hättenschwiler and Gasser 2005; Jonsson and Wardle 2008). Further, there is increasing recent recognition that species richness manipulation experiments may not realistically represent how ecosystem processes are affected by loss of species in real ecosystems, because species are not lost from real communities randomly (Wardle 1999; Leps 2004; Ridder 2008). In this chapter, we therefore argue that an improved understanding of the aboveground and belowground consequences of biodiversity loss requires experiments to be done in real ecosystems, and that this is best achieved through 'removal experiments' in which species are experimentally removed from real ecosystems so as to represent real loss of species or functional groups resulting from changes in abiotic and biotic extrinsic factors caused by human activity (Díaz et al. 2003; Wardle and Zackrisson 2005; Suding et al. 2008; Ward et al. 2009). We have highlighted the utility of this approach by presenting examples of how experimental removals of taxa of plants, soil microbes and fauna, and aboveground consumers can affect key ecosystem processes both above and belowground. Such studies, together with investigations of ecosystem effects of actual species losses (e.g. Zimov et al. 1995; Ripple and Betscha 2006), highlight that ecosystem processes are likely to be influenced primarily by differences in key characteristics or traits between species that are lost and those that remain.

There has been much recent work, and some significant advances, in understanding how gains of species through invasion in real ecosystems affect aboveground--belowground linkages and ecosystem processes. In particular, several studies comparing coexisting native and invasive species or floras have demonstrated important differences between them in terms of functionally important traits (Baruch and Goldstein 1999; Leishman et al. 2007), effects on decomposition and mineralization processes (Ehrenfeld 2003), and feedback interactions involving the soil biota (Klironomos 2002). Recent studies have also highlighted that invasive plants may undergo different interactions with both soil biota and other plant species in their new habitat than in their native range (Reinhart et al. 2003; Wolfe and Klironomos 2005), which in turn can contribute to the invader's success. While the majority of studies on invader effects on aboveground--belowground linkages have focused on invasive plants, in this chapter we highlight several recent examples that reveal the capacity of invasive consumers to transform ecosystems when they have key functional capabilities that are lacking in the native biota. Some of the most spectacular examples we describe involve invasive herbivores such as beavers, deer and gypsy moths, invasive predators such as ship rats, foxes, and ants, and invasive soil animals such as earthworms and flatworms. Further, we highlight that global climate change is not only an important driver of species gains and losses in ecosystems through inducing shifts in phenlogy and interaction strengths, but also by causing range expansion of many taxa towards higher latitudes and elevations. Although comparatively few examples have considered range-expanding species from an aboveground--belowground perspective, recent studies have shown that that such species can have important effects on ecosystem properties and potentially carbon-cycle feedbacks (e.g. Knapp et al. 2008; Kurz et al. 2008), and may show different interactions with their antagonists to related native species in the new or expanded range (Engelkes et al. 2008).

Despite much recent effort on understanding how species gains and losses affect ecosystems, our knowledge of how species loss affects ecosystem processes remains limited, largely because relatively few studies have considered the impacts of non-random species losses in real ecosystems and how they vary in different environment contexts. We argue that the most productive way forward would be the widespread use of experimental or theoretical approaches that directly investigate how loss of species with functionally important traits directly impact upon other organisms and ecosystem processes, and whether traits that predispose species to become extinct from a community are also those that drive ecosystem processes. With regard to invasive biota, although we have a reasonable understanding of how ingress of new plant species affects aboveground and belowground processes, much remains unknown about the mechanisms involved, and how these mechanisms may contribute to the spread of the invader. While interesting mechanistic theories (e.g. the novel weapons and enemy escape hypotheses) have been suggested, our understanding of their ecological significance remains limited. The issue of how invasive plants exert ecological effects through not just their per-capita effects but also the rate and extent to which they increase their range and abundance remains little explored, even though such information is crucial to predicting the ecosystem-level consequences of biological invasions over time. Further, while a growing number of studies have recognized the impacts of animal invasions both aboveground and belowground, we have yet to move away from accumulating a collection of intriguing examples to developing strong predictive or conceptual frameworks for understanding how invasive animals affect ecosystems. Finally, we are only beginning to understand the ecological consequences of range-expanding species in new habitats resulting from global change (e.g. global warming), and an important challenge will be to understand how range expansion and its ecological consequences are driven by species traits and interactions, as well as the biotic and abiotic characteristics of the environment.

6

Underlying themes and ways forward

6.1 Introduction

Our aim in this book was to draw attention to the importance of biotic interactions between aboveground and belowground communities as fundamental drivers of community dynamics and ecosystem functioning. Although having a long history, this area of research has seen tremendous amount of activity over the past two decades and has yielded several recent conceptual advances, as apparent in the preceding chapters. These developments have arisen through increasing recognition that the aboveground and belowground subsystems do not work in isolation, but rather that they are strongly interdependent. Feedbacks between the aboveground and belowground subsystems emerge through plants interacting both with decomposers via the indirect pathway and with root-associated organisms (e.g. pathogens, root feeders, and mycorrhizal fungi) via the direct pathway, with these interactions in turn being influenced by fauna that feed on plants, microbes, and on each other (Fig. 2.1). While it is certainly true that historically there have been ecologists who have fully acknowledged that understanding aboveground processes requires explicit recognition of those that occur belowground, and vice versa (e.g. Müller 1884; Handley 1954; Vitousek and Walker 1989), over the past few years recognition of this issue has been the norm rather than the exception. This recognition has facilitated a great deal of recent research that has contributed to better understanding of the feedbacks and interactions that take place in the plant–soil system and the ecosystem processes that they drive; this book provides an up-to-date synthesis of this understanding.

We have addressed the issue of aboveground–belowground feedbacks by highlighting four key components: biotic interactions in the soil (Chapter 2); plant community influences (Chapter 3); the role of aboveground consumers (Chapter 4); and the influence of species gains and losses (Chapter 5). While these topics remain as discrete entities around which the book is structured, there are a number of cross-cutting themes that run throughout those chapters; in this chapter we focus on these themes. These themes fall into three distinct but complementary categories that operate at different spatial and temporal scales. The first involves biotic interactions and feedbacks that occur at local scales (i.e. the scale of the individual organism) and their consequences for ecosystem processes. The second occurs at larger spatial and temporal scales, and involves characterizing variation both aboveground and belowground over space and time, as well as what drives differences among contrasting ecosystems. The third involves considering linkages between the aboveground and belowground subsystems in relation to human-induced global change phenomena, such as climate change. Through considering these cross-cutting themes in this chapter, we aim to explicitly highlight areas of significant recent development, as well as those that we believe are productive avenues for future research.

6.2 Biotic interactions, feedbacks, and ecosystem processes

6.2.1 Linkages and feedbacks between the aboveground and belowground subsystems

It has long been known that the type of plant species present affects the soil community, and that components of the soil community in turn affect the plant community. As such, feedbacks between the plant and soil communities are inevitable. A core element of this book has been to highlight the positive and negative effects that plant and soil communities can have on each other, and therefore to develop insights about feedbacks between these communities. As emphasized throughout the book, and as depicted in Fig. 2.1, soil communities consist of a myriad of organisms that interact with plants both via the indirect pathway (i.e. decomposers) and the direct pathway (including root-associated pathogens, root herbivores, and mycorrhizal fungi). A large and growing number of recent studies have used standard feedback-experiment methodology (Chapter 3) to investigate the net strength and direction of feedbacks between plants and their associated soil biota. As highlighted in Chapters 2, 3, and 5, such experiments have greatly advanced our understanding of how these feedbacks may influence plant species replacement (Packer and Clay 2000) and succession (Kardol et al. 2006), success of invasive plants (Klironomos 2002; Reinhart et al. 2003), and plant range expansion under global change (Engelkes et al. 2008). Such studies emphasize that the strength and direction of feedbacks between plants and their soil biota may differ greatly among plant species with different attributes, and highlight that the soil community is a major driver of plant species dynamics and the balance of interactions between coexisting plant species.

Although many plant–soil feedback studies point to negative feedbacks (Kulmatiski et al. 2008), we suggest in Chapter 3 that current methodology is more likely to encourage soil biota that undergo negative rather than positive associations with the soil biota. Further, the majority of feedback studies have focused on early succes-sional and herbaceous plant species which are more likely to show negative associ-ations with soil organisms than are late successional and woody species. As we discuss in Chapters 2 and 3, mycorrhizal fungi frequently enter into positive associ-ations with components of the plant community, and this may be especially true for forests dominated by one or a few species of ectomycorrhizal tree species. Further, if plant species do indeed promote decomposer communities that preferentially miner-alize their own plant litter as opposed to that of other plant species (including those with which they coexist; e.g. Vivanco and Austin 2008), then this makes positive feedbacks between plant species and their decomposer biota likely and probably widespread. Determining the net magnitude of feedback of the diverse components of the soil community (including those not favoured in current soil-feedback methodologies) with their associated plant species, and understanding how this contributes to plant community dynamics and ultimately ecosystem processes, remains one of the main unresolved challenges in this field.

Both plants and those soil organisms with which they directly interact are consumed by fauna, and trophic relationships that involve fauna are important drivers of aboveground–belowground associations. As such, several studies have shown the indirect effects of soil fauna and faunal community structure on plants and their aboveground consumers (Chapter 2), and of foliar herbivores on belowground processes (Chapter 4). A growing number of studies have also characterized belowground responses to losses of herbivore species, invasive herbivores, and herbivore range expansion (Chapter 5). Recent advances have involved not only quantifying the aboveground or belowground effects of herbivorous fauna, but also the mechanistic basis through which their effects occur (e.g. Mikola et al. 2009; Sørensen et al. 2009) and feedback to plant growth (Hamilton and Frank 2001; Hamilton et al. 2008). However, studies of this type are still at their infancy, and further research is needed to determine the relative importance of different herbivore-driven mechanisms and feedbacks across contrasting ecosystems. Further, there is growing recognition that upper-level consumers (i.e. predators of herbivorous and microbe-feeding fauna), both aboveground and belowground, have the potential to induce trophic cascades that can indirectly influence organisms on the other side of the aboveground–belowground interface, sometimes considerably. In this light, recent studies have provided spectacular examples of how invasions of islands by aboveground predators, such as ship rats (Fukami et al. 2006), foxes (Maron et al. 2006), and ants (O'Dowd et al. 2003), have exerted cascading effects that profoundly affect the plant community, soil fertility, and feedbacks to plant growth (Chapter 5). However, studies on how top predators (including predator invasion and extinction) affect ecosystem functioning are still relatively few (see Chapters 2, 4, and 5), and much remains unknown about the extent to which recent examples in the literature represent special cases as opposed to widespread phenomena. As such, there is still a significant need for a more fully developed multitrophic perspective about how interactions on both sides of the aboveground–belowground interface affect the functioning of terrestrial ecosystems.

Central to understanding aboveground–belowground interactions and feedbacks is the need to know the extent to which plant community structure serves as a determinant of soil community structure and vice versa. Over the past decade, several studies have made some headway in determining the degree to which the structure of the two communities is associated (Chapters 2 and 3), as well as the extent to which their diversities are coupled (reviewed by Wardle 2006). However, one major challenge that continues to impair progress on this topic involves difficulties in adequately quantifying soil community structure, particularly at taxonomically fine levels of resolution. There are obvious taxonomic problems in characterizing the community structure of much of the soil biota. Indeed, over 90% of microfloral and microfaunal species probably remain undescribed, and over the past few years there has been an increasing dearth of taxonomists that are well qualified to identify mesofaunal and smaller organisms. These impediments serve as a major reason as to why many ecologists and soil scientists over the past decades have continued to describe the soil community as a 'black box'. One way forward is the recent development of molecular approaches that have the potential to describe soil communities in a manner that has not existed historically. Traditional PCR-based molecular techniques for assessing microbial communities (e.g. denaturing-gradient gel electrophoresis (DGGE), terminal restriction fragment length polymorphism (TRFLP)) have provided lists of what soil organisms are present in soil, but generally not their relative abundances, limiting their utility for addressing ecological questions or their sensitivity for detecting microbial community responses to environmental factors (Ramsey et al. 2006). However, new techniques based on 454 pyrosequencing (Goldberg et al. 2006; Hudson 2008) offer much potential for determining how microbial community composition may respond to biotic and abiotic factors at much finer levels of resolution than has been the case to present. Such approaches have recently been used to reveal an unexpectedly high diversity of microbes in terrestrial ecosystems, for instance of fungi in soil (Buée et al. 2009) and in the phyllosphere of temperate trees (Jumpponen and Jones 2009), and contrasting levels of bacterial diversity between rhizosphere and bulk soil (Uroz et al. 2010). Such approaches, and the vast quantity of data that they generate, provide ecologists with opportunities to describe microbial communities in greater detail than has been the case previously, and could greatly advance our understanding of how plant and soil communities are linked.

6.2.2 Organism traits as ecological drivers

There is a long history in plant ecology of characterizing how plant traits differ among species with different ecological strategies (Grime 1977, 1979). As discussed in Chapter 3, it is now well recognized that among plant species there is a primary axis of evolutionary specialization ranging from those with traits associated with rapid resource capture, to those that have traits linked to resource conservation (Grime et al. 1997; Díaz et al. 2004). Of particular relevance to this book is the recognition that these traits (and trait differences among species) are important drivers of ecosystem processes and properties. This is because they determine litter and foliar quality, the decomposability of root, shoot, and stem litter (Chapter 3), the palatability and consumption of foliage (Chapter 4), and, ultimately, nutrient cycling and feedbacks between the aboveground and belowground subsystems (Chapters 3 and 4). As such, they serve as the agents through which differences between dominant species influence how ecosystems function. Further, as discussed in Chapter 5, invasive plant species can differ greatly in key functional traits from the native species of the invaded community, with potentially important ecosystem-level consequences. The same may also be true for plant species that are lost from a community, although as discussed in Chapter 5 this issue is less well understood. One important recent advance in this topic is the recognition that there can be tremendous variation in key functional traits even at very local scales (Hättenschwiler et al. 2008), which can be almost as great as the magnitude of variation observed globally (Richardson et al. 2008). As such, there is evidence that rates of litter decomposition are influenced more by trait differences among species, even at very local scales, than by macroclimatic variation across biomes (Cornwell et al. 2008). Further, a number of recent publications are now recognizing that plant species trait differences can be important predictors of differences in ecosystem functioning across communities (e.g. Vile et al. 2006; Fortunel et al. 2009) and of how human activities may affect ecosystems and the delivery of ecosystem services (e.g. Díaz et al. 2007; De Deyn et al. 2008).

Despite the considerable interest in the ecological significance of plant traits, there are several important issues which remain little understood, and which are only now beginning to gain attention. Here we highlight three issues raised in Chapter 3 that deserve particular emphasis. First, most ecosystem-centred studies have focused on foliar traits, and much remains unknown about how traits of other tissues that may represent a large proportion of ecosystem NPP, such as roots, stem tissues, and (in forested areas) wood, can drive ecological processes either aboveground or belowground (but see Weedon et al. 2009; Cornwell et al. 2009). Second, most studies have focused on traits of higher plants, yet in many ecosystems worldwide mosses and ferns also serve as important drivers, and in many boreal forests and the tundra mosses can represent a substantial proportion of total NPP. However, the issue of how moss and fern traits might affect ecological processes has been addressed only in a few studies, and mostly in relation to litter decomposition (e.g. Wardle et al. 2002 and Amantangelo and Vitousek 2008 for forest ferns; Dorrepaal et al. 2005 and Lang et al. 2009 for sub-Arctic mosses). Third, the vast majority of studies on plant traits have been at either the species or functional-group level of resolution. As discussed in Chapter 3, some plant species show considerable intraspecific phenotypic or genotypic variation of both foliar and litter quality, but little is known about whether functionally important traits vary within species in a similar way to that often found across species. Large differences among genotypes within species have been shown for both litter decomposability and foliage palatability (e.g. Classen et al. 2007; Silfver et al. 2007), and it is therefore likely that there is an ecologically meaningful level of variability of traits within some species, especially those that occupy vastly contrasting environments. However, rigorous testing of this idea remains minimal.

A major issue that remains poorly understood is how plant traits influence consumer organisms, their community structure, and their ecosystem impacts both belowground and aboveground. At a coarse level, it is recognized that bacterial-based and fungal-based energy channels in soil food webs vary greatly in their effects on nutrient cycling (Chapter 2), and that plant trait combinations can influence which of the two energy channels dominate (Chapter 3; Fig. 4.3). Further, animal body size distribution can vary greatly among ecosystems (Mulder and Elser 2009) which should in turn influence soil processes (Chapter 2), although almost nothing is known about how this is affected by traits of dominant plant species. At a finer resolution, coexisting fungi differ greatly in traits that determine their metabolic capacity, and for major faunal groups, such as Nematoda and Acari, functionally important traits relating to feeding structures and mouthparts vary greatly among species. While these traits are undoubtedly crucial in driving belowground interactions and processes, very little is known about how plant effect traits select for soil community structure, or determine the trait distribution of the soil community. However, an enhanced knowledge of this is essential for advancing our understanding of how aboveground–belowground linkages at the community level drives ecosystem functioning. Although aboveground plant traits are increasingly being recognized as important determinants of how plant communities interact with herbivores (Díaz et al. 2006), as discussed in Chapter 4, the issue of how plant traits select for herbivore communities and herbivore trait distributions remains essentially unexplored. In sum, much remains to be determined as to how suites of plant traits select for consumer traits (either aboveground or belowground) that may in turn be important in driving ecosystem processes and have consequences for higher trophic levels.

One issue for where a trait-based approach could usefully inform is in terms of understanding ecosystem processes for multiple-species communities. As outlined in Chapter 3, weighted trait average approaches (in which trait values of each species in a community are weighted in terms of their relative abundance) have recently been used to predict both plant productivity (Vile et al. 2006) and rates of decomposition (Quested et al. 2007; Fortunel et al. 2009) at the community level. However, as we discuss in Chapter 3, such approaches have limitations for understanding the effects of non-additive interactions among component species on ecosystem processes. To date, this issue has attracted limited attention, although some studies have focused on quantifying how ecosystem-level processes can be influenced by the magnitude of trait variation among coexisting species of live plants (Hooper et al. 2005), decomposer invertebrates (Heemsbergen et al. 2004), and litters of different qualities (Wardle et al. 1997a; Quested et al. 2005). Such approaches have the potential to assist our understanding of how ecosystems may be affected by gains (through invasion) or losses (through extinction) of species, especially if these gains or losses involve shifts in the range of trait values in the community. However, as highlighted in Chapter 5, our knowledge of how species losses in real communities affect ecosystem functioning remains limited, largely because most studies on this issue have relied on a single approach which involves assembling artificial communities in which organism diversity is experimentally varied, often by drawing random assemblages of species from a pool of species (Wardle and Jonsson 2010). We argue in Chapter 5 that alternative approaches based on understanding which functional traits affect ecosystem processes, and whether these traits also predispose species to being lost from a community, represent a more fruitful and direct means for understanding how biodiversity loss in real ecosystems influences ecosystem functioning.

6.3 Drivers of spatial and temporal variability

6.3.1 *Drivers of variation over time*

Relationships between the aboveground and belowground subsystems operate over a spectrum of temporal scales, ranging from days to seasons to centuries and ultimately to millions of years. Those which occur over the shortest timescales (days to years) often involve temporal variation in resource availability (Bardgett et al. 2005). For example, in Chapter 2 we emphasized through examples in alpine ecosystems that seasonal dynamics of the soil microbial community can drive temporal patterns of nutrient availability for plants, leading to microbes and plants partitioning nutrients across seasons. Further, as discussed in Chapter 4, foliar herbivory can lead to a pulse in root exudation, which often promotes soil microbial activity and nutrient mineralization in the root zone, and ultimately benefits plant growth. Many other examples have been reported of ephemeral resource pulses influencing the belowground and aboveground subsystems, and these can arise from both plants (e.g. seasonal litter fall, storm damage, mast seeding events, pollen inputs; Chapter 3) and animals (e.g. faecal deposits and carcasses; Chapter 4). One important example described in Chapter 4 involves adult cicadas, which emerge as adults only every 17 years. The carcasses of cicadas serve as a major resource pulse when they die, with wide-ranging consequences both above and below ground (Yang 2004). Despite a growing number of examples of resource pulses in ecosystems (Yang et al. 2008), drivers of temporal variability of aboveground–belowground linkages remain poorly understood, especially at the inter-annual level. As such, while we know that ecosystem NPP varies greatly across years in response to macroclimatic variation, and that the magnitude of this variation depends on site factors such as vegetation type (Knapp and Smith 2001), much remains unknown about how this in turn influences inter-annual variation in the belowground subsystem or feedbacks aboveground. Further, as discussed in Chapter 4, inter-annual variation in rainfall has been shown to drive variation in how grazing herbivores affect plants and nutrient cycling (e.g. Augustine and McNaughton 2006), again suggesting likely effects of inter-annual climatic variation on aboveground–belowground linkages.

With regard to longer time-scales (decades to centuries), ecologists have long recognized that aboveground and belowground properties change in predictable directions during ecosystem succession and development, as discussed in Chapter 3. While it is reasonably well understood that particular characteristics of the aboveground and belowground subsystems often shift in parallel over successional time, the mechanistic basis of how these subsystems interact during succession is less well understood. However, some recent plant–soil feedback experiments in herbaceous systems have provided evidence that negative feedbacks between early successional plant species and their associated soil biota may hasten their replacement during succession (e.g. De Deyn et al. 2003; Kardol et al. 2006). Further, as described in Chapter 3, types of soil organisms that may not thrive in plant–soil feedback experiments, such as mycorrhizal fungi and decomposer organisms, are also involved in feedbacks with the plant community, which influence plant species replacement, although this has been seldom investigated. In particular, little attention has been given to understanding how plant species feedbacks with the decomposer subsystem influence soil nutrient supply and availability, and ultimately vegetation change (but see Berendse 1998; Bowman et al. 2004; Meier et al. 2008). Further, as discussed in Chapter 5, some recent studies are beginning to provide evidence that plant–soil feedbacks are relevant for understanding plant invasions and the extent to which invaders may be able to dominate the plant community and replace native species (Van der Putten et al. 2007).

It is widely recognized that foliar herbivores have important effects on vegetation change and ecosystem succession, and that depending on context they can both accelerate and retard plant succession. As discussed in Chapter 4, this often has major consequences for the functional composition of vegetation, the quantity and quality of resources that plants return to the soil, and ultimately soil properties and nutrient availability. However, a recent and growing body of literature also highlights the potential for herbivores to cause ecosystems to rapidly undergo large stepwise transitions to alternative 'stable states'; as highlighted in Chapter 4, this is especially apparent when overgrazing has led to soil degradation and an irreversible collapse of existing vegetation (Rietkerk et al. 1997; Rietkerk and van de Koppel 1997). In these circumstances, mosaics of vastly contrasting patches can result, with the status of each patch depending on whether the herbivores have driven it past a critical transition. From Chapter 5, it is apparent that these rapid changes to alternative stable states can also occur when functionally important herbivores either invade a new ecosystem (e.g. invasions of beavers and subsequent tree felling in southern South America; Fig. 5.16) or become extinct (e.g. the Pleistocene extinctions of mega-herbivores in the sub-Arctic; Zimov et al. 1995). These types of sudden transitions have been studied mostly from an aboveground perspective, but are also likely to have large and rapid effects on the belowground subsystem, especially given the large changes in plant-derived resource input and soil disturbance which they frequently involve.

Relationships also occur between aboveground and belowground communities over much longer timescales, in the order of millennia and beyond. As outlined in Chapter 3, in the long-term absence of catastrophic disturbance during succession, a decline or retrogressive phase can occur which results from a prolonged decline in the availability of nutrients (notably phosphorus), which eventually leads to declines in plant production and biomass, litter quality, abundance of soil organisms, and decomposition and nutrient mineralization (Vitousek 2004; Wardle et al. 2004b). This retrogressive phase often occurs over geological timescales (e.g. 4.1 million years on the Hawai'ian islands; Vitousek 2004) and is associated with changes in the parent material and soil that occur during pedogenesis. Studies of ecosystem retrogression have greatly advanced our understanding of how variation in nutrient availability at local spatial scales drives community and ecosystem processes. However, retrogression has been comprehensively studied in only a few locations, and the issue of whether the phenomena that have been identified in retrogressed ecosystems also apply to other areas containing nutrient-depleted old soils (e.g. as occurs in much of the tropics) is poorly understood. Before leaving this topic, the nature of relationships between the aboveground and belowground subsystems and the consequences for ecosystem change over even more lengthy geological and evolutionary timescales (in the order of millions of years) has scarcely been explored. However, Berendse and Scheffer (2009) recently proposed that the large increase in angiosperms relative to gymnosperms during the Cretaceous may have resulted from angiosperms producing much higher-quality litter than gymnosperms, leading to greater nutrient supply rates in the soil. This in turn set a positive feedback in motion, leading to a rapid increase in angiosperms (and eventually angiosperm dominance) once they had reached a critical density.

6.3.2 Drivers of variation over space

Variability in aboveground and belowground properties of terrestrial ecosystems occurs across a very broad spectrum of spatial scales (Ettema and Wardle 2002) (Fig. 3.4). Such spatial variation occurs in a nested structure and is driven by a hierarchy of extrinsic and intrinsic factors, leading to patterns of spatial aggregation ranging from square millimetres to hectares (Ettema and Wardle 2002). For example, at large spatial scales, topographic variability has an important role in determining heterogeneity of resources such as soil nutrients and moisture, as well as soil structure, which in turn drives the spatial variability of plants, soil biota, and aboveground consumers. Nested within this at local scales, spatial variability in both the aboveground and belowground subsystems is driven by patchiness of vegetation and activities of larger bodied aboveground and belowground fauna. At the finest spatial scale, spatial variability in microtopography and soil porosity, as well as the spatial arrangement of litter fragments and soil organic matter hotspots, determines the spatial distribution of small-bodied soil organisms such as microbes, nematodes, protozoa, and microarthropods, as well as the decomposition and mineralization processes that they drive. A recurrent theme throughout this book has been the elucidation of how plants, soil organisms, and herbivores can contribute to spatial variability on both sides of the aboveground–belowground interface, and the consequences of this variability for community- and ecosystem-level processes and properties.

As outlined in Chapter 3, coexisting plant species can differ greatly in their key traits, including those that are important in driving belowground communities and their activity. When individuals of different coexisting species (or even different coexisting genotypes within species) are spatially segregated, and when horizontal transport of plant-derived resources is low, then each individual plant has the potential to strongly affect the soil beneath it. When these coexisting plants differ greatly in terms of their belowground effects (e.g. in terms of their litter quality and rhizosphere influences) then this can be a powerful driver of spatial heterogeneity of soil organisms, decomposer processes, and nutrient supply rates from the soil. At its most extreme, this can lead to individual plants creating 'islands of fertility' and 'tree islands', as discussed in Chapter 3. What remains less well understood is how these individual plant effects on belowground spatial variability feedback aboveground, although these effects can conceivably be either positive or negative. As an example of a negative effect, individual plants can promote pathogens in the patches that they occupy, adversely influencing their success, as has been shown for *Prunus serotina* by Packer and Clay (2000). This could in turn reduce or promote spatial variability of the belowground subsystem, depending on what effect that species has on below-ground patchiness. As an example of a positive effect, each individual plant in a multispecies community may control belowground resource availability in the patch that it occupies, with all species potentially being maintained in their own patches. This would in turn maintain vegetation heterogeneity and, potentially, plant species diversity. Although there is theoretical support for such a mechanism (e.g. Huston and De Angelis 1994), direct experimental evidence for it is scarce. However, the ability of some coexisting tree species to select for decomposer communities that preferentially break down (and presumably mineralize) their own litter (Vivanco and Austin 2008) provides some support for this idea. Further knowledge of the mechanisms of how aboveground–belowground feedbacks may contribute or respond to plant patchiness may help our understanding of the factors that structure plant communities and influence interactions involving invasive and range-expanding plant species, as considered in Chapter 5.

Consumer organisms may be an important driver of spatial variability both above-ground and belowground, especially when they drive the spatial redistribution of resources. As we revealed through several examples in Chapter 2, belowground invertebrates can greatly promote spatial heterogeneity at a range of scales, determined in part by animal body size. Detritivorous microarthropods promote heterogeneity at very local spatial scales by producing faecal pellets that serve as hotspots for microbial activity and nutrient release. At larger scales, earthworms form casts and middens that influence microbes and smaller-bodied soil organisms, and in grasslands promote the establishment of some plant species over others. Social soil animals such as termites and ants concentrate resources into the nests that they construct, which in turn eventually become colonized by some plant species and not others, often contributing to vegetation heterogeneity in the landscape and driving plant succession Further, as discussed in Chapter 4, aboveground animals, notably grazing herbivores, can contribute greatly to belowground and vegetation heterogeneity, through patchy returns of resources such as dung and urine (e.g. Schütz et al. 2006; Jewell et al. 2007), and even their carcasses (e.g. Bump et al. 2009a, 2009c; Parmenter and MacMahon 2009). An important development is the recognition that herbivores can also promote landscape heterogeneity through causing alternative stable states of vegetation, leading to vegetation mosaics (Rietkerk et al. 1997; Rietkerk and van de Koppel 1997). However, most studies on this issue have focused above ground, and the consequences for belowground organisms and processes have been seldom explored. Further, a number of studies have highlighted the ability of consumer organisms to reorganize the spatial distribution of resources in the landscape, either through consuming resources in one part of the landscape and depositing them elsewhere, or through moving resources across ecosystem boundaries such as from aquatic to terrestrial. Finally, as highlighted in Chapter 5, ecosystems may be subjected to particularly significant transformations when animals that drive the redistribution of resources are either lost from the ecosystem or invade it. For example, invasive predators of seabirds that thwart the transfer of resources by these birds from the ocean to land can cause spectacular transformations in both aboveground and belowground properties, and the functioning of the ecosystem (e.g. Fukami et al. 2006; Maron et al. 2006).

6.3.3 Differences across ecosystems

Much of the material in this book has focused on comparisons among contrasting ecosystems. Ecosystems can vary greatly even over small geographic scales, as a result of fundamental differences in either soil fertility (driven by parent material and abiotic soil properties) or macroclimate, although at local spatial scales more attention has focused on the former. As such, there are widely recognized fundamental biotic differences between fertile and infertile ecosystems that are driven by soil properties and fertility (Fig. 4.3). As discussed in Chapter 3, fertile habitats are dominated by plant species that are adapted for rapid resource acquisition rather than resource conservation, and also produce litter that is of much lower quality when compared with plant species characteristic of infertile habitats. Further, as we describe in Chapter 2, fertile habitats favour bacterial-based food webs that promote rapid cycling of nutrients, rather than fungal-based food webs that are more adapted for conserving nutrients. And, as outlined in Chapter 4, herbivores consume a much higher proportion of total plant productivity in fertile than in infertile systems, and are likely to retard rather than promote plant succession. As such, in infertile ecosystems, plants, decomposers, and herbivores will interact in such a way as to enhance nutrient retention and conservation of limiting nutrients, while in fertile systems these groups tend to interact in a manner that promotes rapid nutrient acquisition and use. As emphasized through numerous examples throughout this book, interactions between these three groups of organisms involve a range of feedback mechanisms which both positive and negative outcomes for both aboveground and belowground ecosystem processes. Much remains to be discovered, however, about how these feedback mechanisms vary between fertile and infertile ecosystems, or how they may themselves enhance or reduce soil fertility.

A key element of how ecosystems differ involves 'context dependency' of species effects on community and ecosystem properties. While much research has focused on how species affect ecosystems, greater emphasis is now focusing on how these effects can be driven by environmental context. As noted in Chapter 2, soil animals can vary greatly in their effects on microbes and decomposer processes depending on what habitat they are in. For example, the indirect effect of spiders on decomposition can depend on habitat moisture regime (Lensing and Wise 2006), and the indirect effect of predatory mites on soil microflora depends on soil fertility (Lenoir et al. 2007). Similarly, as outlined in Chapter 4, foliar herbivores can indirectly influence belowground biota and processes through a range of mechanisms, some positive and some negative. Different mechanisms dominate in different contexts, for instance along gradients of soil fertility or topography, and herbivores can therefore have either net positive or negative effects depending on which ecosystem or part of an ecosystem they are in. The nature of plant–soil feedbacks may also be context-dependent, and as outlined in Chapter 5 the magnitude and even the direction of feedback that some invasive plant species have with their associated soil biota depends on whether they are growing in their native range or in a newly invaded area. The issue of context dependency has also arisen in the literature of how changes in biodiversity affect the functioning of ecosystems; as discussed in Chapter 5, a growing number of studies show that environmental factors that differ across ecosystems helps determine whether and how community diversity of plants and their consumers may affect key aboveground and belowground processes. One of the challenges in studying the ecology of aboveground–belowground relationships continues to be to understand not just the basis through which species affect community and ecosystem properties, but also how and why these effects may differ in different environmental contexts and across contrasting ecosystems.

Other attributes that characterize differences among ecosystems are ratios of major nutrients (Elser et al. 2000). As outlined in Chapter 3, the nitrogen-to-phosphorus ratio of ecosystem components, such as plants, microbes, animals, and soil organic materials, are linked and have the potential to vary greatly across ecosystems. For example, ecosystems that have undergone retrogression have much lower availability of phosphorus relative to nitrogen than those that have not. This diminished availability of phosphorus greatly elevates the nitrogen-to-phosphorus ratio of biological tissues, and leads to reduced plant productivity, and rates of decomposition and nutrient supply from the soil (Wardle et al. 2004b). Additional changes include increased domination of the fungal-based relative to the bacterial-based energy channel, and reductions in the abundance of most groups of soil invertebrates. Recent work by Mulder and Elser (2009) and Reuman et al. (2009) also provides strong evidence for community-level effects of variation in nitrogen-to-phosphorus ratios in soil food webs in Dutch agricultural fields. Here, increases in soil nitrogen-to-phosphorus ratios across fields alter the allometric slope between faunal abundance and body size, in part through favouring small-bodied organisms associated with the bacterial-based energy channel at the expense of larger-bodied organisms associated with the fungal-based energy channel. As is apparent through Chapter 2, this should in turn have important implications for differences across ecosystems in soil nutrient cycling, and feedbacks aboveground. However, studies of how variation in the nitrogen-to-phosphorus ratio across ecosystems affects community and ecosystem properties are still few, and much remains unknown about how this ratio drives interactions between decomposers, plants, and their consumers.

6.3.4 Global-scale contrasts

As is apparent in the preceding chapters, aboveground and belowground biota, and the processes that they drive, vary greatly among biomes and therefore at the sub-global to global scale. Variation in major groups of soil biota at the global scale has been characterized by various authors, both qualitatively (e.g. Swift et al. 1979) and quantitatively (e.g. Wardle 1992; Fierer et al. 2009). These is also some recognition that the functional capabilities and importance of major groups of soil organisms varies greatly across different climatic zones; for example, earthworm species from lower latitudes have the enzymatic capability to digest more recalcitrant substrates than do those from higher latitudes (Lavelle et al. 1995). As such, while we have a reasonable knowledge as to how the main groups of soil biota vary sub-globally or globally, there have been few robust tests of how the different components of the soil biota vary in their functional role at these broad scales, or the consequences of this for ecosystem processes and interactions between the aboveground and belowground subsystems. However, some recent studies have utilized a manipulative approach to investigate the role of soil fauna in driving litter decomposition across ecosystems at these larger scales (e.g. Gonzalez and Seastedt 2001; Wall et al. 2008; Powers et al. 2009). These have involved placement of litter bags in combination with treatments to allow or exclude access by decomposer fauna, and with findings being interpreted in terms of the interactive effects of fauna and macroclimate (temperature and/or precipitation) on rates of decomposition.

One major issue that has attracted surprisingly little attention is the relative role of macroclimate versus soil fertility in explaining sub-global- or global-scale variation in belowground and aboveground biota, or the ecosystem processes that they drive. This is reflective of the macroecological literature in general, which has often paid scant attention to soil properties and/or soil fertility as a driver of patterns at these scales (but see Huston 1993). Tropical soils are generally less fertile than those at higher latitudes, because they are more depleted in organic matter, and are much older, more weathered, and contain less phosphorus. Therefore, at a global scale, macroclimate and soil fertility are not independent, and as such there are problems in determining whether climate drives sub-global or global variation of organisms and processes directly (through direct effects of temperature and precipitation) or indirectly (via long-term effects on soil fertility). However, studies at the sub-global to global scale suggest that both are probably important. For example, soil fertility has shown to play an important role in explaining variation among biomes in soil biota (Wardle 1992; Fierer et al. 2009), and macroclimatic variables serve as useful predictors of both the rate of litter decomposition (Berg et al. 1993) and the role of soil animals in driving decomposition at sub-global scales (Wall et al. 2008). However, some variables may be driven much more by soil fertility than by climate. For example, a recent analysis by Huston and Wolverton (2009) provides compelling evidence that global variation in NPP in forests is driven largely by soil fertility rather than temperature, with tropical forests on average being less productive than temperate ones because they grow on less fertile soils. This has important implications for the belowground subsystem, which as discussed in Chapter 3 is driven in part by ecosystem productivity. Further, Cornwell et al. (2008) provided evidence that local variation in plant traits, which is likely to be driven primarily by soil fertility, is a more powerful determinant of rates of litter decomposition than is sub-global to global variation in macroclimate. However, direct evidence about the role of macro-climate versus soil fertility in explaining global variation in aboveground or below-ground biota and the processes they drive remains scarce.

6.4 Global change phenomena

A topic that we have emphasized throughout this book involves the impact of human-induced global change (including carbon dioxide enrichment, climate change, nitrogen deposition, invasion of alien species into new territories, and losses of native species) on aboveground and belowground biota and the processes that they drive. Moreover, we have illustrated how these effects of global change can have far-reaching consequences for ecosystem-level properties and feedbacks to the Earth-system. One important development in this area has been the recognition that global change phenomena have both direct and indirect effects on the belowground subsystem, and that such effects create feedbacks that influence aboveground biota, ecosystem nutrient and carbon cycling, and the flux of carbon dioxide and other greenhouse gases from land to the atmosphere. This understanding is perhaps most developed in the context of climate change and its consequences for ecosystem carbon dynamics and carbon-cycle feedbacks. For example, as detailed in Chapters 2 and 3, many examples now exist in the literature showing that climate change can impact directly on soil organisms and ecosystem carbon cycling through changes in temperature and precipitation, and extreme climatic events, and indirectly via climate-driven changes in plant productivity and community composition. This in turn alters soil physicochemical conditions, the supply of carbon to soil, and the structure and activity of soil communities involved in decomposition processes and carbon loss from soil (Fig. 2.19). Likewise, much litera-ture exists demonstrating that nitrogen enrichment, which has increased substantially as a result of anthropogenic activity, can significantly affect the belowground subsystem and ecosystem-level processes through strong direct effects on soil organisms and through indirect effects on vegetation change.

Although understanding in these areas has increased dramatically in recent years, significant gaps nevertheless remain. For instance, very little is known about the relative contributions of direct and indirect effects of global change phenomena on ecosystem properties, although recent manipulative experiments are beginning to shed light on this issue (e.g. Manning et al. 2006; Suding et al. 2008). Also, there are considerable gaps in our understanding of the response of soil respiration to climate change, which in part stems from the fact that it is regulated by a myriad of factors, including resource quality (Davidson and Janssens 2006) and a variety of interactions and feedbacks between climate, plants, their herbivores and symbionts, and free-living soil organisms (Wolters et al. 2000; Högberg and Read 2006; Bardgett et al. 2008). Moreover, very little is known about the consequences of extreme events associated with climate change, such as droughts and soil freezing, for aboveground and below-ground biota and ecosystem processes, or of the role that simultaneous global change drivers play in regulating community and ecosystem-level processes. Indeed, as highlighted in Chapters 2 and 3, it is becoming increasingly apparent that our ability to predict future responses to global change requires a greater understanding of the simultaneous effects of multiple global change drivers on soil biological communities (Bardgett et al. 2008; Tylianakis et al. 2008). As we illustrate in Chapter 2, there is much potential for interactions between global change drivers to amplify, suppress or even neutralize climate change driven effects on the belowground subsystem and their feedback to carbon exchange. Studies in this area are still few, and we therefore suggest that experiments which simultaneously vary two or more global change drivers offer much promise for improving our understanding of how aboveground–belowground interactions and their influence on ecosystem-level processes may respond to current global change scenarios. This represents a major research challenge for the future.

An important but rarely considered issue that emerges from this book is that aboveground consumers have the potential to significantly alter the impacts of global change on the functioning of terrestrial ecosystems. One important development in this topic is the recent recognition that aboveground herbivores not only strongly affect net ecosystem carbon exchange (e.g. Welker et al. 2004; Ward et al. 2007; Van der Wal et al. 2007), and hence the source–sink activity of terrestrial ecosystems, but they can also alter the influence of climate itself on land–atmosphere carbon fluxes (Polley et al. 2008). Moreover, recent experimental studies are beginning to show that grazing can strongly modulate the effects of climate change on the carbon balance of terrestrial ecosystems. For example, studies in the Arctic point to interactive effects of warming and grazing on plant community composition (Post and Pedersen 2008), belowground processes (Rinnan et al. 2009) and ecosystem carbon-sink strength (Sjögersten et al. 2008). Further, evidence from long-term studies suggest that woody plant expansion in response to warming is able to be constrained by large herbivores (Post et al. 1999; Olofsson et al. 2009). As we stress in Chapter 4, much remains to be learned about the response of herbivores (and other trophic groups such as predators) to climate change and the consequences of this for community- and ecosystem-level processes, including ecosystem carbon exchange and carbon-cycle feedbacks. Given the dearth of studies in this area, we maintain that there is a significant need for more research aimed at an improved understanding of how herbivores, and other biotic factors, modulate ecosystem responses to climate change, and ultimately how herbivore management might be altered to mitigate climate change. Such research is especially important given the need for predictive models to accommodate biotic factors such as grazing in order to accurately simulate carbon dynamics in terrestrial ecosystems.

As we highlight in Chapter 5, global climate change is an important driver of species gains and losses in ecosystems, with potentially far reaching consequences for aboveground and belowground properties of ecosystems, and climate feedbacks. An issue that we highlight in Chapter 5 involves invasive and range-expanding species, and their potential to influence community- and ecosystem-properties properties. Indeed, evidence is now mounting that species range shifts are widespread across taxonomic groups and geographic regions (e.g. Hickling et al. 2006;

Menendez et al. 2007; Lenoir et al. 2008), and that climate warming plays a significant role in these shifts (Walther et al. 2002; Parmesan and Yohe 2003). Although a relatively understudied topic, there is now mounting evidence that range shifts can have important effects on ecosystem properties and potentially on carbon-cycle feedbacks (e.g. Knapp et al. 2008; Kurz et al. 2008; Takatsuki 2009), and that range-expanding species may differ from related native species in the nature of their interactions with their antagonists in their new or expanded range (Engelkes et al. 2008). However, as we highlight in Chapter 5, it is apparent that the scale of impacts of range-expanding species will vary greatly depending on many factors, including the traits of those species that are shifting in their range and the environmental characteristics of the habitats into which they invade, which may also be affected by climate change. A major challenge for the future, therefore, will be to understand the mechanisms by which climate change simultaneously impacts on terrestrial ecosystems through range shifts and through its direct and indirect effects on resident aboveground and belowground communities.

As we highlighted at the start of this book, the aboveground and belowground components of ecosystems have traditionally often been considered in isolation from one another. However, as illustrated in this book, the last two decades, and especially the last 5 years, have witnessed a proliferation of studies that have explored the influence that these components exert on each other, and the role that aboveground–belowground interactions play in controlling the structure and functioning of terrestrial ecosystems and their interactions with drivers of global change. This recent activity has led to major advances in the field of ecosystem ecology, especially through demonstrating the over-riding role that organism traits and biotic interactions, both among and within trophic levels and on both sides on the aboveground–belowground interface, play as terrestrial ecosystem drivers. Moreover, we now know much more about the role that aboveground–belowground interactions have in regulating ecosystem responses to global change, and in controlling carbon-cycle feedbacks with consequences for the Earth-system. However, as we highlight throughout this book, many challenges remain, and most notably there is a significant need to use our advancing knowledge of aboveground–belowground interactions to improve their representation in current concepts relating to global-scale patterns and processes. Indeed, while many detailed empirical studies have been carried out on different types of aboveground–belowground interactions, we still lack a strong and unifying theoretical basis for the subject, and this limits our ability to make predictions about their role in regulating terrestrial ecosystem processes and their response to global change. Likewise, while our understanding of plant and soil microbial processes involved in carbon cycling has advanced greatly, this is not well represented in global carbon models. Finally, the science of ecosystem ecology is becoming increasingly directed towards understanding and predicting the effects of global environmental change on the delivery of ecosystem services upon which humans depend. This approach presents many new challenges for ecosystem ecologists, including the need to better understand the role of aboveground–belowground interactions as drivers of ecosystem services and human well-being.

References

Aarssen, L. 1997. High productivity in grassland ecosystems: effected by species diversity or productive species? *Oikos* **80**, 183–184.

Abdul-Fattah, H.A. and F.A. Bazzaz. 1979. The biology of *Ambrosia trifida* L. 1. Influence of species removal on the organization of the plant community. *New Phytologist* **83**, 813–816.

Aber, J.D., K.J. Nadelhoffer, P. Steudler, et al. 1989. Nitrogen saturation in northern forest ecosystems: excess nitrogen from fossil fuel combustion may stress the biosphere. *Bioscience* **39**, 378–386.

Aber, T., D.E. Bignell and M. Higashi. 2000. *Termites: Evolution, Sociality, Symbioses, Ecology.* Kluwer, Dordrecht.

Aberdeen, J.E.C. 1956. Factors influencing the distribution of fungi and plant roots. Part I. Different host species and fungal interactions. *Papers of the Department of Botany of the University of Queensland* **3**, 113–124.

Aerts, R. and F.S. Chapin. 2000 The mineral nutrition of wild plants revisited: a re-evaluation of processes and patterns. *Advances in Ecological Research* **30**, 1–67.

Afzal, M. and W.A. Adams. 1992. Heterogeneity of soil mineral nitrogen in pasture grazed by cattle. *Soil Science Society of America Journal* **56**, 1160–1166.

Agrawal, A.A., S. Tuzun and E. Bent. 1999. *Induced Plant Defenses Against Pathogens and Herbivores: Biochemistry, Ecology and Agriculture.* APS Press, Saint Paul, MN.

Ågren, G.I. and M. Knecht. 2001. Simulation of soil carbon and nutrient development under *Pinus sylvestris* and *Pinus contorta. Forest Ecology and Management* **141**, 117–129.

Allen, R.B. and W.G. Lee (eds). 2006. *Biological Invasions in New Zealand.* Springer, Berlin.

Allen-Morley, C.R. and D.C. Coleman. 1989. Resilience of soil biota in various food webs to freezing perturbations. *Ecology* **70**, 1127–1141.

Allison, S.D. and P.M. Vitousek. 2004. Rapid nutrient cycling in leaf litter from invasive plants in Hawai'i. *Oecologia* **141**, 612–619.

Allison, S.D., C.I. Czimczik and K.K. Treseder. 2008. Microbial activity and soil respiration under nitrogen addition in Alaskan boreal forest. *Global Change Biology* **14**, 1156–1168.

Alphei, J., M. Bonkowski and S. Scheu. 1996. Protozoa, Nematoda and Lumbricidae in the rhizosphere of *Hordelymus europaeus* (Poaceae): faunal interaction, response of microorganisms and effects on plant growth. *Oecologia* **106**, 111–126.

Amatangelo, K.L. and P.M. Vitousek. 2008. Stoichiometry of ferns in Hawaii: implications for nutrient cycling. *Oecologia* **157**, 619–627.

Amundson, R. 2001. The carbon budget in soils. *Annual Review of Earth and Planetary Sciences* **29**, 535–562.

Anderson, C.B., C.R. Griffith and A.D. Rosemond. 2006. The effects of invasive North American beavers on riparian plant communities in Cape Horn, Chile—Do exotic beavers engineer differently in sub-Antarctic ecosystems? *Biological Conservation* **128**, 467–474.

Anderson, C.B., G.M. Pastur, M.V. Lencinas, et al. 2009. Do introduced North American beavers *Castor canadensis* engineer differently in southern South America? An overview with implications for restoration. *Mammal Review* **39**, 33–52.

Anderson, J.M. and P. Ineson. 1984. Interactions between microorganisms and soil invertebrates in nutrient flux in pathways of forest ecosystems. In: *Invertebrate-Microbial Interactions* (Anderson, J.M., A.D.M. Rayner and D.W.H. Walton, eds), pp. 59–88. Cambridge University Press, Cambridge.

Anderson, J.M., P. Ineson and S.A. Huish. 1983. Nitrogen and cation mobilization by soil fauna feeding on leaf litter and soil organic matter from deciduous woodlands. *Soil Biology and Biochemistry* **15**, 463–467.

Anderson, R.V., D.C. Coleman and C.V. Cole. 1981. Effects of saprophytic grazing on net mineralization. In: *Terrestrial Nitrogen Cycles* (F.E. Clark and T. Rosswall, eds), pp. 201–216. Ecological Bulletin 33. Swedish National Science Research Council, Stockholm.

Anderson, W.B. and G.A. Polis. 1999. Nutrient fluxes from water to land: seabirds affect plant nutrient status of Gulf of California islands. *Oecologia* **118**, 324–332.

Anderson, W.B., D.A. Wait and P. Stapp. 2008. Resources from another place and time: responses to pulses in a spatially subsidized system. *Ecology* **89**, 660–670.

Andresen, E. and D.J. Levey. 2004. Effects of dung and seed size on secondary dispersal, seed predation, and seedling establishment of rain forest trees. *Oecologia* **139**, 45–54.

Angel, A., R.M. Wanless and J. Cooper. 2009. Review of impacts of the introduced house mouse on islands in the Southern Ocean: are mice equivalent to rats? *Biological Invasions* **11**, 1743–1754.

Anisimov, O.A., F.E. Nelson and A.V. Pavlov. 1999. Predictive scenarios of permafrost development under conditions of global climate change in the XXI century. *Earth Cryology* **3**, 15–25.

Anser, G.P., S.R. Levick, T. Kennedy-Bowdoin, et al. 2009. Large-scale impacts of herbivores on the structural diversity of African savannas. *Proceedings of the National Academy of Sciences USA* **106**, 4947–4952.

Archer, S., D.S. Schimel and E.A. Holland. 1995. Mechanisms of shrubland expansion: land use, climate, or CO_2? *Climatic Change* **29**, 91–99.

Armesto, J.J. and S.T.A. Pickett. 1985. Experiments on disturbance in old-field plant communities: impacts on species richness and abundance. *Ecology* **66**, 230–240.

Arndt, E. and J. Perner. 2008. Invasion patterns of ground-dwelling arthropods in Canarian laurel forests. *Acta Oecologica* **34**, 202–213.

Arnold, J.F. 1955. Plant life-form classification and its use in evaluating range conditions and trend. *Journal of Range Management* **8**, 176–181.

Arrow, G.J. 1931. *The Fauna of the British India, including Ceylon and Burma*. Coleoptera, Lamellicornia Part III. Coprinae. Taylor and Francis, London.

Augustine, D.J. and S.J. McNaughton. 1998. Ungulate effects on the functional species composition of plant communities: herbivore selectivity and plant tolerance. *Journal of Wildlife Management* **62**, 1165–1183.

Augustine, D.J. and Frank, D.A. 2001. Effects of migratory grazers on spatial heterogeneity of soil nitrogen properties in a grassland ecosystem. *Ecology* **82**, 3149–3162.

Augustine, D.J. and S.J. McNaughton. 2006. Interactive effects of ungulate herbivores, soil fertility, and variable rainfall on ecosystem processes in a semi-arid savanna. *Ecosystems* **9**, 1242–1256.

Austin, M.P. and B.O. Austin. 1980. Behavior of experimental plant communities along a nutrient gradient. *Journal of Ecology* **68**, 891–918.

Avni, Y., N. Porat, J. Plakht, et al. 2006. Geographic changes leading to natural desertification versus anthropogenic land conservation in an environment, the Negev Highlands, Israel. *Geomorphology* **82**, 177–200.

Ayres, E., J. Heath, M. Possell, et al. 2004. Tree physiological responses to above-ground herbivory directly modify below-ground processes of soil carbon and nitrogen cycling. *Ecology Letters* **7**, 469–479.

Ayres, E., K.M. Dromph and R.D. Bardgett. 2006. Do plant species encourage soil biota that specialize in the rapid decomposition of their litter? *Soil Biology and Biochemistry* **38**, 183–186.

Ayres, E., N.J. Ostle, R. Cook, et al. 2007. The influence of below-ground herbivory and defoliation of a legume on nitrogen transfer to neighbouring plants. *Functional Ecology* **21**, 256–263.

Ayres, E., H. Steltzer, S. Berg, et al. 2009. Soil biota accelerate decomposition in high-elevation forests by specializing in the breakdown of litter produced by the plant species above them. *Journal of Ecology* **97**, 901–912.

Bailey, D.W. and F.D. Provenza. 2008. Mechanisms determining large-herbivore distribution. In: *Resource Ecology: Spatial and Temporal Dynamics of Foraging* (H.H.T. Prins and F. van Langevelde, eds), pp. 7–28. Springer, Amsterdam.

Bailey, J.K., J.A. Schweitzer, F. Ubeda, et al. 2009. From genes to ecosystems: a synthesis of the effects of plant genetic factors across levels of organization. *Philosophical Transactions of the Royal Society of London Series B Biological Sciences* **364**, 1607–1616.

Bais, H.P., T.S. Walker, F.R. Stermitz, et al. 2003. Allelopathy and invasive plants: from genes to invasion. *Science* **301**, 1377–1380.

Bakker, E.S., H. Olff, M. Boekhoff, et al. 2004. Impact of herbivores on nitrogen cycling: contrasting effects of small and large species. *Oecologia* **138**, 91–101.

Bakker, E.S., M.E. Ritchie, H. Olff, et al. 2006. Herbivore impact on grassland plant diversity depends on habitat productivity and herbivore size. *Ecology Letters* **9**, 780–787.

Baldwin, I.T., R.K. Olson and W.A. Reiners. 1983. Protein binding phenolics and the inhibition of nitrification in subalpine balsam fir soils. *Soil Biology and Biochemistry* **15**, 419–423.

Ball, B.A., M.A. Hunter, J.S. Kominoski, et al. 2008. Consequences of non-random species loss for decomposition dynamics: experimental evidence for additive and non-additive effects. *Journal of Ecology* **96**, 303–313.

Ball, B.A., M.A. Bradford, D.C. Coleman, et al. 2009. Linkages between below and above-ground communities: decomposer responses to simulated tree species loss are largely additive. *Soil Biology and Biochemistry* **41**, 1155–1163.

Ballinger, A. and P.S. Lake. 2006. Energy and nutrient fluxes from rivers and streams into terrestrial food webs. *Marine and Freshwater Research* **57**, 15–28.

Balvanera, P., A.B. Pfisterer, N. Buchmann, et al. 2006. Quantifying the evidence for biodiversity effects on ecosystem functioning and services. *Ecology Letters* **9**, 1146–1156.

Bardgett, R.D. 2005. *The Biology of Soil: A Community and Ecosystem Approach*. Oxford University Press, Oxford.

Bardgett, R.D. and K.F. Chan. 1999. Experimental evidence that soil fauna enhance nutrient mineralization and plant nutrient uptake in montane grassland ecosystems. *Soil Biology and Biochemistry* **31**, 1007–1014.

Bardgett, R.D. and E. McAlister. 1999. The measurement of soil fungal:bacterial biomass ratios as an indicator of ecosystem self-regulation in temperate meadow grasslands. *Biology and Fertility of Soils* **29**, 282–290.

Bardgett, R.D. and A. Shine. 1999. Linkages between plant litter diversity, soil microbial biomass and exosystem function in temperate grasslands. *Soil Biology and Biochemistry* **31**, 317–321.

Bardgett, R.D. and D.A. Wardle. 2003. Herbivore mediated linkages between above-ground and below-ground communities. *Ecology* **84**, 2258–2268.

Bardgett, R.D. and L.R. Walker. 2004. Impact of coloniser plant species on the development of decomposer microbial communities following deglaciation. *Soil Biology and Biochemistry* **36**, 555–559.

Bardgett, R.D., J.C. Frankland and J.B. Whittaker. 1993a. The effects of agricultural practices on the soil biota of some upland grasslands. *Agriculture, Ecosystems and Environment* **45**, 25–45.

Bardgett, R.D., J.B. Whittaker and J.C. Frankland. 1993b. The diet and food preferences of *Onychiurus procampatus* (Collembola) from upland grassland soils. *Biology and Fertility of Soils* **16**, 296–298.

Bardgett, R.D., J.B. Whittaker and J.C. Frankland. 1993c. The effect of collembolan grazing on fungal activity in differently managed upland pastures—a microcosm study. *Biology and Fertility of Soils* **16**, 255–262.

Bardgett, R.D., D.K. Leemans, R. Cook, et al. 1997. Seasonality of the soil biota of grazed and ungrazed hill grasslands. *Soil Biology and Biochemistry* **29**, 1285–1294.

Bardgett, R.D., S. Keiller, R. Cook, et al. 1998a. Dynamic interactions between soil fauna and microorganisms in upland grassland soils: a microcosm experiment. *Soil Biology and Biochemistry* **30**, 531–539.

Bardgett, R.D., D.A. Wardle and G.W. Yeates. 1998b. Linking above-ground and below-ground food webs: how plant responses to foliar herbivory influence soil organisms. *Soil Biology and Biochemistry* **30**, 1867–1878.

Bardgett, R.D., C.S. Denton and R. Cook. 1999a. Below-ground herbivory promotes soil nutrient transfer and root growth in grassland. *Ecology Letters* **2**, 357–360.

Bardgett, R.D., E. Kandeler, D. Tscherko, et al. 1999b. Below-ground microbial community development in a high temperature world. *Oikos* **85**, 193–203.

Bardgett, R.D., J.L. Mawdsley, S. Edwards, et al. 1999c. Plant species and nitrogen effects on soil biological properties of temperate upland grasslands. *Functional Ecology* **13**, 650–660.

Bardgett, R.D., J.M. Anderson, B. Behan-Pelletier, et al. 2001a. The role of soil biodiversity in the transfer of materials between terrestrial and aquatic systems. *Ecosystems* **4**, 421–429.

Bardgett, R.D., A.C. Jones, D.L. Jones, et al. 2001b. Soil microbial community patterns related to the history and intensity of grazing in sub-montane ecosystems. *Soil Biology and Biochemistry* **33**, 1653–1664.

Bardgett, R.D., T.C. Streeter, L. Cole, et al. 2002. Linkages between soil biota, nitrogen availability, and plant nitrogen uptake in a mountain ecosystem in the Scottish Highlands. *Applied Soil Ecology* **19**, 121–134.

Bardgett, R.D., T.C. Streeter and R. Bol. 2003. Soil microbes compete effectively with plants for organic nitrogen inputs to temperate grasslands. *Ecology* **84**, 1277–1287.

Bardgett, RD, W.D. Bowman, R. Kaufmann, et al. 2005. A temporal approach to linking aboveground and belowground ecology. *Trends in Ecology and Evolution* **20**, 634–641.

Bardgett, R.D., R.S. Smith, R.S. Shiel, et al. 2006. Parasitic plants indirectly regulate below-ground properties in grassland ecosystems. *Nature* **439**, 969–972.

Bardgett, R.D., A. Richter, R. Bol, et al. 2007a. Heterotrophic microbial communities use ancient carbon following glacial retreat. *Biology Letters* **3**, 487–490.

Bardgett, R.D., R. Van der Wal, I.S. Jõnsdõttir, et al. 2007b. Temporal variability in plant and soil nitrogen pools in a high Arctic ecosystem. *Soil Biology and Biochemistry* **39**, 2129–2137.

Bardgett, R.D., C. Freeman and N.J. Ostle. 2008. Microbial contributions to climate change through carbon-cycle feedbacks. *The ISME Journal* **2**, 805–814.

Bardgett, R.D., G.B. De Deyn and N.J. Ostle. 2009. Plant-soil interactions and the carbon cycle. *Journal of Ecology* **97**, 838–839.

Barford, C.C., S.C. Wofsy, M.L. Goulden, et al. 2001. Factors controlling long- and short-term sequestration of atmospheric CO_2 in a mid-latitude forest. *Science* **294**, 1688–1691.

Barrett, J.E., R.A. Virginia, D.H. Wall, et al. 2004. Variation in biogeochemistry and soil biodiversity across spatial scales in a polar desert ecosystem. *Ecology* **85**, 3105–3118.

Baruch, Z. and G. Goldstein. 1999. Leaf construction cost, nutrient concentration and net CO_2 assimilation rate of native and invasive species in Hawaii. *Oecologia* **121**, 183–192.

Bauhus, J. and R. Barthel. 1995. Mechanisms for carbon and nutrient release and retention in beech forest gaps. II. The role of soil microbial biomass. *Plant and Soil* **168**, 585–592.

Beare, M.H., R.W. Parmelee, P.H. Hendrix, et al. 1992. Microbial and faunal interactions and effects on litter nitrogen and decomposition in agroecosystems. *Ecological Monographs* **62**, 569–591.

Beggs, J.R. and J.S. Rees. 1999. Restructuring of Lepidoptera communities by introduced *Vespula* wasps in a New Zealand beech forest. *Oecologia* **119**, 565–571.

Bellingham, P.J., L.R. Walker and D.A. Wardle. 2001. Differential facilitation by a nitrogen fixing shrub during primary succession influences relative performance of canopy tree species. *Journal of Ecology* **89**, 861–875.

Belnap, J. 2003. Factors influencing nitrogen fixation and nitrogen release in biological soil crusts. In: *Biological Soil Crusts. Structure, Function, and Management* (J. Belnap and O.L. Lange, eds), pp. 241–262. Springer, Berlin.

Belnap, J. and O.L. Lange. 2003. *Biological Soil Crusts: Structure, Function, and Management*. Springer, Berlin.

Belnap, J., S.L. Phillips, S.K. Sherrod, et al. 2005. Soil biota can change after exotic plant invasion: does this affect ecosystem processes? *Ecology* **86**, 3007–3017.

Belovsky, G.E. and J.B. Slade. 2000. Insect herbivory accelerates nutrient cycling and increases plant production. *Proceedings of the National Academy of Sciences USA* **97**, 14412–14417.

Belsky, A.J. 1986. Population and community processes in a mosaic grassland in the Serengeti, Tanzania. *Journal of Ecology* **74**, 841–856.

Bengtsson, G. and S. Rundgren. 1983. Respiration and growth of a fungus, *Mortierella isabellina*, in response to grazing by *Onychiurus armatus* (Collembola). *Soil Biology and Biochemistry* **15**, 469–473.

Bengtsson, J., D.W. Zheng, G.I. Ågren, et al. 1995. Food webs in soil: an interface between population and ecosystem ecology. In: *Linking Species and Ecosystems* (C.G. Jones and J.H. Lawton, eds), pp. 159–165. Chapman and Hall, London.

Berendse, F. 1998. Effects of dominant plant species on soils during succession in nutrient poor ecosystems. *Biogeochemistry* **42**, 73–88.

Berendse, F. and M. Scheffer. 2009. The angiosperm radiation revisited, an ecological explanation for Darwin's 'abominable mystery'. *Ecology Letters* **12**, 865–872.

Berg, B. and G. Ekbohm. 1991. Litter mass loss and decomposition patterns in some leaf and litter types. Long term decomposition in a Scots pine forest. *Canadian Journal of Botany* **69**, 1449–1456.

Berg, B. and C. McClaugherty. 2003. *Plant Litter: Decomposition, Humus Formation and Carbon Sequestration*. Springer, Berlin.

Berg, B., M.P. Berg, P. Bottner, et al. 1993. Litter mass loss rates in pine forests of Europe and eastern United States – some relationships with climate and litter quality. *Biogeochemistry* **20**, 127–159.

Berg, M.P. and J. Bengtsson. 2007. Temporal and spatial variability in soil food web structure. *Oikos* **116**, 1789–1804.

Beschta, R.L. and W.J. Ripple. 2008. Wolves, trophic cascades, and rivers in the Olympic National Park, USA. *Ecohydrology* **1**, 118–130.

Bever, J.D. 1994. Feedback between plants and their soil communities in an old field community. *Ecology* **75**, 1965–1977.

Bever, J.D. 2003. Soil community feedback and the coexistence of competitors: conceptual frameworks and empirical tests. *New Phytologist* **157**, 465–473.

Bever, J.D., K.M. Westover and J. Antonovics. 1997. Incorporating the soil community into plant population dynamics: the utility of the feedback approach. *Journal of Ecology* **85**, 561–573.

Bezemer, T.M. and W.H. Van der Putten. 2007. Diversity in plant communities. *Nature* **446**, E6–E7.

Bezemer, T.M., R. Wagenaar, N.M. Van Dam, et al. 2004. Interactions between above- and belowground insect herbivores as mediated by plant defense system. *Oikos* **101**, 555–562.

Bezemer, T.M., C.D. Lawson, K. Hedlund, et al. 2006. Plant species and functional group effects on abiotic and microbial soil properties and plant-soil feedback responses in two grasslands. *Journal of Ecology* **94**, 893–904.

Billes, G., H. Rouhier and P. Bottner. 1993. Modifications of the carbon and nitrogen allocations in the plant (*Triticum-Aestivum* L) soil system in response to increased atmospheric CO_2 concentration. *Plant and Soil* **157**, 215–225.

Binet, F., L. Fayolle and M. Pussard. 1998. Significance of earthworms in stimulating soil microbial activity, *Biology and Fertility of Soils* **27**, 79–84.

Binkley, D. and C. Giardina (1998) Why do trees affect soils? The Warp and Woof of tree-soil interactions. *Biogeochemistry* **42**, 89–106.

Birch, H. 1958. The effect of soil drying on humus decomposition and nitrogen availability. *Plant and Soil* **10**, 9–31.

Blair, A.C., B.D. Hanson, G.R. Brunk, et al. 2005. New techniques and findings in the study of a candidate allelochemical implicated in invasion success. *Ecology Letters* **8**, 1039–1047.

Blair, J.M., R.W. Parmelee and M.H. Beare. 1990. Decay rates, nitrogen fluxes and decomposer communities in single and mixed-species foliar litter. *Ecology* **71**, 1976–1985.

Blanka, V., J. Raabová, T. Kyncl, et al. 2009. Ants accelerate succession from mountain grassland towards spruce forest. *Journal of Vegetation Science* **20**, 577–587.

Bloemers G.F., M. Hodda, P.J.D. Lambshead, et al. 1997. The effects of forest disturbance on diversity of tropical soil nematodes. *Oecologia* **111**, 575–558.

Boag, B. 2000. The impact of the New Zealand flatworm on earthworms and moles in agricultural land in western Scotland. *Aspects of Applied Biology* **62**, 79–84.

Boag, B. and G.W. Yeates. 2001. The potential impact of the New Zealand flatworm, a predator of earthworms, in western Europe. *Ecological Applications* **11**, 1276–1286.

Bobbink, R. 1991. Effects of nutrient enrichment in Dutch chalk grassland. *Journal of Applied Ecology* **28**, 28–41.

Bobbink, R. and L.P.M. Lamers. 2002. Effects of increased nitrogen deposition. In: *Air Pollution and Plant Life* (J.N.D. Bell and M. Treshow, eds), 2nd edn, pp. 201–345. John Wiley and Sons, Chichester.

Boddey, R.M., R. Macedo, R.M. Tarre, et al. 2004. Nitrogen cycling in *Brachiaria* pastures: the key to understanding the process of pasture decline. *Agriculture, Ecosystems and Environment* **103**, 389–403.

Boddy, L. 1999. Saprotrophic cord-forming fungi: meeting the challenge of heterogeneous environments. *Mycologia* **91**, 13–32.

Bodmer, R.E., J.F. Eisenberg and K.H. Redford. 1997. Hunting and the likelihood of extinction of Amazonian mammals. *Conservation Biology* **11**, 460–466.

Boege, K. 2004. Induced responses in three tropical dry forest plant species—direct and indirect effects on herbivory. *Oikos* **107**, 541–548.

Bohlen, P.J., R.W. Parmelee and J.M. Blair. 1995. Efficacy of methods for manipulating earthwork populations in large scale field experiments in agroecosystems. *Soil Biology and Biochemistry* **27**, 993–999.

Bohlen, P.J., P.M. Groffman, T.J. Fahey, et al. 2004a. Ecosystem consequences of exotic earthworm invasion of north temperate forests. *Ecosystems* **7**, 1–12.

Bohlen, P.J., S. Scheu, C.M. Hale, et al. 2004b. Non-native invasive earthworms as agents of change in northern temperate forests. *Frontiers in Ecology and Evolution* **2**, 427–435.

Bonanomi, G., F. Giannino and S. Mazzoleni. 2005. Negative plant-soil feedback and species coexistence. *Oikos* **111**, 311–321.

Bonkowski, M. 2004. Protozoa and plant growth: the microbial loop revisited. *New Phytologist* **162**, 617–631.

Bonkowski, M., I.E. Geoghegan, A.N.E. Birch, et al. 2001. Effects of soil decomposer invertebrates (protozoa and earthworms) on an above-ground phytophagous insect (cereal aphid) mediated through changes in the host plant. *Oikos* **95**, 441–450.

Boone, R.B., S.B. Burnsilver, J.S. Worden, et al. 2008. Large-scale movements of large herbivores. In: *Resource Ecology: Spatial and Temporal Dynamics of Foraging* (H.H.T. Prins and F. van Langevelde, eds), pp. 187–206. Springer, Amsterdam.

Bowker, M.A., F.T. Maestre and C. Escolar. 2010. Biodiversity of biological crusts influences ecosystem function: A review and reanalysis. *Soil Biology and Biochemistry* **42**, 405–417.

Bowman, W.D. 2000. Biotic controls over ecosystem response to environmental change in alpine tundra of the Rocky Mountains. *Ambio* **29**, 396–400.

Bowman, W.D., T.A. Theodose, J.C. Schardt, et al. 1993. Constraints of nutrient availability on primary production in two alpine communities. *Ecology* **74**, 2085–2098.

Bowman, W.D., T.A. Theodose and M.C. Fisk. 1995. Physiological and production responses of plant growth forms to increases in limiting resources in alpine tundra: Implications for differential community response to environmental change. *Oecologia* **101**, 217–227.

Bowman, W.D., H. Steltzer, T.N. Rosenstiel, et al. 2004. Litter effects of two co-occurring alpine species on plant growth, microbial activity and immobilization of nitrogen. *Oikos* **104**, 336–344.

Bradford, M.A., T.H. Jones, R.D. Bardgett, et al. 2002. Impacts of soil faunal community composition on model grassland ecosystems. *Science* **298**, 615–618.

Bradford, M.A., C.D. Davies, S.D. Frey, et al. 2008. Thermal adaptation of soil microbial respiration to elevated temperature. *Ecology Letters* **11**, 1316–1327.

Bradley, B.A., R.A. Houghton, J.F. Mustard, et al. 2006. Invasive grass reduces aboveground carbon stocks in shrublands of the western U.S. *Global Change Biology* **12**, 1815–1822.

Brais, S., C. Camire, Y. Bergeron, et al. 1995. Changes in nutrient availability and forest floor characteristics in relation to stand age and forest composition in the southern part of the boreal forests of northwest Quebec. *Forest Ecology and Management* **76**, 181–189.

Bremer, D.J., J.M. Ham, C.E. Owensby, et al. 1998. Response of soil respiration to clipping and grazing in a tallgrass prairie. *Journal of Environmental Quality* **27**, 1539–1548.

Bret-Harte, M.S., G.R. Shaver and F.S. Chapin. 2002. Primary and secondary stem growth in arctic shrubs: implications for community response to environmental change. *Journal of Ecology* **90**, 251–267.

Breznak, J.A. and A. Brune. 1994. Role of microorganisms in the digestion of lignocellulose by termites. *Annual Review of Entomology* **39**, 453–487.

Brinkman, E.P., H. Duyts and W.H. Van der Putten. 2005. Consequences of variation in species diversity in a community of root-feeding herbivores for nematode dynamics and host plant biomass. *Oikos* **110**, 417–427.

Briones, M.J.I., P. Ineson and J. Poskitt. 1998. Climate change and *Cognettia sphagnetorum*: effects on carbon dynamics in organic soils. *Functional Ecology* **12**, 528–535.

Brooker, R. and R. Van der Wal. 2003. Can soil temperature direct the composition of high Arctic plant communities? *Journal of Vegetation Science* **14**, 535–542.

Brooks, M.L., C.M. D'Antonio, D.M. Richardson, et al. 2004. Effects of invasive alien plants on fire regimes. *BioScience* **54**, 677–688.

Brown, V.K. and A.C. Gange. 1989. Differential effects of above-ground and below-ground insect herbivory during early plant succession. *Oikos* **54**, 67–76.

Brown, V.K. and A.C. Gange. 1990. Insect herbivory below ground. *Advances in Ecological Research* **20**, 1–58.

Brown V.K. and A.C. Gange. 1992. Secondary plant succession—how is it modified by insect herbivory? *Vegetatio* **101**, 3–13.

Bruno, J.F. and B.J. Cardinale. 2008. Cascading effects of predator richness. *Frontiers in Ecology and the Environment* **6**, 539–546.

Brussaard, L., V.M. Behan-Pelletier, D.E. Bignell, et al. 1997. Biodiversity and ecosystem functioning in soil. *Ambio* **26**, 563–570.

Buckland, S.M. and J.P. Grime. 2000. The effects of trophic structure and soil fertility on the assembly of plant communities: a microcosm experiment. *Oikos* **91**, 336–352.

Buée, M., M. Reich, C. Murat, et al. 2009. 454 pyrosequencing analyses of forest soils reveal an unexpectedly high fungal diversity. *New Phytologist* **184**, 449–456.

Bump, J.K., R.O. Peterson and J.A. Vucetich. 2009a. Wolves modulate soil nutrient heterogeneity and foliar nitrogen by configuring the distribution of ungulate carcasses. *Ecology* **90**, 3159–3167.

Bump, J.K., K.B. Tischler, A.J. Schrank, et al. 2009b. Large herbivores and aquatic-terrestrial links in southern boreal forests. *Journal of Animal Ecology* **78**, 338–345.

Bump, J.K., C.R. Webster, J.A. Vucetich, et al. 2009c. Ungulate carcasses perforate ecological filters and create bighgoechemical hotsopts in forest herbaceous layers allowing trees a competitive advantage. *Ecosystems* **12**, 996–1007.

Bunker, D.E., F. De Clerck, J.C. Bradford, et al. 2005. Species loss and aboveground carbon storage in a tropical forest. *Science* **310**, 1029–1031.

Burke, M.J.W. and J.P. Grime. 1996. An experimental study of plant community invasibility. *Ecology* **77**, 776–790.

Cadisch, G. and Giller, K.E. (eds) 1997. *Driven by Nature—Plant Litter Quality and Decomposition*, pp. 107–124. CAB International, Wallingford.

Callaway, R.M. and W.M. Ridenour. 2004. Novel weapons: invasive success and the evolution of increased competitive ability. *Frontiers in Ecology and the Environment* **2**, 436–443.

Callaway, R.M., G.C. Thelan, A. Rodriguez, et al. 2004. Soil biota and exotic plant invasion. *Nature* **427**, 731–737.

Callaway, R.M., D. Cippolini, K. Barto, et al. 2008. Novel weapons: invasive plant suppresses fungal mutualists but not in its native Europe. *Ecology* **89**, 1043–1055.

Cameron, G.N. and S.R. Spencer. 1989. Rapid leaf decay and nutrient release in a Chinese tallow forest. *Oecologia* **80**, 222–228.

Cardillo, M., G.M. Mace, K.E. Jones, et al. 2005. Multiple cause of high extinction risk in large mammal species. *Science* **309**, 1239–1241.

Cardinale, B.J., D.S. Srivastava, J.E. Duffy, et al. 2006. Effects of biodiversity on the functioning of trophic groups and ecosystems: a meta-analysis. *Nature* **443**, 989–992.

Carline, K.A., H.E. Jones and R.D. Bardgett. 2005. Large herbivores affect the stoichiometry of nutrients in a regenerating woodland ecosystem. *Oikos* **110**, 453–460.

Carpenter, S.R., J.F. Kitchell and J.R. Hodgson. 1985. Cascading trophic interactions and lake productivity. *BioScience* **35**, 634–639.

Carpenter, S.R., H.A. Mooney, J. Agard, et al. 2009. Science for managing ecosystem services: Beyond the Millennium Ecosystem Assessment. *Proceedings of the National Academy of Sciences USA* **106**, 1305–1312.

Carreiro, M.M., R.L. Sinsabaugh, D.A. Repert, et al. 2000. Microbial enzyme shifts explain litter decay responses to simulated nitrogen deposition. *Ecology* **81**, 2359–2365.

Certini, G. 2005. Effects of fire on properties of forest soils: a review. *Oecologia* **143**, 1–10.

Chaneton, E.J., J.H. Lemcoff and R.S. Lavado. 1996. Nitrogen and phosphorus cycling in grazed and ungrazed plots in a temperate subhumid grassland in Argentina. *Journal of Applied Ecology* **33**, 291–302.

Chapin, F.S. and A.M. Starfield. 1997. Time lags and novel ecosystems in response to transient climatic change in Arctic Alaska. *Climate Change* **35**, 449–461.

Chapin, F.S., Walker, L.R., C. Fastie, et al. 1994. Mechanisms of post-glacial primary succession at Glacier Bay, Alaska. *Ecological Monographs* **64**, 149–175.

Chapin, F.S., M. Sturm, M.C. Serreze, et al. 2005. Role of land-surface changes in arctic summer warming. *Science* **310**, 657–660.

Chapin, F.S., J. McFarland, A.D. McGuire, et al. 2009. The changing global carbon cycle: linking plant–soil carbon dynamics to global consequences. *Journal of Ecology* **97**, 840–850.

Chapman, K., J.B. Whittaker and O.W. Heal. 1988. Metabolic and faunal activity in litter mixtures compared with pure stands. *Agriculture, Ecosystems and Environment* **34**, 65–73.

Chapuis-Lardy, L., S. Vanderhoeven, N. Dassonville, et al. 2006. Effect of the exotic invasive plant *Solidago gigantean* on soil phosphorus status. *Biology and Fertility of Soils* **42**, 373–378.

Chauvel, M., E. Grimaldi, E. Barros, et al. 1999. Pasture damage by an Amazonian earthworm, *Nature* **398**, 32–33.

Christensen, S. and J.M. Tiedje. 1990. Brief and vigorous N_2O production by soil at spring thaw. *Journal of Soil Science* **41**, 1–4.

Christie, P., E.I. Newman and R. Campbell. 1974. Grassland plant species can influence the abundance of microbes on each other's roots. *Nature* **250**, 570–571.

Christie, P., E.I. Newman and R. Campbell. 1978. The influence of neighboring grassland plants on each others' endomycorrhizas and root surface microorganisms. *Soil Biology and Biochemistry* **10**, 521–527.

Clarholm, M. 1985. Interactions of bacteria, protozoa and plants leading to mineralization of soil nitrogen. *Soil Biology and Biochemistry* **17**, 181–187.

Classen, A.T., S.K. Chapman, T.G. Whitham, et al. 2007. Genetic-based plant resistance and susceptibility traits to herbivory influence needle and root litter nutrient dynamics. *Journal of Ecology* **95**, 1181–1194.

Cleveland, C.C. and D. Liptzin. 2007. C : N : P stoichiometry in soil: is there a "Redfield ratio" for the microbial biomass? *Biogeochemistry* **85**, 235–252.

Cleveland, C.C., A.R. Townsend, D.S. Schimel, et al. 1999. Global patterns of terrestrial biological nitrogen (N-2) fixation in natural ecosystems. *Global Biogeochemical Cycles* **13**, 623–645.

Cole, L., R.D. Bardgett, P. Ineson, et al. 2002a. Relationships between enchytraeid worms (Oligochaeta), temperature, and the release of dissolved organic carbon from blanket peat in northern England. *Soil Biology and Biochemistry* **34**, 599–607.

Cole, L., R.D. Bardgett, P. Ineson, et al. 2002b. Enchytraeid worm (Oligochaeta) influences on microbial community structure, nutrient dynamics, and plant growth in blanket peat subjected to warming. *Soil Biology and Biochemistry* **34**, 83–92.

Cole, L., P.L. Staddon, D. Sleep, et al. 2004. Soil animals influence microbial abundance, but not plant-microbial competition for soil organic. *Functional Ecology* **18**, 631–640.

Cole, L., M.A. Bradford, P.J.A. Shaw, et al. 2006. The abundance, richness and functional role of soil meso- and macrofauna in temperate grassland—a case study. *Applied Soil Ecology* **33**, 186–198.

Cole, L., S.M. Buckland and R.D. Bardgett. 2008. Influence of disturbance and nitrogen addition on plant and soil animal diversity in grassland. *Soil Biology and Biochemistry* **40**, 505–514.

Coleman, D.C. 1985. Through a ped darkly: an ecological assessment of root-soil-microbial-faunal interactions. In: *Ecological Interactions in Soil: Plants, Microbes and Animals* (A.H. Fitter, D. Atkinson, D.J. Read, et al., eds), pp. 1–21. British Ecological Society Special Publication Number 4. Blackwell, Oxford.

Coleman, D.C. and D.A. Crossley. 1995. *Fundamentals of Soil Ecology*. Academic Press, San Diego, CA.

Coleman, D.C., C.V. Cole, R.V. Anderson, et al. 1977. Analysis of rhizosphere-saprophage interactions in terrestrial ecosystems. In: *Soil Organisms as Components of Ecosystems* (U. Lohm and T. Persson, eds), pp. 299–309. Ecological Bulletin 25, Swedish National Science Research Council, Stockholm.

Coleman, D.C., C.P.P. Reid and C.V. Cole. 1983. Biological strategies of nutrient cycling in soil systems. *Advances in Ecological Research* **13**, 1–51.

Coley, P.D., J.P. Bryant and F.S. Chapin. 1985. Resource availability and plant antiherbivore defense. *Science* **230**, 895–899.

Conant, R.T., K. Paustian and E.T. Elliott. 2001. Grassland management and conversion into grassland: effects on soil carbon. *Ecological Applications* **11**, 343–355.

Conant, R.T., R.A. Drijber, M.L. Haddix, et al. 2008. Sensitivity of organic matter decomposition to warming varies with its quality. *Global Change Biology* **14**, 868–877.

Conen, F., J. Leifeld, B. Seth and C. Alewell. 2006. Warming mobilises young and old soil carbon equally. *Biogeosciences* **3**, 515–519.

Connell, J.H. and R.O. Slatyer. 1977. Mechanisms of succession in natural communities and their role in community stability and organization. *American Naturalist* **111**, 1119–1144.

Connell, J.H. and M.D. Lowman. 1989. Low-diversity tropical rain forests: some possible mechanisms for their existence. *American Naturalist* **134**, 88–119.

Convey, P., P. Greenslade, R.J. Arnold, et al. 1999. Collembola of sub-Antarctic South Georgia. *Polar Biology* **22**, 1–6.

Coomes, D.A., R.B. Allen, W.A. Bently, et al. 2005. The hare, the tortoise, and the crocodile: the ecology of angiosperm dominance, conifer persistence and fern filtering. *Journal of Ecology* **93**, 918–935.

Cooper, E. and Wookey, P. 2001. Field measurements of the growth rates of forage lichens, and the implications of grazing by Svalbard reindeer. *Symbiosis* **31**, 173–186.

Cornelissen, J.H.C. 1996. An experimental comparison of leaf decomposition rates in a wide range of temperate plant species and types. *Journal of Ecology* **84**, 573–582.

Cornelissen, J.H.C. and K. Thompson. 1997. Functional leaf attributes predict litter decomposition rate in herbaceous plants. *New Phytologist* **135**, 109–114.

Cornelissen, J.H.C., P.C. Dies and R. Hunt. 1996. Seedling growth, allocation and leaf attributes in a wide range of woody plants and types. *Journal of Ecology* **84**, 755–765.

Cornelissen, J.H.C., N. Perez-Harguindeguy, S. Díaz, et al. 1999. Leaf structure and defence control litter decomposition rate across species and life forms in regional floras in two continents. *New Phytologist* **143**, 191–200.

Cornelissen, J.H.C., R. Aerts, B. Cerabolini, et al. 2001a. Carbon cycling traits of plant species are linked with mycorrhizal strategy. *Oecologia* **129**, 611–619.

Cornelissen, J.H.C., T.V. Callaghan, J.M. Alatalo, et al. 2001b. Global change and arctic ecosystems: is lichen decline a function of increases in vascular plant biomass? *Journal of Ecology* **89**, 984–994.

Cornelissen, J.H.C., H.M. Quested, D. Gwynn-Jones, et al. 2004. Leaf digestibility and litter decomposability are related in a wide range of subarctic plant species and types. *Functional Ecology* **18**, 779–786.

Cornelissen, J.H.C., P.M. Van Bodegom, R. Aerts, et al. 2007. Global negative vegetation feedback to climate warming responses of leaf litter decomposition rates in cold biomes. *Ecology Letters* **10**, 619–627.

Cornwell, W.K., J.H.C. Cornelissen, K. Amatangelo, et al. 2008. Plant traits are the predominant control of litter decomposition within biomes worldwide. *Ecology Letters* **11**, 1065–1071.

Cornwell, W.K., J.H.C. Cornelissen, S.D. Allison, et al. 2009. Plant traits and wood fates across the globe; rotted, burned or consumed? *Global Change Biology* **15**, 2431–2449.

Cortez, J., R. Hameed and M.B. Bouché. 1989. C and N transfer in soil with or without earthworms fed with ^{14}C- and ^{15}N labelled wheat straw. *Soil Biology and Biochemistry* **21**, 491–497.

Côte, S.D., T.P. Rooney, J.-P. Tremblay, et al. 2004. Ecological impacts of deer overabundance. *Annual Reviews of Ecology and Systematics* **35**, 113–147.

Cottingham, K.L., B.L. Brown and J.L. Lennon. 2001. Biodiversity may regulate the temporal variability of ecological systems. *Ecology Letters* **4**, 72–85.

Coughenour, M.B. 1991. Biomass and N responses to grazing of upland steppe on Yellowstone's northern winter range. *Journal of Applied Ecology* **28**, 71–82.

Coulson, S.J., H.P. Leinass, R.A. Ims, et al. 2000. Experimental manipulation of the winter surface ice layer: the effects on Arctic soil microarthropod community. *Ecography* **23**, 299–306.

Coûteaux. M.M., C. Kurz, P. Bottner, et al. 1999. Influence of increased atmospheric CO_2 concentration on quality of plant material and litter decomposition. *Tree Physiology* **19**, 301–311.

Cox, P.M., R.A. Betts, C.D. Jones, et al. 2000. Acceleration of global warming due to carbon cycle feedbacks in a coupled climate model. *Nature* **408**, 184–187.

Crafford, J.E. 1990. The role of feral house mice in ecosystem functioning on Marion Island. In: *Antarctic Ecosystems: Ecological Change and Conservation* (K.R. Kerry and G. Hempel, eds), pp. 359–364. Springer, Berlin.

Cragg, R.G. and R.D. Bardgett. 2001. How changes in soil faunal diversity and compostion within a trophic group influence decomposition processes. *Soil Biology and Biochemistry* **33**, 2073–2081.

Craine, J.M., D.A. Wedin and F.S. Chapin. 1999. Predominance of ecophysiological controls on soil CO_2 flux in a Minnesota grassland. *Plant and Soil* **207**, 77–86.

Craine, J.M., D. Tilman, D. Wedin, et al. 2002. Functional traits, productivity and effects on nitrogen cycling of 33 grassland species. *Functional Ecology* **16**, 563–574.

Craine, J.M., C. Morrow and N. Fierer. 2007. Microbial nitrogen limitation increases decomposition. *Ecology* **88**, 2105–2113.

Creel, S. and D. Christianson. 2009. Wolf presence and increased willow consumption by Yellowstone elk: implications for trophic cascades. *Ecology* **90**, 2454–2466.

Creel, S., J. Winnie, B. Maxwell, et al. 2005. Elk alter habitat selection as an antipredator response to wolves. *Ecology* **86**, 3387–3397.

Crews, T.E., K. Kitayama, J.H. Fownes, et al. 1995. Changes in soil phosphorus fractions and ecosystem dynamics across a long chronosequence in Hawaii. *Ecology* **76**, 1407–1424.

Crocker, R.L. and J. Major. 1955. Soil development in relation to vegetation and surface age at Glacier Bay, Alaska. *Journal of Ecology* **43**, 427–448.

Croll, D.A., J.L. Maron, J.A. Estes, et al. 2005. Introduced predators transform subarctic islands from grassland to tundra. *Science* **307**, 1959–1961.

Crutsinger, G.M., W.N. Reynolds, A.T. Classen, et al. 2008. Disparate effects of plant genotypic diversity on foliage and litter arthropod communities. *Oecologia* **158**, 65–75.

Curtis, P.S. and X.Z. Wang, X.Z. 1998. A meta-analysis of elevated CO_2 effects on woody plant mass, form, and physiology. *Oecologia* **113**, 299–313.

Curtis T.P., W.T. Sloan and J.W. Scannell. 2002. Estimating prokaryotic diversity and its limits. *Proceedings of the National Academy of Sciences USA* **99**, 10494–10499.

Daily, G.C. and P.A. Matson. 2008. Ecosystem services: from theory to implementation. *Proceedings of the National Academy of Sciences USA* **105**, 9455–9456.

Danby, R.K. and D.S. Hik. 2007. Variability, contingency and rapid change in recent subarctic alpine tree line dynamics. *Journal of Ecology* **95**, 352–363.

D'Antonio, C.M. and P.M. Vitousek. 1992. Biological invasions by exotic grasses, the grass/fire cycle and global change. *Annual Review of Ecology and Systematics* **23**, 63–87.

Davidson, E.A. and I.A. Janssens. 2006. Temperature sensitivity of soil carbon decomposition and feedbacks to climate change. *Nature* **440**, 165–173.

Davis, M.A., J.P. Grime and K. Thompson. 2000. Fluctuating resources in plant communities: a general theory of invasability. *Journal of Ecology* **88**, 528–534.

DeAngelis, D.L. and W.M. Post. 1991. Positive feedback and ecosystem organization. In: *Theoretical Studies of Ecosystems: The Network Perspective* (M. Higashi and T.P. Burns, eds), pp. 155–178. Cambridge University Press, Cambridge.

De Deyn, G.B. and W.H. Van der Putten. 2005 Linking aboveground and belowground ecology. *Trends in Ecology and Evolution* **20**, 625–633.

De Deyn, G.B., C.E. Raaijmakers, H.R. Zoomer, et al. 2003. Soil invertebrate fauna enhances grassland succession and diversity. *Nature* **422**, 711–713.

De Deyn, G.B., C.E. Raaijmakers, J. van Ruijven, et al. 2004. Plant species identity and diversity effects on different trophic levels of nematodes in the soil food web. *Oikos* **106**, 576–586.

De Deyn, G.B., J. van Ruijven, E. Ciska, et al. 2007. Above- and belowground insect herbivores differentially affect soil nematode communities in species-rich plant communities. *Oikos* **116**, 923–930.

De Deyn, G.B., H.C. Cornelissen and R.D. Bardgett. 2008. Plant functional traits and soil carbon sequestration in contrasting biomes. *Ecology Letters* **11**, 516–531.

De Deyn, G.B., H. Quirk, Y. Zou, et al. 2009. Vegetation composition promotes carbon and nitrogen storage in model grassland communities of contrasting soil fertility. *Journal of Ecology* **97**, 864–875.

Degens, B. 1998. Decreases in microbial functional diversity do not result in corresponding changes in decomposition under different moisture regimes. *Soil Biology and Biochemistry* **30**, 1989–2000.

De Graaff, M.A., K.J. van Groenigen, J. Six, et al. 2006. Interactions between plant growth and soil nutrient cycling under elevated CO_2: a meta-analysis. *Global Change Biology* **12**, 2077–2091.

De Graaff, M.A., J. Six and C. van Kessel. 2007. Elevated CO_2 increases ntrogen rhizodeposition and microbial immobilization of root-derived nitrogen. *New Phytologist* **173**, 778–786.

DeLuca, T.H., M.-C. Nilsson and O. Zackrisson. 2002a. Nitrogen mineralization and phenol accumulation along a fire chronosequence in northern Sweden. *Oecologia* **133**, 206–214.

DeLuca, T.H., O. Zackrisson, M.-C. Nilsson, et al. 2002b. Quantifying nitrogen-fixation in feather moss carpets of boreal forests. *Nature* **419**, 917–920.

De Mazancourt, C., M. Loreau and L. Abbadie. 1999. Grazing optimization and nutrient cycling: potential impact of large herbivores in a savannah system. *Ecological Applications* **9**, 784–797.

Denton, C.S., R.D. Bardgett, R. Cook, et al. 1999. Low amounts of root herbivory positively influence the rhizosphere microbial community in a temperate grassland soil. *Soil Biology and Biochemistry* **31**, 155–165.

De Rooij-van der Goes, P.C.E.M. 1995. The role of plant – parasitic nematodes in the decline of *Ammophila arenaria* L. Link. *New Phytologist* **129**, 661–669.

de Ruiter, P.C., A.-M. Nuetel and J.C. Moore. 1995. Energetics, patterns of interactions strengths, and stability in real ecosystems. *Science* **269**, 1257–1260.

Desprez-Loustau, M.L., C. Robin, M. Buee, et al. 2007. The fungal dimension of biological invasions. *Trends in Ecology and Evolution* **22**, 472–480.

de Vries, F.T., E. Hoffland, N. van Eekeren, et al. 2006. Fungal/bacterial ratios in grasslands with contrasting nitrogen management. *Soil Biology and Biochemistry* **38**, 2092–2103.

Diamond, J.M. 2005. *Collapse: How Societies Choose to Fail or Succeed.* Penguin Books, London.

Díaz, S., J.P. Grime, J. Harris, et al. 1993. Evidence of a feedback mechanism limiting plant-response to elevated carbon-dioxide. *Nature* **364**, 616–617.

Díaz, S., F.S. Chapin III, A. Symstad, et al. 2003. Functional diversity revealed through removal experiments. *Trends in Ecology and Evolution* **18**, 140–146.

Díaz, S., J.G. Hodgson, K. Thompson, et al. 2004. The plant traits that drive ecosystems: evidence from three continents. *Journal of Vegetation Science* **15**, 295–304.

Díaz, S., S. Lavorel, S. McIntyre, et al. 2006. Plant trait responses to grazing—a global synthesis. *Global Change Biology* **13**, 313–341.

Díaz, S., S. Lavorel, F. de Bello, et al. 2007. Incorporating plant functional diversity effects in ecosystem service assessments. *Proceedings of the National Academy of Sciences USA* **104**, 20684–20689.

Díaz, S., A. Hector and D.A. Wardle. 2009. Biodiversity in forest carbon sequestration initiatives: not just a side benefit. *Current Opinion in Environmental Sustainability* **1**, 55–60.

Dickie, I.A., R.T. Koide and K.C. Steiner. 2002. Influences of established trees on mycorrhizas, nutrition, and growth of *Quercus rubra* seedlings. *Ecological Monographs* **72**, 505–521.

Dijkstra, F.A. and W.X. Cheng. 2007. Interactions between soil and tree roots accelerate long-term soil carbon decomposition. *Ecology Letters* **10**, 1046–1053.

Dillon, R.J. and V.M. Dillon. 2004. The gut bacteria of insects: nonpathogenic interactions. *Annual Review of Entomology* **49**, 71–92.

Doblas-Miranda, E., D.A. Wardle, D.A. Peltzer, et al. 2008. Changes in the community structure and diversity of soil invertebrates across the Franz Josef Glacier chronosequence. *Soil Biology and Biochemistry* **40**, 1069–1081.

Donnison, L.M., G.S. Griffith, J. Hedger, et al. 2000. Management influences on soil microbial communities and their function in botanically diverse haymeadows of northern England and Wales. *Soil Biology and Biochemistry* **32**, 253–263.

Dormaar, J.F. 1990. Effect of active roots on the decomposition of soil organic matter. *Biology and Fertility of Soils* **10**, 121–126.

Dorrepaal, E., J.H.C. Cornelissen, R. Aerts, et al. 2005. Are growth forms consistent predictors of leaf litter quality and decomposability across peatlands along a latitudinal gradient? *Journal of Ecology* **93**, 817–828.

Dorrepaal, E., S. Toet, R.S.P. van Logtestijn, et al. 2009. Carbon respiration from subsurface peat accelerated by climate warming in the subarctic. *Nature* **460**, 616–619.

Dowrick, D.J., S. Hughes, C. Freeman, et al. 1999. Nitrous oxide emissions from a gully mire in mid-Wales UK, under simulated summer drought. *Biogeochemistry* **44**, 151–162.

Dromph, K., R. Cook, N.J. Ostle, et al. 2006. Root parasite induced nitrogen transfer between plants is density dependent. *Soil Biology and Biochemistry* **38**, 2495–2498.

Duffy, J.E. 2009. Why biodiversity is important to the functioning of real world ecosystems. *Frontiers in Ecology and the Environment* **8**, 437–444.

Duffy, J.E., B.J. Cardinale, K.E. France, et al. 2007. The functional role of biodiversity in ecosystems: incorporating trophic complexity. *Ecology Letters* **10**, 522–538.

Duke, S.O., A.C. Blair, F.E. Dayan, et al. 2009. Is (-)-Catechin a novel weapon of spotted knapweed (*Centaurea stoebe*)? *Journal of Chemical Ecology* **35**, 141–153.

Duncan, R.P. and J.R. Young. 2000. Determinants of plant extinction and rarity 145 years after European settlement of Auckland, New Zealand. *Ecology* **81**, 3048–3061.

Dunham, A.E. 2008. Above and below ground impacts of terrestrial mammals and birds in a tropical forest. *Oikos* **117**, 571–579.

Dunn, R., J. Mikola, R. Bol, et al. 2006. Influence of microbial activity on plant-microbial competition for organic and inorganic nitrogen. *Plant and Soil* **289**, 321–334.

Dyer, H.C., L. Boddy and C.M. Preston-Meek. 1992. Effect of the nematode *Panagrellus redivivus* on growth and enzyme production by *Phanerochaete velutina* and *Stereum hirsutum*. *Mycological Research* **96**, 1019–1028.

Dyer, L.A. and D. Letourneau. 2003. Top down and bottom up diversity cascades in detrital vs. living food webs. *Ecology Letters* **6**, 60–68.

Dyksterhuis, E.J. 1949. Condition and management of rangelands based on quantitative ecology. *Journal of Range Management* **2**, 104–115.

Edwards, C.A. 2004. *Earthworm Ecology*, 2nd edn. CRC Press, Boca Raton, FL.

Edwards, C.A. and P.J. Bohlen. 1996. *Biology and Ecology of Earthworms*, 3rd edn. Chapman and Hall: London.

Edwards, P.J. and S. Hollis. 1982. The distribution of excreta on New Forest grassland used by cattle, ponfies and deer. *Journal of Applied Ecology* **19**, 953–964.

Egerton-Warburton, L.M. and E.B. Allen. 2000. Shifts in arbuscular mycorrhizal communities along an anthropogenic nitrogen deposition gradient. *Ecological Applications* **10**, 484–496.

Ehrenfeld, J.G. 2003. Effects of exotic plant invasions on soil nutrient cycling processes. *Ecosystems* **6**, 503–523.

Ehrlich, P.R. and H.A. Mooney. 1983. Extinction, substitution and ecosystem services. *BioScience* **33**, 248–254.

Eisenhauer, N. and S. Scheu. 2008. Earthworms as drivers of the competition between grasses and legumes. *Soil Biology and Biochemistry* **40**, 2650–2659.

Eldridge, D.J. 1993. Effects of ants on sandy soils in semiarid eastern Australia: local distribution of nest entrances and their effect in infiltration of water. *Australian Journal of Soil Research* **31**, 509–518.

Ellis, J.C. 2005. Marine birds on land; a review of plant biomass, species richness and community composition in seabird colonies. *Plant Ecology* **181**, 227–241.

Ellis, J.C., J.M. Farina and J.D. Whitman. 2006. Nutrient transfer from sea to land: the case of gulls and cormorants in the Gulf of Maine. *Journal of Animal Ecology* **75**, 565–574.

Ellison, A.M., M.S. Bank, B.D. Clinton, et al. 2005. Loss of foundation species: consequences for the structure and dynamics of forested ecosystems. *Frontiers in Ecology and the Environment* **3**, 479–486.

Elser, J.J., R.W. Sterner, E. Gorokhova, et al. 2000. Biological stoichiometry from genes to ecosystems. *Ecology Letters* **3**, 540–550.

Elton, C.S. 1958. *The Ecology of Invasions by Animals and Plants*. Methuen, London.

Engelbrecht, B.M., L.S. Comita, R. Condit, et al. 2007. Drought sensitivity shapes species distribution patterns in tropical forests. *Nature* **447**, 80–82.

Engelkes, T., E. Morrien, K.J.F. Verhoeven, et al. 2008 Successful range-expanding plants experience less above-ground and below-ground enemy impact. *Nature* **456**, 946–948.

Enríquez, S., C.M. Duarte and K. Sandjensen. 1993. Patterns in decomposition rates among photosynthetic organisms: the importance of detritus C:N:P content. *Oecologia* **94**, 457–471.

Epstein, H.E., W.A. Beringer, A.H. Gould, et al. 2004. The nature of spatial transitions in the Arctic. *Journal of Biogeography* **31**, 1917–1933.

Ettema, C. and D.A. Wardle. 2002. Spatial soil ecology. *Trends in Ecology and Evolution* **17**, 177–183.

Ettema, C.H., R. Lowrance, D.C. Coleman, et al. 1999. Riparian soil response to surface nitrogen input: the indicator potential of free living soil nematode populations. *Soil Biology and Biochemistry* **31**, 1625–1638.

Euskirchen, E.S., A.D. McGuire, D.W. Kicklighter, et al. 2006. Importance of recent shifts in soil thermal dynamics on growing season length, productivity, and carbon sequestration in terrestrial high-latitude ecosystems. *Global Change Biology* **12**, 731–750.

Fang, C.M., P. Smith, J.B. Moncrieff, et al. 2005. Similar response of labile and resistant soil organic matter pools to changes in temperature. *Nature* **433**, 57–59.

Fastie, C.L. 1995. Causes and ecosystem consequences of multiple pathways of primary succession at Glacier Bay, Alaska. *Ecology* **76**, 1899–1916.

Feeley, K.J. and J.W. Terborgh. 2005. The effects of herbivore density on soil nutrients and tree growth in tropical forest fragments. *Ecology* **86**, 116–124.

Fenner, N., Freeman, C., Lock, M.A., et al. 2007a. Interactions between elevated CO_2 and warming could amplify DOC exports from peatland catchments. *Environmental Science and Technology* **41**, 3146–3152.

Fenner, N., N.J. Ostle, N. McNamara, et al. 2007b. Elevated CO_2 Effects on peatland plant community carbon dynamics and DOC production. *Ecosystems* **10**, 635–647.

Fierer, N and J.P. Schimel. 2002. Effects of drying-rewetting frequency on soil carbon and nitrogen transformations. *Soil Biology and Biochemistry* **34**, 777–787.

Fierer, N., J.M. Craine, K. McLauchlan, et al. 2005. Litter quality and the temperature sensitivity of decomposition. *Ecology* **86**, 320–326.

Fierer, N., M.S. Strickland, D. Liptzin, et al. 2009. Global patterns in belowground communities. *Ecology Letters* **12**, 1238–1249.

Findlay, S., M. Carreiro, V. Krischik, et al. 1996. Effects of damage to living plants on leaf litter quality. *Ecological Applications* **6**, 269–275.

Finlay, B.J. 2002. Global dispersal of free-living microbial eukaryote species. *Science* **296**, 1061–1063.

Finlay, R.D. 1985. Interactions between soil micro-arthropods and endomycorrhizal associations of higher plants. In: *Ecological Interactions in Soil—Plants, Microbes and Animals* (A.H. Fitter, ed.), pp. 319–331. Blackwell Scientific Publications, Oxford.

Finzi, A.C. and S.T. Berthrong. 2005. The uptake of amino acids by microbes and trees in three cold-temperate forests. *Ecology* **86**, 3345–3353.

Finzi, A.C., E.H. DeLucia, R.G. Hamilton, et al. 2002. The nitrogen budget of a pine forest under free-air CO_2 enrichment. *Oecologia* **132**, 567–578.

Fisk, M.C., D.R. Zak and T.R. Crow. 2002. Nitrogen storage and cycling in old- and second-growth northern hardwood forests. *Ecology* **83**, 73–87.

Flanagan, L.B., L.A. Wever and P.J. Carlson. 2002. Seasonal and interannual variation in carbon dioxide exchange and carbon balance in a northern temperate grassland. *Global Change Biology* **8**, 599–615.

Floate, M.J.S. 1970a. Mineralization of nitrogen and phosphorus from organic materials of plant and animal origin and its significance in the nutrient cycle in grazed upland hills and soils. *Journal of the British Grassland Society* **25**, 295–302.

Floate, M.J.S. 1970b. Decomposition of organic materials from hill soils and pastures II. Comparative studies on the mineralization of carbon, nitrogen and phosphorus from plant materials and sheep faeces. *Soil Biology and Biochemistry* **2**, 173–185.

Floate, M.J.S. 1981. Effects of grazing by large herbivores on N cycling in agricultural ecosystems. In: *Terrestrial Nitrogen Cycles – Processes, Ecosystem Strategies and Management Impacts* (F.E. Clark and T. Rosswalls, eds), pp. 585–597. Ecological Bulletins, Stockholm.

Foissner, W., A. Chao, L.A. Katz, et al. 2008. Diversity and geographic distribution of ciliates (Protista: Ciliophora). *Biodiversity and Conservation* **17**, 345–363.

Fontaine S. and S. Barot. 2005. Size and functional diversity of microbe populations control plant persistence and long-term soil carbon accumulation. *Ecology Letters* **8**, 1075–1087.

Fornara, D.A. and J.T. Du Toit. 2007. Browsing lawns? Response of *Acacia nigrescens* to ungulate browsing in an African savanna. *Ecology* **88**, 200–209.

Fornara, D.A. and J.T. Du Toit. 2008. Browsing-induced effects of leaf litter quality and decomposition in a southern African savanna. *Ecosystems* **11**, 238–249.

Fornara, D.A. and D. Tilman. 2008. Plant functional composition influences rates of soil carbon and nitrogen accumulation. *Journal of Ecology* **96**, 314–322.

Fortunel, C., E. Garnier, R. Joffre, et al. 2009. Leaf traits capture the effects of land use and climate on litter decomposability of herbaceous communities across Europe. *Ecology* **90**, 598–611.

Fox, L.R., S.P., Ribeiro, V.K. Brown, et al. 1999. Direct and indirect effects of climate change on St John's wort, *Hypericum perforatum* L. (Hypericaceae). *Oecologia* **120**, 113–122.

Fox, A.D., Madsen, J., Boyd, H. et al. 2005. Effects of agricultural change on abundance, fitness components and distribution of two arctic-nesting goose populations. *Global Change Biology* **11**, 881–893.

Francis, R. and D.J. Read. 1995. Mutualism and antagonism in the mycorrhizal symbiosis, with special reference to impacts on plant community structure. *Canadian Journal of Botany* **73**, S1301–S1309.

Frank, A.B. 2002. Carbon dioxide fluxes over a grazed prairie and seeded pasture in the Northern Great Plains. *Environmental Pollution* **116**, 397–403.

Frank, A.B., D.L. Tanaka, L., Hofmann, et al. 1995. Soil carbon and nitrogen of Northern Great Plains grasslands as influenced by long-term grazing. *Journal of Range Management* **48**, 470–474.

Frank, D.A. 2008. Evidence for top predator control of a grazing system. *Oikos* **117**, 1718–1724.

Frank, D.A. and P.M. Groffman. 2009. Plant rhizospheric N processes: what we don't know and why we should care. *Ecology* **90**, 1512–1519.

Frank, D.A. and S.J. McNaughton. 1992. The ecology of plants, large mammalian herbivores, and drought in Yellowstone National Park. *Ecology* **73**, 2043–2058.

Freckman, D.W. and R. Mankau. 1986. Abundance, distribution, biomass and energetics of soil nematodes in a northern Mojave desert. *Pedobiologia* **29**, 129–142.

Freckman, D.W. and R.A. Virginia. 1997. Low-diversity Antarctic soil nematode communities: distribution and response to disturbance. *Ecology* **78**, 363–369.

Freeman, C., G.B. Nevison, H. Kang, et al. 2002. Contrasted effects of simulated drought on the production and oxidation of methane in a mid-Wales wetland. *Soil Biology and Biochemistry* **34**, 61–67.

Freeman C., N.J. Ostle, N. Fenner, et al. 2004a. A regulatory role for phenol oxidase during decomposition in peatlands. *Soil Biology and Biochemistry* **36**, 1663–1667.

Freeman, C., N. Fenner, N.J. Ostle, et al. 2004b. Dissolved organic carbon export from peatlands under elevated carbon dioxide levels. *Nature* **430**, 195–198.

Frelich, L.E., C.M. Hale, S. Scheu, et al. 2006. Earthworm invasion into previously earthworm-free temperate and boreal forests. *Biological Invasions* **8**, 1235–1245.

Frenot, Y., S.L. Chown, J. Whinam, et al. 2005. Biological invasions in the Antarctic: extent, impacts and implications. *Biological Reviews* **80**, 45–72.

Freschet, G.T., J.H.C. Cornelissen, L.S.P. van Longtestijn, et al. 2010. Evidence of the plant ecnomics spectrum in a subarctic flora. *Journal of Ecology* **98**, 362–373.

Frey, S.D., M. Knorr, J.L. Parrent, et al. 2004. Chronic nitrogen enrichment affects the structure and function of the soil microbial community in temperate hardwood and pine forests. *Forest Ecology and Management* **196**, 159–171.

Fridley, J.D. 2002. Resource availability dominates and alters the relationship between species diversity and ecosystem productivity in experimental plant communities. *Oecologia* **132**, 271–277.

Fridley, J.D., J.J. Stachowicz, S. Naeem, et al. 2007. The invasion paradox: reconciling pattern and process in species invasion. *Ecology* **88**, 3–17.

Friedlingstein, P., P. Cox, R. Betts, et al. 2006. Climate-carbon cycle feedback analysis: results from the (CMIP)-M-4 model intercomparison. *Journal of Climate* **19**, 3337–3353.

Frost, C.J. and M.D. Hunter. 2004. Insect canopy herbivory and frass deposition affect soil nutrient dynamics and export in oak mesocosms. *Ecology* **85**, 3335–3347.

Frost, C.J. and M.D. Hunter. 2007. Recycling of nitrogen in herbivore feces: plant recovery, herbivore assimilation, soil retention, and leaching losses. *Oecologia* **151**, 42–53.

Frost, C.J. and M.D. Hunter. 2008a. Insect herbivores and their frass affect *Quercus rubra* leaf quality and initial stages of subsequent litter decomposition, *Oikos* **117**, 13–22.

Frost, C.J. and M.D. Hunter. 2008b. Herbivore-induced shifts in carbon and nitrogen allocation in red oak seedlings. *New Phytologist* **178**, 835–845.

Frouz, J., D. Elhottová, V. Sustr, et al. 2002. Preliminary data about the compartmentalization of the gut of the saprophagous dipteran larvae *Penthetria holosericea* (Bibionidae). *European Journal of Soil Biology* **38**, 47–51.

Frouz, J., V. Kristufek, X. Li, et al. 2003. Changes in the amount of bacteria during gut passage of leaf litter and during coprophagy in three species of bibionidae (Diptera) larvae. *Folia Microbiology* **48**, 535–542.

Fukami, T., D.A. Wardle, P.J. Bellingham, et al. 2006. Above- and below-ground impacts of introduced predators in seabird-dominated island ecosystems. *Ecology Letters* **9**, 1299–1307.

Funk, J.L. and P.M. Vitousek. 2007. Resource-use efficiency and plant invasions in low-resource systems. *Nature* **446**, 1079–1081.

Gadgill, R.L. 1971. The nutritional role of *Lupinus arboreus* in coastal sand dune forestry. I. The potential influence of undamaged lupin plants on nitrogen uptake by *Pinus radiata*. *Plant and Soil* **34**, 357–367.

Galloway, J.N., F.J. Dentener, D.G. Capone, et al. 2004. Nitrogen cycles: past, present, and future. *Biogeochemistry* **70**, 153–226.

Galloway, J.N., A.R. Townsend, J. Willem Erisman, et al. 2008. Transformation of the nitrogen cycle: recent trends, questions, and potential solutions. *Science* **320**, 889–892.

Gange, A.C. and V.K. Brown. 1989. Effects of root herbivory by an insect on a folier-feeding species, mediated by changes in the host plant. *Oecologia* **81**, 38–42.

Gange, A.C. and H.M. West. 1994. Interactions between arbuscular mycorrhizal fungi and foliar-feeding insects in *Plantago lanceolata* L. *New Phytologist* **128**, 79–87.

Gange, A.C., V.K. Brown and G.S. Sinclair. 1993. Vesicular-arbuscular mycorrhizal fungi: a determinant of plant community structure in early succession. *Functional Ecology* **7**, 616–622.

Gange, A.C., E.G. Gange, T.H. Sparks, et al. 2007. Rapid and recent changes in fungal fruiting patterns. *Science* **316**, 71.

Garnier, E., J. Cortez, G. Billes, et al. 2004. Plant functional markers capture ecosystem properties during secondary succession. *Ecology* **85**, 2630–2637.

Garrettson, M., J.F. Stetzel, B.S. Halpern, et al. 1998. Diversity and abundance of understory plants on active and abandoned nests of leaf-cutting ants (Atta cephalotes) in a Costa Rican rain forest. *Journal of Tropical Ecology* **14**, 17–26.

Gartner, B and Z. Cardon. 2004. Decomposition dynamics in mixed species leaf litter. *Oikos* **104**, 230–246.

Gaston, K.J., A.G. Jones, C. Hanel, et al. 2003. Rates of species introduction to a remote oceanic island. *Proceedings of the Royal Society of London Series B Biological Sciences* **270**, 1091–1098.

Geden, K.N. and M.D. Bertness. 2009. Experimental warming causes rapid loss of plant diversity in New England salt marshes. *Ecology Letters* **12**, 842–848.

Gehring C.A. and T.G. Whitham. 1994. Interactions between aboveground herbivores and the mycorrhizal mutualists of plants. *Trends in Ecology and Evolution* **9**, 251–255.

Gehring C.A. and T.G. Whitham. 2002. Mycorrhizae–herbivory interactions: population and community consequences. In: *Mycorrhizal Ecology* (M.G.A. Van der Heijden and I.R. Sanders, eds), pp. 295–320. Springer, Berlin.

Gende, S.M., A.E. Miller and E. Hood. 2007. The effects of salmon carcases on soil nitrogen pools in riparian forest of southeastern Alaska. *Canadian Journal of Forest Research* **37**, 1194–1202.

Giller, K.E., E. Witter and Sp.P. McGrath. 1998. Toxicity of heavy metals to microorganisms and microbial processes in agricultural soils—a review. *Soil Biology and Biochemistry* **30**, 1389–1414.

Goldberg, S.M.D., J. Johnson, D. Busam, et al. 2006. A Sanger/pyrosequencing hybrid approach for the generation of high-quality draft assemblies of marine microbial genomes. *Proceedings of the National Academy of Sciences USA* **103**, 11240–11245.

Golluscio, R.A., A.T. Austin, G.C. García Martínez, et al. 2009. Sheep grazing decreases organic carbon and nitrogen pools in the Patagonian Steppe: combination of direct and indirect effects. *Ecosystems* **12**, 686–697.

Gonzalez, G. and T.R. Seastedt. 2001. Soil fauna and plant litter decomposition in tropical and subalpine forests. *Ecology* **82**, 955–964.

Gonzalez, G., C.Y. Huang, X. Zou, et al. 2006. Earthworm invasion in the tropics. *Biological Invasions* **8**, 1247–1256.

Gordon, H., P.M. Haygarth and R.D. Bardgett. 2008. Drying and rewetting effects on soil microbial community composition and nutrient leaching. *Soil Biology and Biochemistry* **40**, 302–311.

Gotelli, N.J. and A.E. Arnett. 2000. Biogeographic effects of red fire ant invasion. *Ecology Letters* **3**, 257–261.

Gough, L., P.A. Wookey and G.R. Shaver. 2002. Dry heath arctic tundra responses to long-term nutrient and light manipulation. *Arctic, Antarctic and Alpine Research* **34**, 211–218.

Gough, L., E.A. Ramsey and D.R. Johnson. 2007. Plant-herbivore interactions in Alaskan arctic tundra change with soil nutrient availability. *Oikos* **116**, 407–418.

Grace, J.B., T.M. Anderson, M.D. Smith, et al. 2007. Does species diversity limit productivity in natural grassland communities? *Ecology Letters* **10**, 680–689.

Gratton, C. and J. Vander Zanden. 2009. Flux of aquatic insect productivity to land: comparison of lentic and lotic systems. *Ecology* **90**, 2689–2699.

Gratton, C., J. Donaldson and M.J. Vander Zanden. 2008. Ecosystem linkages between lakes and the surrounding terrestrial landscape in northeast Iceland. *Ecosystems* **11**, 764–774.

Grayston, S.J., S.Q. Wang, C.D. Campbell, et al. 1998. Selective influence of plant species on microbial diversity in the rhizosphere. *Soil Biology and Biochemistry* **30**, 369–378.

Green, P.T., D.J. O'Dowd and P.S. Lake. 2008. Recruitment dynamics in a rainforest seedling community: context-independent impact of a keystone consumer. *Oecologia* **156**, 373–385.

Greenfield, L.G. 1999. Weight loss and release of mineral nitrogen from decomposing pollen. *Soil Biology and Biochemistry* **31**, 353–361.

Griffiths, B.S., K. Ritz, R.D. Bardgett, et al. 2000. Ecosystem response of pasture soil communities to fumigation-induced microbial diversity reductions: an examination of the biodiversity-ecosystem function relationship. *Oikos* **90**, 279–294.

Griffiths, B.S., K. Ritz, R. Wheatley, et al. 2001. An examination of the biodiversity-ecosystem function relationship in arable soil microbial communities. *Soil Biology and Biochemistry* **33**, 1713–1722.

Grime, J.P. 1977. Evidence for the existence of three primary strategies of plants and its relevance to ecological and evolutionary theory. *American Naturalist* **111**, 1169–1194.

Grime, J.P. 1979. *Plant Strategies and Vegetation Processes*. John Wiley, Chichester, UK.

Grime, J.P. 1998. Benefits of plant diversity to ecosystems: immediate, filter and founder effects. *Journal of Ecology* **86**, 902–910.

Grime, J.P. 2001. *Plant Strategies, Vegetation Processes and Ecosystem Functioning*. Wiley, Chichester.

Grime, J.P., J.M.L. Mackey, S.N. Hillier, et al. 1987. Floristic diversity in a model system using experimental microcosms. *Nature* **328**, 420–422.

Grime, J.P., J.H.C. Cornelissen, K. Thompson, et al. 1996. Evidence of a causal connection between anti-herbivore defence and the decomposition rate of leaves. *Oikos* **77**, 489–494.

Grime, J.P., K. Thompson, R. Hunt, et al. 1997. Integrated screening validates primary axes of specialization in plants. *Oikos* **79**, 259–281.

Groffman, P.M., C.T. Driscoll, T.J. Fahey, et al. 2001. Colder soils in a warmer world: A snow manipulation study in a northern hardwood forest ecosystem. *Biogeochemistry* **56**, 135–150.

Grundmann, G.L. and D. Debouzie. 2000. Geostatistical analysis of the distribution of NH4 and NO2 oxidizing bacteria and serotypes at the millimeter scale along a soil transect. *FEMS Microbiology Ecology* **34**, 57–62.

Güsewell, S. and M.O. Gessner. 2009. N:P ratios influence litter decomposition and colonization by fungi and bacteria in microcosms. *Functional Ecology* **23**, 211–219.

Güsewell, S., W. Koerselman and T.A. Verhoeven. 2003. Biomass N : P ratios as indicators of nutrient limitation for plant populations in wetlands. *Ecological Applications* **13**, 372–384.

Güsewell, S., G. Jacobs and E. Weber. 2006. Native and introduced populations of *Solidago gigantea* differ in shoot production but not in leaf traits or litter decomposition. *Functional Ecology* **20**, 575–584.

Guitian, R. and R.D. Bardgett. 2000. Plant and soil microbial responses to defoliation in temperate semi-natural grassland. *Plant and Soil* **220**, 271–277.

Gundale, M.J. 2002. Influence of exotic earthworms on the soil organic horizon and the rare fern *Botrychium mormo*. *Conservation Biology* **16**, 1555–1561.

Gundale, M.J., D.A. Wardle and M.-C. Nilsson. 2010. Vascular plant removal effects on biological N-fixation vary across a boreal forest island gradient. *Ecology*, in press.

Gunn, A., F.L. Miller, S.J. Barry, et al. 2006. A near-total decline in caribou on Prince of Wales, Somerset and Russell Islands, Canadian Arctic. *Arctic* **59**, 1–13.

Guthrie, R.D. 2003. Rapid body size decline in Alaskan horses before extinction. *Nature* **426**, 169–171.

Hairston, N.G., F.E. Smith and L.B. Slobodkin. 1960. Community structure, population control and competition. *American Naturalist* **94**, 421–425.

Halaj, J. and D.H. Wise. 2001. Terrestrial trophic cascades: how much do they trickle? *American Naturalist* **157**, 262–281.

Hale, C.M., L.E. Frelich and P.B. Reich. 2005. Exotic European earthworm invasion dynamics in northern hardwood forests of Minnesota, USA. *Ecological Applications* **15**, 848–860.

Hale, C.M., L.E. Frelich and P.B. Reich. 2006. Changes in hardwood forest understory plant communities in response to European earthworm invasions. *Ecology* **87**, 1637–1649.

Halvorson, J.J., J.L. Smith and E.H. Franz. 1991. Lupine influence on soil carbon, nitrogen and microbial activity in developing ecosystems at Mt St Helens. *Oecologia* **87**, 162–170.

Hamilton, E.W. and D.A. Frank. 2001. Can plants stimulate soil microbes and their own nutrient supply? Evidence from a grazing tolerant grass. *Ecology* **82**, 239–244.

Hamilton, E.W., D.A. Frank, P.M. Hinchey, et al. 2008. Defoliation induces root exudation and triggers positive rhizospheric feedbacks in a temperate grassland. *Soil Biology and Biochemistry* **40**, 2865–2873.

Han, G., X. Hao, M. Zhao, et al. 2008. Effect of grazing intensity on carbon and nitrogen in soil and vegetation in a meadow steppe in Inner Mongolia. *Agriculture, Ecosystems and Environment* **125**, 21–32.

Handley, W.R.C. 1954. *Mull and Mor in Relation to Forest Soils*. Her Majesty's Stationery Office, London.

Handley, W.R.C. 1961. Further evidence for the importance of residual leaf protein complexes in litter decomposition and the supply of nitrogen for plant growth. *Plant and Soil* **15**, 37–73.

Hanley, M.E., S. Trofmov and G. Taylor. 2004. Species-level effects more important than functional group-level responses to elevated CO_2: evidence from simulated turves. *Functional Ecology* **18**, 304–313.

Hanlon, R.D.G. and J.M. Anderson. 1979. Effects of Collembola grazing on microbial activity in decomposing leaf litter. *Oecologia* **38**, 93–99.

Hansen, R.A. 1999. Red oak litter promotes a microarthropod functional group that accelerates its decomposition. *Plant and Soil* **209**, 37–45.

Hansen, R.A. 2000. Effect of habitat complexity and composition on a diverse litter microarthropod assemblage. *Ecology* **81**, 1120–1132.

Hanski, I. and Y. Cambefort. 1991. *Dung Beetle Ecology*. Princeton University Press, Princeton, NJ.

Harley, J.L. 1969. *The Biology of Mycorrhiza*. Leonard Hill, London.

Harper, J.L. 1977. *Population Biology of Plants*. Academic Press, London.

Harris, W.V. 1964. *Termites: Their Recognition and Control*. Longman, London.

Harrison, K.A. and R.D. Bardgett. 2004. Browsing by red deer negatively impacts on soil nitrogen availability in regenerating woodland. *Soil Biology and Biochemistry* **36**, 115–126.

Harrison, K.A., R. Bol and R.D. Bardgett. 2007. Preferences for uptake of different nitrogen forms by co-existing plant species and soil microbes in temperate grasslands. *Ecology* **88**, 989–999.

Harrison, K.A., R. Bol and R.D. Bardgett. 2008. Do plant species with different growth strategies vary in their ability to compete with soil microbes for chemical forms of nitrogen? *Soil Biology and Biochemistry* **40**, 228–237.

Harsch, M.A., P.E. Hulme, M.S. McGlone, et al. 2009. Are treelines advancing? A global meta-analysis of treeline response to climate warming. *Ecology Letters* **12**, 1040–1049.

Hartley, I.P., D.W. Hopkins, M.H. Garnett, et al. 2008. Soil microbial respiration in arctic soil does not acclimate to temperature. *Ecology Letters* **11**, 1092–1100.

Hartley, S.E. and C.J. Jones. 1997. Plant chemistry and herbivory, or why is the world green? In: *Plant Ecology* (M.J. Crawley, ed.), pp. 284–324. Blackwell Science, Oxford.

Hartnett, D.C. and G.W.T. Wilson. 1999. Mycorrhizae influence plant community structure and diversity in tallgrass prairie. *Ecology* **80**, 1187–1195.

Hassall, M., J.G. Turner and M.R.W. Rands. 1987. Effects of terrestrial isopods on the decomposition of woodland leaf litter. *Oecologia* **72**, 597–604.

Hättenschwiler, S. and P. Gasser. 2005. Soil animals alter plant litter effects on decomposition. *Proceedings of the National Academy of Sciences USA* **102**, 1519–1524.

Hättenschwiler, S. and P.M. Vitousek. 2000. The role of polyphenols in terrestrial ecosystem nutrient cycling. *Trends in Ecology and Evolution* **15**, 238–243.

Hättenschwiler, S., A.V. Tiunov and S. Scheu. 2005. Biodiversity and litter decomposition in terrestrial ecosystems. *Annual Reviews of Ecology, Evolution and Systematics* **36**, 191–218.

Hättenschwiler, S., B. Aeschlimann, M.-M. Coûteaux, et al. 2008. High variation in foliage and leaf litter chemistry among 45 tree species in a neotropical rainforest community. *New Phytologist* **179**, 165–175.

Hawes, C., A.J.A. Stewart and H.F. Evans. 2002. The impact of wood ants (*Formica rufa*) on the distribution and abundance of ground beetles (Coleoptera: Carabidae) in a Scots pine plantation. *Oecologia* **131**, 612–619.

Haystead, A., N. Malajczuk and T.S. Grove. 1988. Underground transfer of nitrogen between pasture plants infected with arbuscular mycorrhizal fungi. *New Phytologist* **108**, 417–423.

He, N., L. Wu, Y. Wang, et al. 2009. Changes in carbon and nitrogen in soil particle-size fractions along a grassland restoration chronosequence in northern China. *Geoderma* **150**, 302–308.

He, Z., T.J. Gentry, C.W. Schadt, et al. 2007. GeoChip: a comprehensive microarray for investigating biogeochemical, ecological and environmental processes. *The ISME Journal* **1**, 67–77.

Heath. J., E. Ayres, M. Possell, et al. 2005. Rising atmospheric CO_2 reduces soil carbon sequestration. *Science* **309**, 1711–1713.

Hector, A., B. Schmid, C. Beierkuhnlein, et al. 1999. Plant diversity and productivity experiments in European grasslands. *Science* **286**, 1123–1127.

Hedlund, K. and M.S. Öhrn. 2000. Tritrophic interactions in a soil community enhance decomposition rates. *Oikos* **88**, 585–591.

Hedlund, K., L. Boddy and C.M. Preston. 1991. Mycelial responses of the soil fungus *Mortierella isabellina* to grazing by *Onychiurus armatus* (Collembola). *Soil Biology and Biochemistry* **23**, 361–366.

Hedlund, K., I.S. Regina, W.H. Van der Putten, et al. 2003. Plant species diversity, plant biomass and responses of the soil community on abandoned land across Europe: idiosyncracy or above-belowground time lags. *Oikos* **103**, 45–58.

Heemsbergen, D.A., M.P. Berg, M. Loreau, et al. 2004. Biodiversity effects on soil processes explained by interspecific functional dissimilarity. *Science* **306**, 1019–1020.

Heimann, M. and M. Reichstein. 2008. Terrestrial ecosystem carbon dynamics and climate feedbacks. *Nature* **451**, 289–292.

Helfield, J.M. and R.J. Naiman. 2002. Salmon and alder as niutrofgen sources to riparian forests in a boreal Alaska watershed. *Oecologia* **133**, 573–582.

Helfield, J.M. and R.J. Naiman. 2006. Keystone interactions; salmon and bear in riparian forests in Alaska. *Ecosystems* **9**, 167–180.

Hendriksen, N.B. 1990. Leaf litter selection by detritivore and geophagous earthworms. *Biology and Fertility of Soils* **10**, 17–21.

Hendrix, P.F., R.W. Parmelee, D.A. Crossley, et al. 1986. Detritus food webs in conventional and no-tillage agroecosystems. *BioScience* **36**, 374–380.

Hendrix, P.F., M.A. Callaham, J.M. Drake, et al. 2008. Pandora's Box contained bait: the global problem of introduced earthworms. *Annual Review of Ecology and Systematics* **39**, 593–613.

Heneghan, L., D.C. Coleman, X. Zou, et al. 1999. Soil microarthropod contributions to decomposition dynamics: Tropical-temperate comparisons of a single substrate. *Ecology* **80**. 1873–1882.

Henry, H., J.D. Juarez, C.B. Field, et al. 2005. Interactive effects of elevated CO_2, N deposition and climate change on extracellular enzyme activity and soil density fractionation in a Californian annual grassland. *Global Change Biology* **11**, 1808–1815.

Herbert, D.A., J.H. Fownes and P.M. Vitousek. 1999. Hurricane damage to a Hawaiian forest: Nutrient supply rate affects resistance and resilience. *Ecology* **80**, 908–920.

Herms, D.A. and W.J. Mattson. 1992. The dilemma of plants: to grow or defend. *Quarterly Review of Biology* **67**, 283–335.

Hickling, R., D.B. Roy, J.K. Hill, et al. 2006. The distributions of a wide range of taxonomic groups are expanding polewards. *Global Change Biology* **12**, 450–455.

Hill, J.K., Y.C. Collingham, C.D. Thomas, et al. 2001. Impacts of landscape structure on butterfly range expansion. *Ecology Letters* **4**, 313–321.

Hobbie, J.E. and E.A. Hobbie. 2006. N-15 in symbiotic fungi and plants estimates nitrogen and carbon flux rates in Arctic tundra. *Ecology* **87**, 816–822.

Hobbie, S.E. 1992. Effects of plant species on nutrient cycling. *Trends in Ecology and Evolution* **7**, 336–339.

Hobbie, S.E. and P.M. Vitousek. 2000. Nutrient limitation of decomposition in Hawaiian forests. *Ecology* **81**, 1867–1877.

Hobbie, S.E., A. Shevtsova and F.S. Chapin. 1999. Plant responses to species removal and experimental warming in Alaskan tussock tundra. *Oikos* **84**, 417–434.

Hobbie, S.E., J.P. Schimel, S.E. Trumbore, et al. 2001. Controls over carbon storage and turnover in high-latitude soils. *Global Change Biology* **6**, 196–210.

Hobbie, S.E., L. Gough and G.R. Shaver, G.R. 2005. Species compositional differences on different-aged glacial landscapes drive contrasting responses of tundra to nutrient addition. *Journal of Ecology* **93**, 770–782.

Hobbie, S.E., P.B. Reich, J. Oleksyn, et al. 2006. Tree species effects on decomposition and forest floor dynamics in a common garden. *Ecology* **87**, 2288–2297.

Hocking, M.D. and T.E. Reimchen. 2002. Salmon-derived nitrogen in terrestrial invertebrates from coniferous forests of the Pacific Northwest. *BMC Ecology* **2**, 4.

Hodge, A., C.D. Campbell and A.H. Fitter. 2001. An arbuscular mycorrhizal fungus accelerates decomposition and acquires nitrogen directly from organic material. *Nature* **413**, 297–299.

Hodkinson, I.D., S.J. Coulson, N.R. Webb, et al. 1996. Can high Arctic soil microarthropods survive elevated summer temperatures? *Functional Ecology* **10**, 314–321.

Hodkinson, I.D., S.J. Coulson, N.R. Webb, et al. 2004. Invertebrate community structure across proglacial chronosequences in the high Arctic. *Journal of Animal Ecology* **73**, 556–568.

Högberg, M.N., Y. Chen and P. Högberg. 2007. Gross nitrogen mineralisation and fungi-to-bacteria ratios are negatively correlated in boreal forests. *Biology and Fertility of Soils* **44**, 363–366.

Högberg, P. 1992. Root symbioses of trees in African dry tropical forests. *Journal of Vegetation Science* **3**, 393–400.

Högberg, P. and D.J. Read. 2006. Towards a more plant physiological perspective on soil ecology. *Trends in Ecology and Evolution* **21**, 548–554.

Högberg, P., A. Nordgren, N. Buchmann, et al. 2001. Large-scale forest girdling shows that current photosynthesis drives soil respiration. *Nature* **411**, 789–792.

Högberg, P., M.N. Högberg, S.G. Gottlicher, et al. 2008. High temporal resolution tracing of photosynthate carbon from the tree canopy to forest soil microorganisms. *New Phytologist* **177**, 220–228.

‌

Hokka, V., J. Mikola, M. Vestberg, et al. 2004. Interactive effects of defoliation and an AM fungus on plants and soil organisms in experimental legume-grass communities. *Oikos* **106**, 73–84.

Holland, E.A. and J.K. Detling. 1990. Plant response to herbivory and below-ground nitrogen cycling. *Ecology* **71**, 1040–1049.

Holland, E.A., F.J. Dentener and B.H. Braswell, et al. 1999. Contemporary and pre-industrial global reactive nitrogen budgets. *Biogeochemistry* **46**, 7–43.

Holland J.N., W. Cheng and D.A. Crossley Jr. 1996. Herbivore-induced changes in plant carbon allocation: assessment of below-ground C fluxes using carbon-14. *Oecologia* **107**, 87–94.

Hölldobler, B. and E. Wilson. 1990. *The Ants*. Springer-Verlag, Berlin.

Holtgrieve, G.W., D.E. Schindler and P.K. Jewett. 2009. Large predators and biogeochemical hotspots: brown bear (*Ursus arctos*) predation on salmon alters nitrogen cycling in riparian soils. *Ecological Research* **24**, 1125–1135.

Hoogerkamp, M., H. Rogaar and H.J.P. Eijsackers. 1983. Effect of earthworms on grassland ion recently reclaimed polder soils in the Netherlands. In: *Earthworm Ecology: From Darwin to Vermiculture* (J.E. Satchell, ed.), pp. 85–105. Chapman and Hall, London.

Hooper, D.U. and J.S. Dukes. 2004. Overyielding among plant functional groups in a long-term experiment. *Ecology Letters* **7**, 95–105.

Hooper, D.U. and P.M. Vitousek. 1997. The effects of plant composition and diversity on ecosystem processes. *Science* **277**, 1302–1305.

Hooper, D.U. and P.M. Vitousek. 1998. Effects of plant composition and diversity on nutrient cycling. *Ecological Monographs* **68**, 121–149.

Hooper, D.U., D.E. Bignell, V.K. Brown, et al. 2000. Interactions between aboveground and belowground biodiversity in terrestrial ecosystems: patterns, mechanisms and feedbacks. *BioScience* **50**, 1049–1061.

Hooper, D.U., F.S. Chapin, J.J. Ewel, et al. 2005. Effects of biodiversity on ecosystem functioning: a consensus of current knowledge and needs for future research. *Ecological Monographs* **75**, 3–35.

Hoorens, B., R. Aerts and M. Strogenga. 2003. Does initial leaf chemistry explain litter mixture effects on decomposition? *Oecologia* **442**, 578–586.

Hopkins, A. and R. Wilkins. 2006. Temperate grassland: key developments in the last century and future perspectives. *Journal of Agricultural Science* **144**, 503–523.

Hopp, H. and Slater, C.S. 1948. Influence of earthworms on soil productivity. *Soil Science* **66**, 421–428.

Houghton, J.T., L.G. Meira Filho, B.A. Callender. 1996. *Climate Change 1995. The Science of Climate Change*. Intergovernmental Panel on Climate Change. Cambridge University Press, Cambridge.

Houlton, B.Z., D.M. Sigman and L.O. Hedin. 2006. Isotopic evidence for large gaseous nitrogen losses from tropical rainforests. *Proceedings of the National Academy of Sciences USA* **103**, 8745–8750.

Houlton, B.Z., Y.P. Wang, P.M. Vitousek, et al. 2008. A unifying framework for dinitrogen fixation in the terrestrial biosphere. *Nature* **454**, 327–334.

Hu, S., F.S. Chapin, M.K. Firestone, et al. 2001. Nitrogen limitation of microbial decomposition in a grassland under elevated CO_2. *Nature* **409**, 188–191.

Hudson, M.E. 2008. Sequencing breakthroughs for genomic ecology and evolutionary biology. *Molecular Biology Resources* **8**, 3–17.

Hungate, B.A., E.A. Holland, R.B. Jackson, et al. 1997. On the fate of carbon in grasslands under carbon dioxide enrichment. *Nature* **388**, 576–579.

Hunt, H.W. and D.H. Wall. 2002. Modelling the effects of loss of soil biodiversity on ecosystem function. *Global Change Biology* **8**, 33–50.

Hunt, H.W., D.C. Coleman, E.R. Ingham, et al. 1987. The detrital food web in shortgrass prairie. *Biology and Fertility of Soils* **3**, 57–68.

Hunt, H.W., E.R. Ingham, D.C. Coleman, et al. 1988. Nitrogen limitation of production and decomposition in prairie, mountain meadow and pine forest. *Ecology* **69**, 1009–1016.

Hunter, M.D. and R.E. Forkner. 1999. Hurricane damage influences foliar nutrient phenolics and subsequent herbivory on surviving trees. *Ecology* **80**, 2676–2682.

Huntly, N. 1991. Herbivores and the dynamics of communities and ecosystems. *Annual Review of Ecology and Systematics* **22**, 477–503.

Huston, M.A. 1993. Biological diversity, soils and economics. *Science* **262**, 1676–1680.

Huston, M.A. 1994. *Biological Diversity. The Coexistence of Species on Changing Landscapes*. Cambridge University Press, Cambridge.

Huston, M.A. 1997. Hidden treatments in ecological experiments: re-evaluating the ecosystem function of biodiversity. *Oecologia* **110**, 449–460.

Huston, M.A. and D.L. De Angelis. 1994. Competition and coexistence: the effects of resource transport and supply rates. *American Naturalist* **144**, 854–877.

Huston, M.A. and S. Wolverton. 2009. The global distribution of net primary production: resolving the paradox. *Ecological Monographs* **79**, 343–377.

Hyodo, F. and D.A. Wardle. 2009. Effect of ecosystem retrogression on stable nitrogen and carbon isotopes of plants, soils and consumer organisms in boreal forest islands. *Rapid Communications in Mass Spectrometry* **23**, 1892–1898.

Hyodo, F., T. Inour, J.I. Azuma, et al. 2000. Role of the mutualistic fungus in lignin degradation of the soil fungus *Macromycetes gilvus* (Isoptera: Macrotermitin ae). *Soil Biology and Biochemistry* **32**, 653–658.

Ilieva-Makulec, K. and G. Makulec. 2002. Effect of the earthworm *Lumbricus rubellus* on the nematode community in a peat meadow soil. *European Journal of Soil Biology* **38**, 59–62.

Ilmarinen, K., J. Mikola, M. Nieminen, et al. 2005. Does plant growth phase determine the response of plants and soil organisms to defoliation? *Soil Biology and Biochemistry* **37**, 433–443.

Ingham, E.R., J.A. Trofymow, R.N. Ames, et al. 1986. Trophic interactions and nitrogen cycling in a semiarid grassland soil. 2. System responses to removal of different groups of soil microbes or fauna. *Journal of Applied Ecology* **23**, 615–630.

Ingham, R.E., J.A. Trofymow, E.R. Ingham, et al. 1985. Interactions of bacteria, fungi, and their nematode grazers: effects on nutrient cycling and plant growth. *Ecological Monographs* **55**, 119–140.

Insam, H. and K. Haselwandter. 1989. Metabolic quotient of the soil microflora in relation to plant succession. *Oecologia* **79**, 174–178.

Insam, H., C.C. Mitchell and J.F. Dormaar. 1991. Relationship of soil microbial biomass and activity with fertilization practice and crop yield of three Ultisols. *Soil Biology and Biochemistry* **23**, 459–464.

IPCC. 2007 *Climate Change 2007: The Physical Science Basis, Contribution of Working Group I to the Fourth Assessment Report of the Intergovernmental Panel on Climate Change*. Cambridge University Press, Cambridge.

Ishida, T.A., K. Nara, M. Tanaka, et al. 2008. Germination and infectivity of ectomycorrhizal fungal spores in relation to their ecological traits during primary succession. *New Phytologist* **180**, 491–500.

Iversen, C.M., J. Ledford and R.J. Norby. 2008. CO_2 enrichment increases carbon and nitrogen input from fine roots in a deciduous forest. *New Phytologist* **179**, 837–847.

Ives, A.R. and S.R. Carpenter. 2007. Stability and diversity of ecosystems. *Science* **317**, 58–62.

Ives, A.R., A. Einarsson, V.A.A. Jansen, et al. 2008. High amplitude fluctuations and alternative dynamical states of midges in Lake Myvatn. *Nature* **452**, 84–87.

Jackson, L.E., D.S. Schimel and M.K. Firestone. 1989. Short-term partitioning of ammonium and nitrate between plants and microbes in an annual grassland. *Soil Biology and Biochemistry* **21**, 409–415.

Jackson, R.B., J. Canadell, J.R. Ehleringer, et al. 1996. A global analysis of root distributions for terrestrial biomes. *Oecologia* **108**, 389–411.

Jackson, R.B., J.L. Banner, E.G. Jobbagy, et al. 2002. Ecosystem carbon loss with woody plant invasion of grasslands. *Nature* **418**, 623–626.

Jackson, R.B., C.W. Cook, J.S. Pippen, et al. 2009. Increased belowground biomass and soil CO_2 fluxes after a decade of carbon dioxide enrichment in a warm-temperate forest. *Ecology* **90**, 3352–3366.

Jaeger, C.H., R.K. Monson, M.C. Fisk, et al. 1999. Seasonal partitioning of nitrogen by plants and soil microorganisms in an alpine ecosystem. *Ecology* **80**, 1883–1891.

Jefferies, R.L. 1998. Pattern and process in arctic coastal vegetation in response to foraging by lesser snow geese. In: *Plant Form and Vegetation Structure: Adaptation, Plasticity and Relationships with Herbivory* (M.P.J. Werger, P.J.M. Van der Aart, H.J. During, et al., eds), pp. 281–300. SPB Publishing, The Hague.

Jenkinson, D.S., D.E. Adams and A. Wild. 1991. Model estimates of CO2 emissions from soil in response to global warming. *Nature* **351**, 304–306.

Jenny, H. 1941. *Factors of Soil Formation*. McGraw Hill, New York.

Jensen, R.A., J. Madsen, M. O'Connell, et al. 2008. Prediction of the distribution of Arctic nesting pink-footed geese under a warmer climate scenario. *Global Change Biology* **14**, 1–10.

Jentschke, G., M. Bonkowski, D.L. Godbold, et al. 1995. Soil protozoa and plant growth:nonnutritional effects and interaction with mycorrhizas. *Biology and Fertility of Soils* **20**, 263–269.

Jewell, P.L., D. Käuferle, N.R. Berry, et al. 2007. Redistribution of phosphorus by cattle on a traditional mountain pasture in the Alps. *Agriculture, Ecosystems and Environment* **122**, 377–386.

Johnson, D.J.R. Leake, N. Ostle, et al. 2002. In situ $^{13}CO_2$ pulse labelling of upland grassland demonstrates a rapid pathway of carbon flux from arbuscular mycorrhizal mycelia to soil. *New Phytologist* **153**, 327–334.

Johnson, D., M. Krsek, E.M. Wellington, et al. 2005. Soil invertebrates disrupt carbon flow through fungal networks. *Science* **309**, 1047.

Johnson, K.H. 2000. Trophic-dynamic considerations relating species diversity to ecosystem resilience. *Biological Reviews* **75**, 347–376.

Johnson, L.C. and J.R. Matchett. 2001. Fire and grazing regulate belowground processes in tall grass prairie. *Ecology* **82**, 3377–3389.

Johnson, N.C., J.H. Graham and F.A. Smith. 1997. Functioning of mycorrhizal associations along the mutualism-parasitism continuum. *New Phytologist* **135**, 575–586.

Jonasson, S., A. Michelsen and I.K. Schmidt. 1999. Coupling of nutrient cycling and carbon dynamics in the Arctic; integration of soil microbial and plant processes. *Applied Soil Ecology* **11**, 135–146.

Jones, C.G. and J.H. Lawton (eds). 1995. *Linking Species and Ecosystems*. Chapman and Hall, New York.

Jones, D.L. and K. Kielland. 2002. Soil amino acid turnover dominates the nitrogen flux in permafrost-dominated taiga forest soils. *Soil Biology and Biochemistry* **34**, 209–219.

Jones, D.L., A. Hodge and Y. Kuzyakov. 2004. Plant and mycorrhizal regulation of rhizodeposition. *New Phytologist* **163**, 459–480.

Jones, D.L., J.R. Healey, V.B. Willett, et al. 2005. Dissolved organic nitrogen uptake by plants – An important N uptake pathway? *Soil Biology and Biochemistry* **37**, 413–423.

Jones, T.H., L.J. Thompson, J.H. Lawton, et al. 1998. Impacts of rising atmospheric carbon dioxide on model terrestrial ecosystems. *Science* **280**, 441–443.

Jónsdóttir, I.S., B. Magnússon, J. Gudmundsson, et al. 2005. Variable sensitivity of plant communities in Iceland to experimental warming. *Global Change Biology* **11**, 553–563.

Jonsson, L.M., M.-C. Nilsson, D.A. Wardle, et al. 2001. Context dependent effects of ectomycorrhizal species richness on tree seedling productivity. *Oikos* **93**, 353–364.

Jonsson, M. and D.A. Wardle. 2008. Context dependency of litter-mixing effects on decomposition and nutrient release across a long-term chronosequence. *Oikos* **117**, 1674–1682.

Jonsson, M. and D.A. Wardle. 2009. The influence of freshwater-lake subsidies on invertebrates occupying terrestrial vegetation. *Acta Oecologica* **35**, 698–704.

Jonsson, M. and D.A. Wardle. 2010. Structural equation modelling reveals plant-community drivers of carbon storage in boreal forest ecosystems. *Biology Letters* **6**, 116–119.

Jonsson, M., O. Dangles, B., Malmqvist, et al. 2002. Simulating species loss following perturbation: assessing the effects on process rates. *Proceedings of the Royal Society of London Series B Biological Sciences* **269**, 1047–1052.

Jumpponen, A. and K.L. Jones. 2009. Massively parallel 454 sequencing indicates hyperdiverse fungal communities in temperate *Quercus macrocarpa* phyllosphere. *New Phytologist* **184**, 438–448.

Jurgensen, M.F., L. Finér, T. Domisch, et al. 2008. Organic mound-building ants: their impact on soil properties in temperate and boreal forests. *Journal of Applied Entomology* **132**, 266–275.

Kahmen, A., A. Renker, S.B. Unsicker, et al. 2006. Niche complementarity for nitrogen: an explanation for the biodiversity and ecosystem functioning relationship. *Ecology* **87**, 1244–1255.

Kaiser, J. 2000. Rift over biodiversity divides ecologists. *Science* **289**, 1282–3.

Kajak, A., K. Chmielewski, M. Kaczmarek, et al. 1993. Experimental studies on the effect of epigeic predators on matter decomposition process in managed peat grassland. *Polish Ecological Studies* **17**, 289–310.

Kampichler, C and A. Bruckner. 2009. The role of microarthropods in terrestrial decomposition: a meta-analysis of 40 years of litterbag studies. *Biological Reviews* **84**, 375–389.

Kandeler, E., D. Tscherko, R.D. Bardgett, et al. 1998. The response of soil microorganisms and roots to elevated CO_2 and temperature in a terrestrial model ecosystem. *Plant and Soil* **202**, 251–262.

Kardol, P., T.M. Bezemer and W.H. Van der Putten. 2006. Temporal variation in plant-soil feedback controls succession. *Ecology Letters* **9**, 1080–1088.

Kardol, P., N.J. Cornips, M.M.L. van Kempen, et al. 2007. Microbe-mediated plant-soil feedback causes historical contingency effects in plant community assembly. *Ecological Monographs* 77, 147–162.

Kardol, P., M.A. Creggor, C.E. Campany, et al. 2010. Soil ecosystem functioning under climate change: plant species community effects. *Ecology*, in press.

Kaspari, M., M.N. Garcia, K.E. Harms, et al. 2008. Multiple nutrients limit litterfall and decomposition in a tropical forest. *Ecology Letters* 11, 35–43.

Kaufmann R. 2001. Invertebrate succession on an alpine glacier foreland. *Ecology* 82, 2261–2278.

Kauserud, H., L.F. Stige, J.O. Vik, et al. 2008. Mushroom fruiting and climate change. *Proceedings of the National Academy of Sciences USA* 105, 3811–3814.

Kaye, J.P. and S.C. Hart. 1997. Competition for nitrogen between plants and soil microorganisms. *Trends in Ecology and Evolution* 12, 139–143.

Keane, R.M. and M.J. Crawley. 2002. Exotic plant invasions and the enemy release hypothesis. *Trends in Ecology and Evolution* 17, 164–170.

Keeley, J.E. 1988. Allelopathy. *Ecology* 69, 262–263.

Keith, A.M., R. Van der Wal, R.W. Brooker, et al. 2008. Increasing litter species richness reduces variability in a terrestrial decomposer system. *Ecology* 89, 2657–2664.

Kenis, M., M.A. Auger-Rozenberg, A. Roques, et al. 2009. Ecological effects of invasive alien insects. *Biological Invasions* 11, 21–45.

Kielland, K. 1994. Amino acid absorption by arctic plants: Implications for plant nutrition and nitrogen cycling. *Ecology* 75, 2375–2383.

Kielland, K. and J.P. Bryant. 1998. Moose herbivory in Taiga: Effects on biochemistry and vegetation dynamics in primary succession. *Oikos* 82, 377–383.

Killingbeck, K.T. 1996. Nutrients in senesced leaves: keys to the search for potential resorption and resorption proficiency. *Ecology* 77, 1716–1727.

King, L.K. and K.J. Hutchinson. 1976. The effects of sheep stocking intensity on the abundance and distribution of mesofauna in pastures. *Journal of Applied Ecology* 13, 41–55.

King, L.K., K.J. Hutchinson and P. Greenslade. 1976. The effects of sheep numbers on associations of Collembola in sown pastures. *Journal of Applied Ecology* 13, 731–739.

Kirschbaum, M.U.F. 2004. Soil respiration under prolonged soil warming: are rate reductions caused by acclimation or substrate loss? *Global Change Biology* 10, 1870–1877.

Kirschbaum, M.U.F. 2006. The temperature dependence of organic-matter decomposition—still a topic of debate. *Soil Biology and Biochemistry* 38, 2510–2518.

Klanderud, K. and H.J.B. Birks. 2003. Recent increases in species richness and shifts in altitudinal distributions of Norwegian mountain plants. *The Holocene* 13, 1–6.

Kleb, H. and S.D. Wilson. 1997. Vegetation effects on soil resource heterogeneity in prairie and forest. *American Naturalist* 150, 283–298.

Klironomos, J.N. 2002. Feedback with soil biota contributes to plant rarity and invasiveness in communities. *Nature* 417, 67–70.

Klironomos, J.N. 2003. Variation in plant response to native and exotic arbuscular mycorrhizal fungi. *Ecology* 84, 2292–2301.

Klironomos, J.N., M.C. Rillig, M.F. Allen, et al. 1997. Soil fungal-arthropod responses to *Populus tremuloides* grown under enriched atmospheric CO_2 under field conditions. *Global Change Biology* 3, 473–478.

Klironomos, J., M.C. Rillig and M.F. Allen. 1999. Designing belowground field experiments with the help of semi-variance and power analysis. *Applied Soil Ecology* 12, 227–238.

Klironomos, J., J. McCune and P. Moutoglis. 2004. Species of arbuscular mycorrhizal fungi affect mycorrhizal responses to simulated herbivory. *Applied Soil Ecology* **26**, 133–141.

Klopatek, C.C., E.G. O'Neill, D.W. Freckman, et al. 1992. The sustainable biosphere initiative: a commentary from the U.S. Soil Ecology Society. *Bulletin of the Ecological Society of America* **73**, 223–228.

Klumpp, K., J.F. Soussana and R. Falcimagne. 2007. Effects of past and current disturbance on carbon cycling in grassland mesocosms. *Agriculture, Ecosystems & Environment* **121**, 59–73.

Klumpp, K., S. Fontaine and E. Attard. 2009. Grazing triggers soil carbon loss by altering plant roots and their control on soil microbial community. *Journal of Ecology* **97**, 876–885.

Knapp, A.K. and M.D. Smith. 2001. Variation among biomes in temporal dynamics of aboveground primary production. *Science* **291**, 481–484.

Knapp, A.K., S.L. Conrad and J.M. Blair. 1998. Determinants of soil CO_2 flux from a sub-humid grassland: effect of fire and fire history. *Ecological Applications* **8**, 760–770.

Knapp, A.K., J.M. Blair, J.M. Briggs, et al. 1999. The keystone role of bison in North American tallgrass prairie. *BioScience* **49**, 39–50.

Knapp, A.K., J.M., Briggs, S.L. Collins, et al. 2008. Shrub encroachment in North American grasslands: shifts in growth form dominance rapidly alters control of ecosystem carbon inputs. *Global Change Biology* **14**, 615–623.

Knevel, I.C., T. Lans, F.B.J. Menting, et al. 2004. Release from native root herbivores and biotic resistance by soil pathogens in a new habitat both affect the alien *Ammophila arenaria* in South Africa. *Oecologia* **141**, 502–510.

Knight, D., P.W. Elliot, J.M. Anderson, et al. 1992. The role of earthworms in managed, permanent pastures in Devon, England. *Soil Biology and Biochemistry* **24**, 1511–1517.

Knight, T.M., M.W. McCoy, J.M. Chase, et al. 2005. Trophic cascades across ecosystems. *Nature* **437**, 880–883.

Kobe, R.K., C.A. Lepczyk and M. Iyer. 2005. Resorption efficiency decreases with increasing green leaf nutrients in a global data set. *Ecology* **86**, 2780–2805.

Koerselman, W. and A.M.F. Meuleman. 1996. The vegetation N:P ratio: A new tool to detect the nature of nutrient limitation. *Journal of Applied Ecology* **33**, 1441–1450.

Kohler, F., F. Gillet, S. Reust, et al. 2006. Spatial and seasonal patterns of cattle habitat use in a mountain wooded pasture. *Landscape Ecology* **21**, 281–295.

Kohls, S.J., D.D. Baker, C. van Kessel, et al. 2003. An assessment of soil enrichment by actinorhizal N_2 fixation using delta ^{15}N values in a chronosequence of deglaciation at Glacier Bay, Alaska. *Plant and Soil* **254**, 11–17.

Körner, C. and J.A. Arnone. 1992. Responses to elevated carbon dioxide in artificial tropical ecosystems. *Science* **257**, 1672–1675.

Körner, C., M. Diemer, B. Schäppi, et al. 1997. The response of alpine grassland to four seasons of CO_2 enrichment: a synthesis. *Acta Oecologia* **18**, 165–175.

Korthals, G.W., P. Smilauer, C. Van Dijk, et al. 2001. Linking above- and below-ground biodiversity: abundance and trophic complexity in soil as a response to experimental plant communities on abandoned arable land. *Functional Ecology* **15**, 506–514.

Kourtev, P.S., J.G. Ehrenfeld and M. Häggblom. 2002a. Exotic species alter the microbial community structure and function in soil. *Ecology* **83**, 3152–3166.

Kourtev, P.S., J.G. Ehrenfeld and W.Z. Huang. 2002b. Enzyme activity during litter decomposition of two exotic and two native plant species in hardwood forests of New Jersey. *Soil Biology and Biochemistry* **34**, 1207–1218.

Kuikka, K., E. Härmä, A.M. Markkola, et al. 2003. Severe defoliation of Scots pine reduces reproductive investment by ectomycorrhizal symbionts. *Ecology* **84**, 2051–2061.

Kullman, L. 2002. Rapid recent range-margin rise of tree and shrub species in the Swedish Scandes. *Journal of Ecology* **90**, 68–76.

Kullman, L. and L. Öberg. 2009. Post-Little Ice Age tree line rise and climate warming in the Swedish Scandes: a landscape ecological perspective. *Journal of Ecology* **97**, 415–429.

Kulmatiski, A. and P. Kardol. 2008. Getting plant-soil feedback out of the greenhouse: experimental and conceptual approaches. *Progress in Botany* **69**, 449–472.

Kulmatiski, A., K.H. Beard, J.R. Stevens, et al. 2008. Plant–soil feedbacks: a meta-analytical review. *Ecology Letters* **11**, 980–992.

Kurokawa, H. and T. Nakashizuka. 2008. Leaf herbivory and decomposability in a Malaysian tropical rain forest. *Ecology* **89**, 2645–2656.

Kurokawa, H., D.A. Peltzer and D.A. Wardle. 2010. Plant traits, leaf palatability and litter decomposability for coexisting woody species differing in invasion status and nitrogen fixation ability. *Functional Ecology*, in press.

Kurz, W.A., C.C. Dymond, G. Stinson, et al. 2008. Mountain pine beetle and forest carbon feedback to climate change. *Nature* **452**, 987–990.

Kuzyakov, Y. 2006. Sources of CO_2 efflux from soil and review of partitioning methods. *Soil Biology and Biochemistry* **38**, 425–448.

Kytöviita, M.-M., M. Vestberg and J. Tuomi. 2003. A test of mutual aid in common mycorrhizal networks: established vegetation negates benefit in seedlings. *Ecology* **84**, 898–906.

Laakso, J. and H. Setälä. 1997. Nest mounds of red wood ants (*Formica aquilonia*): hot spots for litter-dwelling earthworms. *Oecologia* **111**, 565–569.

Laakso, J. and H. Setälä. 1998. Composition and trophic structure of detrital food web in ant nest mounds of *Formica aquilonia* and in the surrounding forest soil. *Oikos* **81**, 266–278.

Laakso, J. and H. Setälä. 1999a. Sensitivity of primary production to changes in the architecture of belowground food webs. *Oikos* **87**, 57–64.

Laakso, J. and H. Setälä. 1999b. Population- and ecosystem-effects of predation on microbial-feeding nematodes. *Oecologia* **120**, 279–286.

Lagerström, A., M.C. Nilsson, O. Zackrisson, et al. 2007. Ecosystem input of nitrogen through biological fixation in feather mosses during ecosystem retrogression. *Functional Ecology* **21**, 1027–1033.

Lal, R. 2004. Soil carbon sequestration impacts on global change and food security. *Science* **304**, 1623–1627.

Lal, R. 2009. Sequestering carbon in soils of arid ecosystems. *Land Degradation and Development* **20**, 441–454.

Lang, S.I., J.H.C. Cornelissen, T. Klahn, et al. 2009. An experimental comparison of chemical traits and litter decomposition rates in a diverse range of subarctic bryophyte, lichen and vascular plant species. *Journal of Ecology* **97**, 998–900.

Lapied, E. and P. Lavelle. 2003. The peregrine earthworm *Pontoscolex corethrurus* in the East coast of Costa Rica. *Pedobiologia* **47**, 471–474.

Lavelle, P. and A. Martin. 1992. Small-scale and large-scale effects of endogeneic earthworms on soil organic matter dynamics in soil and the humid tropics. *Soil Biology and Biochemistry* **24**, 1491–1498.

Lavelle, P., C. Lattaud, D. Trigo, et al. 1995. Mutualism and biodiversity in soils. *Plant and Soil* **170**, 23–33.

Lavelle, P., D. Bignell, M. Lepage, et al. 1997. Soil function in a changing world: The role of invertebrate ecosystem engineers. *European Journal of Soil Biology* **33**, 159-193.

Lavorel, S. and E. Garnier. 2002. Predicting changes in community composition and ecosystem functioning from plant traits: revisiting the Holy Grail. *Functional Ecology* **16**, 545–556.

Lawton, J.H. 1994. What do species do in ecosystems? *Oikos* **71**, 367–374.

Lawton, J.H. and S. McNeill. 1979. Between the devil and the deep blue sea: on the problems of being a herbivore. In: *Population Dynamics* (R.M. Anderson, B.D. Turner and L.R. Taylor, eds), pp. 223–244. Blackwells, London.

Lawton, J.H. and C.G. Jones. 1995. Linking species and ecosystems – organisms as ecosystem engineers. In: *Linking Species and Ecosystems* (C.G. Jones and J.H. Lawton, eds), pp. 141–150. Chapman and Hall, New York.

Leake, J.R. and D.J. Read. 1997. Mycorrhizal fungi in terrestrial habitats. In: *The Mycota IV. Environmental and Microbial Relationships* (D.T. Wicklow and B. Söderström, eds), pp. 281–301. Springer-Verlag, Heidelberg.

Leake, J.R., D. Johnson, D. Donnelly, et al. 2004. Networks of power and influence: the role of mycorrhizal mycelial networks in controlling plant communities and agroecosystem function. *Canadian Journal of Botany* **82**, 1016–1045.

Lee, K. 1985. *Earthworms: Their Ecology and Relationships with Soils and Land Use.* Academic Press: New York.

Leininger, A., T. Urich, M. Schloter, et al. 2006. Archaea predominate among ammonia-oxidizing prokaryotes in soils. *Nature* **442**, 806–809.

Leishman, M.R., T. Haslehurst, A. Ares, et al. 2007. Leaf trait relationships of native and invasive plants: community- and global-scale comparisons. *New Phytologist* **176**, 635–643.

Lenoir, J., J.-C. Gégout, P.A. Marquet, et al. 2008. A significant upward shift in plant species optimum elevation during the 20th century. *Science* **320**, 1768–1771.

Lenoir, L., T. Persson, J. Bengtsson, et al. 2007. Bottom-up or top-down control in forest soil microcosms? Effects of soil fauna on fungal biomass and C/N mineralization. *Biology and Fertility of Soils* **43**, 281–294.

Lensing, J.R. and D.H. Wise. 2006. Predicted climate change alters the indirect effect of predators on an ecosystem process. *Proceedings of the National Academy of Sciences USA* **103**, 15502–15505.

Leps, J. 2004. What do the biodiversity experiments tell us about consequences of plant species loss in the real world? *Basic and Applied Ecology* **5**, 529–534.

Ley, R.E. and C.M. D'Antonio. 1998. Exotic grass invasion alters potential rates of N fixation in Hawaiian woodlands. *Oecologia* **113**, 179–187.

Liao, C.Z., R.H. Peng, Y.Q. Luo, et al. 2008. Altered ecosystem carbon and nitrogen cycles by plant invasion: a meta-analysis. *New Phytologist* **117**, 706–714.

Liebhold, A.M., W.L. Macdonald, D. Bergdahl, et al. 1995. Invasion by exotic forest pests—a threat to forest ecosystems. *Forest Science* **41**, 1–49.

Liiri, M., H. Setälä, J. Haimi, et al. 2002. Relationship between soil microarthropod species diversity and plant growth does not change when the system is disturbed. *Oikos* **96**, 137–149.

Likens, G.E., F.H. Bormann, R.S. Pierce, et al. 1977. *Biogeochemistry of a Forested Ecosystem.* Springer-Verlag, New York.

Lindeman, R.L. 1942. The trophic-dynamic aspect of ecology. *Ecology* **23**, 399–418.

Lipson, D.A. and S.K. Schmidt. 2004. Seasonal changes in an alpine bacterial community in the Colorado Rocky Mountains. *Applied Environmental Microbiology* **70**, 2867–2879.

Liu, L. and T.L. Greaver. 2009. A review of nitrogen enrichment effects on three biogenic GHGs: the CO_2 sink may be largely offset by stimulated N_2O and CH_4 emission. *Ecology Letters* **12**, 1103–1117.

Logan, J.A. and J.A. Powell. 2001. Ghost forests, global warming, and the mountain pine beetle (Coleoptera: Scolytidae). *American Entomologist* **47**, 160–173.

Loo, J.A. 2009. Ecological impacts of non-indigenous invasive fungi as forest pathogens. *Biological Invasions* **11**, 81–96.

Lovett, G.M. and A.E. Ruesink. 1995. Carbon and nitrogen mineralization from decomposing gypsy moth frass. *Oecologia* **104**, 133–138.

Lovett, G.M., C.D. Canham, M.A. Arthur, et al. 2006. Forest ecosystem responses to exotic pests and pathogens in eastern North America. *BioScience* **56**, 395–405.

Lugo, A.E. 2008. Visible and invisible effects of hurricanes on forest ecosystems: an international review. *Austral Ecology* **33**, 368–398.

Luo, Y., S. Wan and D. Hui. 2001. Acclimization of soil respiration to warming in tall grass prairie. *Nature* **413**, 622–625.

Luo, Y., B. Su, W.S. Currie, et al. 2004. Progressive nitrogen limitation responses to rising atmopsheric carbon dioxide. *BioScience* **54**, 731–739.

Macarthur, R.H. and E.O. Wilson. 1967. *The Theory of Island Biogeography*. Princeton University Press, Princeton, NJ.

MacGillivray, C.W., J.P. Grime, S.R. Band, et al. 1995. Testing predictions of the resistance and resilience of vegetation subjected to extreme events. *Functional Ecology* **9**, 640–649.

Mack, M.C. and C.M. D'Antonio. 1998. Impacts of biological invasions on disturbance regimes. *Trends in Ecology and Evolution* **13**, 195–198.

Mack, M.C., E.A.G. Schuur, M.S. Bret-Harte, et al. 2004. Ecosystem carbon storage in arctic tundra reduced by long-term nutrient fertilization. *Nature* **431**, 440–443.

Madritch, M.D. and M.D. Hunter. 2002. Phenotypic diversity influences ecosystem functioning in an oak sandhill community. *Ecology* **83**, 2084–2090.

Madritch, M.D., S.L. Greene and R.L. Lindroth. 2009. Genetic mosaics of ecosystem functioning across aspen-dominated landscapes. *Oecologia* **160**, 119–127.

Maesako, Y. 1999. Impacts of streaked shearwater (*Calonectris leucomelas*) on tree seedling regeneration in a warm temperate evergreen forest on Kanmurijima Island, Japan. *Plant Ecology* **145**, 183–190.

Maestre, F.T., A. Escudero, I. Martinex, et al. 2005. Does spatial pattern matter to ecosystem functioning? Insights from biological soil crusts. *Functional Ecology* **19**, 566–573.

Maestre, F.T., M.A. Bowker, M.D. Puche, et al. 2009. Shrub encroachment can reverse desertification in semi-arid Mediterranean grasslands. *Ecology Letters* **12**, 930–941.

Maherali, H. and J.N. Klironomos. 2007. Influence of phylogeny on fungal community assembly and ecosystem functioning. *Science* **316**, 1746–1748.

Manning, P., J.E. Newington, H.R. Robson, et al. 2006. Decoupling the direct and indirect effects of nitrogen deposition on ecosystem function. *Ecology Letters* **9**, 1015–1024.

Manseau, M., J. Huot and M. Crete. 1996. Effects of summer grazing by caribou on composition and productivity of vegetation: community and landscape level. *Journal of Ecology* **84**, 503–513.

Maron, J.L., J.A. Estes, D.A. Croll, et al. 2006. An introduced predator alters Aleutian Island plant communities by thwarting nutrient subsidies. *Ecological Monographs* **76**, 3–24.

Martikainen, E. and V. Huhta. 1990. Interactions between nematodes and predatory mites in raw humus soil: a microcosm experiment. *Revue d' et de Biologie du Sol*, **27**, 13–20.

Martinson, H.M., K. Schneider, J. Gilbert, et al. 2008. Detritivory: stoichiometry of a neglected trophic level. *Ecological Research* **23**, 487–491.

Massey, F.P., A.R. Ennos and S.E. Hartley. 2007. Grasses and the resource availability hypothesis: the importance of silica-based defences. *Journal of Ecology* **95**, 414–424.

Masters, G.J. and V.K. Brown. 1992. Plant-mediated interactions between spatially separated insects. *Functional Ecology* **6**, 175–179.

Masters, G.J., V.K. Brown and A.C. Gange. 1993. Plant mediated interactions between above- and below-ground insect herbivores. *Oikos* **66**, 148–151.

Masters, G.J., T.H. Jones and M. Rogers. 2001. Host-plant mediated effects of root herbivory on insect seed predators and their parasitoids. *Oecologia* **127**, 246–250.

Matzner, E. and W. Borken. 2008. Do freeze-thaw events enhance C and N losses from soils of different ecosystems? A review. *European Journal of Soil Science* **59**, 274–284.

Mawdsley, J.L. and R.D. Bardgett. 1997. Continuous defoliation of perennial ryegrass (*Lolium perenne*) and white clover (*Trifolium repens*) and associated changes in the composition and activity of the microbial population of an upland grassland soil. *Biology and Fertility of Soils* **24**, 52–58.

May, R.M. 1973. *Stability and Complexity in Model Ecosystems*. Princeton University Press, Princeton, NJ.

McGlone, M.S. and B.D. Clarkson. 1993. Ghost stories: moa, plant defenses and evolution in New Zealand. *Tuatara* **32**, 1–18.

McGroddy, M.E., T. Daufresne and L.O. Hedin. 2004. Scaling of C : N : P stoichiometry in forests worldwide: Implications of terrestrial redfield-type ratios. *Ecology* **85**, 2390–2401.

McIntoch, P.D., R.B. Allen and N. Scott. 1997. Effects of exclosure and management on biomass and soil nutrient pools in seasonally dry high county, New Zealand. *Journal of Environmental Management* **51**, 169–186.

McIntyre, S., S. Lavorel, J. Landsberg, et al. 1999. Disturbance response in vegetation – towards a global perspective on functional traits. *Journal of Vegetation Science* **10**, 621–630.

McKane, R.B., L.C. Johnson, G.R. Shaver, et al. 2002. Resource-based niches provide a basis for plant species diversity and dominance in arctic tundra. *Nature* **413**, 68–71.

McKinney, M.L. and J.L. Lockwood. 1999., Biotic homogenization: A few winners replacing many losers in the next mass extinction, *Trends in Ecology and Evolution* **14**, 450–453.

McLean, M.A. and D. Parkinson. 2000a. Field evidence of the effects of the epigeic earthworm *Dendrobaena octaedra* on the microfungal community in pine forest floor. *Soil Biology and Biochemistry* **32**, 351–360.

McLean, M.A. and D. Parkinson. 2000b. Introduction of the epigeic earthworm *Dendrobaena octaedra* changes the oribatid community and microarthropod abundances in a pine forest. *Soil Biology and Biochemistry* **32**, 1671–1681.

McNaughton, S.J. 1977. Diversity and stability of ecological communities: a comment on the role of empiricism in ecology. *American Naturalist* **111**, 515–525.

McNaughton, S.J. 1983. Serengeti grassland ecology: the role of composite environmental factors and contingency in community organization. *Ecological Monographs* **53**, 291–320.

McNaughton, S.J. 1985. Ecology of a grazing system: the Serengeti. *Ecological Monographs* **55**, 259–294.

McNaughton, S.J., M. Oesterheld, D.A. Frank, et al. 1989. Ecosystem-level patterns of primary productivity and herbivory in terrestrial habitats. *Nature* **341**, 142–144.

McNaughton, S.J., F.F. Banyikwa and M.M. McNaughton. 1997a. Promotion of the cycling of diet-enhancing nutrients by African grazers. *Science* **278**, 1798–1800.

McNaughton, S.J., G. Zuniga, M.M. McNaughton, et al. 1997b. Ecosystem catalysis: soil urease activity and grazing in the Serengeti ecosystem. *Oikos* **80**, 467–469.

McNaughton, S.J., F.F. Banyikwa and M.M. McNaughton. 1998. Root biomass and productivity in a grazing ecosystem: the Serengeti. *Ecology* **79**, 587–592.

Meier, C.L. and W.D. Bowman. 2008. Links between plant litter chemistry, species diversity, and below-ground ecosystem function. *Proceedings of the National Academy of Sciences USA* **105**, 19780–19785.

Meier, C.L., K.N. Suding and W.D. Bowman. 2008. Carbon flux from plants to soil: roots are a below-ground source of phenolic secondary compounds in an alpine ecosystem. *Journal of Ecology* **96**, 421–430.

Meier, C.L., Keyserling and W.D. Bowman. 2009. Fine root inputs to soil reduce growth of a neighbouring plant via distinct mechanisms dependent on root carbon chemistry. *Journal of Ecology* **97**, 941–949.

Melillo, J.M., P.A. Steudler, J.D. Aber, et al. 2002. Soil warming and carbon-cycle feedbacks to the climate system. *Science* **298**, 2173–2175.

Memmott, J., P.G. Craze, M.N. Waser, et al. 2007. Global warming and the disruption of plant–pollinator interactions. *Ecology Letters* **10**, 710–717.

Menendez, R., A. Gonzalez Megias, J.K. Hill, et al. 2006. Species richness changes lag behind climate change. *Proceedings of the Royal Society of London Series B Biological Sciences* **273**, 1465–1470.

Menendez, R., A. Gonzalez-Megias, Y. Collingham, et al. 2007. Direct and indirect effects of climate and habitat factors on butterfly diversity. *Ecology* **88**, 605–611.

Menendez, R., A. Gonzalez-Megias, O.T. Lewis, et al. 2008 Escape from natural enemies during climate-driven range expansion: a case study. *Ecological Entomology* **33**, 413–421.

Menge, B.A. and J.P. Sutherland. 1976. Species diversity gradients: synthesis of the roles of predation, competition, and temporal heterogeneity. *American Naturalist* **110**, 351–369.

Menge, D.N. and L.O. Hedin. 2009. Nitrogen fixation in different biogeochemical niches along a 120,000-year chronosequence in New Zealand. *Ecology* **90**, 2190–2201.

Menge, D.N.L., S.A. Levin and L.O. Hedin. 2008. Evolutionary tradeoffs can select against nitrogen fixation and thereby maintain nitrogen limitation. *Proceedings of the National Academy of Sciences USA* **105**, 1573–1578.

Mikkelsen, T.N., C. Beier, S. Jonasson, et al. 2008. Experimental design of multifactor climate change experiments with elevated CO_2, warming and drought: the CLIMAITE project. *Functional Ecology* **22**, 185–195.

Mikola, J. and H. Setälä. 1998a. No evidence of trophic cascades in an experimental microbial-based soil food web. *Ecology* **79**, 153–164.

Mikola, J. and H. Setälä. 1998b. Productivity and trophic level biomasses in a microbial-based soil food web. *Oikos* **82**, 158–168.

Mikola, J., G.W. Yeates, G.M. Barker, et al. 2001. Effects of defoliation intensity on soil food-web properties in an experimental grassland community. *Oikos* **92**, 333–343.

Mikola, J., R.D. Bardgett and K. Hedlund. 2002. Biodiversity, ecosystem functioning and soil decomposer food webs. In: *Biodiversity and Ecosystem Functioning: Synthesis and Perspectives* (M. Loreau, S. Naeem and P. Inchausti, eds), pp. 169–180. Oxford University Press, Oxford.

Mikola, J., H. Setälä, P. Virkajärvi, et al. 2009. Defoliation and patchy nutrient return drive grazing effects on plant and soil properties in a dairy cow pasture. *Ecological Monographs* **79**, 221–244.

Milchunas, D.G. and W.K. Lauenroth. 1993. Quantitative effects of grazing on vegetation and soil over a global range of environments. *Ecological Monographs* **63**, 327–366.

Milchunas, D.G., O.E. Sala and W.K. Laurenroth. 1988. A generalized model of the effects of grazing by large herbivores on grassland community structure. *American Naturalist* **132**, 87–106.

Miller, A.E. and W.D. Bowman. 2002. Variation in nitrogen-15 natural abundance and nitrogen uptake traits among co-occuring alpine species: Do species partition by nitrogen form? *Oecologia* **130**, 609–616.

Miller, A.E. and W.D. Bowman. 2003. Alpine plants show species-level difference in the uptake of organic and inorganic nitrogen. *Plant and Soil* **250**, 283–292.

Millennium Ecosystem Assessment. 2005. *Ecosystems and Human Well-being*. Island Press, Washington DC.

Monson, R.K., D.L. Lipson, S.P. Burns, et al. 2006. Winter soil respiration controlled by climate and microbial community composition. *Nature* **439**, 711–714.

Moore, J.C. and P.C. de Ruiter. 2000. Invertebrates in detrital food webs along gradients of productivity. In: *Invertebrates as Webmasters in Ecosystems* (D.C. Coleman and P.F. Hendrix, eds), pp. 161–184. CABI, Oxford.

Moore, J.C. and H.W. Hunt. 1988. Resource compartmentation and the stability of real ecosystems. *Nature* **333**, 261–263.

Moore, J.C., T.V. St. John and D.C. Coleman. 1985. Ingestion of vesicular-arbuscular mycor-rhizal hyphae and spores by soil microarthropods. *Ecology* **66**, 1979–1981.

Moore, J.C., D.E. Walter and H.W. Hunt. 1988. Arthropod regulation of micro- and mesobiota in below-ground detrital food webs. *Annual Review of Entomology* **33**, 419–439.

Moore, J.C., K. McCann, H. Setälä, et al. 2003. Top-down is bottom-up: does predation in the rhizosphere regulate aboveground dynamics? *Ecology* **84**, 846–857.

Moore, J.C., E.L. Berlow, D.C. Coleman, et al. 2004. Detritus, trophic dynamics and biodiversity. *Ecology Letters* **7**, 584–600.

Moritz, C., J.L. Patton, C.J. Conry, et al. 2008. Impact of a century of climate change on small-mammal communities in Yosemite National Park, USA. *Science* **322**, 261–264.

Morris, W.F. and D.M. Wood. 1989. The role of lupine in Mt. St. Helens: facilitation or inhibition? *Ecology* **70**, 697–703.

Mountinho, P., D.C. Nepstad and E.A. Davidson. 2003. Influence of leaf-cutting ant nests on secondary forest growth and soil properties in Amazonia. *Ecology* **84**, 1265–1276.

Mulder, C. and J.A. Elser. 2009. Soil acidity, ecological stoichiometry and allometric scaling in grassland food webs. *Global Change Biology* **11**, 2730–2738.

Mulder, C.P.H. and S.N. Keall. 2001. Burrowing seabirds and reptiles: impacts on seeds, seedlings and soils in an island forest in New Zealand. *Oecologia* **127**, 350–360.

Mulder, C.P.H., J. Koricheva, K. Huss-Danell, et al. 1999. Insects affect relationships between plant species richness and ecosystem processes. *Ecology Letters* **2**, 237–246.

Mulder, C.P.H., M.N. Grant-Hoffman, D.R. Towns, et al. 2009. Direct and indirect effects of rats: will their eradication restore ecosystem functioning of New Zealand seabird islands? *Biological Invasions* **11**, 1671–1688.

Müller, P.E. 1884. Studier over skovjord, som bidrag til skovdyrkningens theori. II. Om muld og mor i egeskove og paa heder. *Tidsskrift for Skovbrug* **7**, 1–232.

Myrold, D.D., P.A. Matson and D.L. Petersson. 1989. Relationships between soil microbial properties and aboveground stand characteristics of coniferous forests in Oregon. *Biogeochemistry* **8**, 265–281.

Naeem, S., L.J. Thompson, S.P. Lawler, et al. 1994. Declining biodiversity can alter the performance of ecosystems. *Nature* **368**, 734–737.

Nara, K. 2006. Pioneer dwarf willow may facilitate tree succession by providing late colonizers with compatible ectomycorrhizal fungi in a primary successional volcanic desert. *New Phytologist* **171**, 187–198.

Nara, K. and T. Hogetsu. 2004. Ectomycorrhizal fungi on established shrubs facilitate subsequent seedling establishment of successional plant species. *Ecology* **85**, 1700–1707.

Nardo, C.D., A. Cinquegrana, S. Papa, et al. 2004. Laccase and peroxidase isoenzymes during leaf litter decomposition of *Quercus ilex* in a Mediterranean ecosystem. *Soil Biology and Biochemistry* **36**, 1539–1544.

Näsholm, T., A. Ekblad, A Nordin, et al. 1998. Boreal forest plants take up organic nitrogen. *Nature* **392**, 914–916.

Negrete-Yankelevich, S., C. Fragoso, A.C. Newton, et al. 2008. Species-specific characteristics of trees can determine the litter invertebrate community and decomposition processes below their canopies. *Plant and Soil* **307**, 83–97.

Nemergut, D.R., S.P. Anderson, C.C. Cleveland, et al. 2007. Microbial community succession in an unvegetated, recently-deglaciated soil. *Microbial Ecology* **53**, 110–122.

Neutel, A.M., J.A.P. Heesterbeek and P.C. de Ruiter. 2002. Stability in real food webs: weak links in long loops. *Science* **296**, 1120–1123.

Neutel, A.M., J.A.P. Heesterbeek, J. van de Koppel, et al. 2007. Reconciling complexity with stability in naturally assembling food webs. *Nature* **449**, 599–601.

Newell, K. 1984a. Interaction between two decomposer basidiomycetes and a collembolan under Sitka spruce: distribution, abundance and selective grazing. *Soil Biology and Biochemistry* **16**, 227–233.

Newell, K. 1984b. Interaction between two decomposer basidiomycetes and a collembolan under Sitka spruce: Grazing and its potential effects on fungal distribution and litter decomposition. *Soil Biology and Biochemistry* **16**, 235–239.

Newsham, K.K., A.R. Watkinson and A.H. Fitter. 1995a. Symbiotic fungi determine plant community structure – changes in a lichen-rich community induced by fungicide application. *Functional Ecology* **9**, 442–447.

Newsham, K.K., A.H. Fitter and A.R. Watkinson. 1995b. Multi-functionality and biodiversity in arbuscular mycorrhizas. *Trends in Ecology and Evolution* **10**, 407–411.

Newton, P.C.D., H. Clark, C.C. Bell, et al. 1995. Plant growth and soil processes in temperate grassland communities at elevated CO_2. *Journal of Biogeography* **22**, 235–240.

Nichols, E., S, Spector, J. Louzada, et al. 2008. Ecological functions and ecosystem services provided by Scarabaenae dung beetles. *Biological Conservation* **141**, 1461–1474.

Nicolai, V. 1988. Phenolic and mineral contents of leaves influences decomposition in European forest systems. *Oecologia* **75**, 575–579.

Niemelä, J., J.R. Spence and H. Carcamo. 1997. Establishment and interactions of carabid populations: an experiment with native and exotic species. *Ecography* **20**, 643–652.

Niklaus, P.A., E. Kandeler, P.W. Leadley, et al. 2001. A link between plant diversity, elevated CO_2 and soil nitrate. *Oecologia* **127**, 540–548.

Nilsson, M.-C. and D.A. Wardle. 2005. Understory vegetation as a forest ecosystem driver: evidence from the northern Swedish boreal forest. *Frontiers in Ecology and the Environment* **3**, 421–428.

Nilsson, M.-C., D.A. Wardle, O. Zackrisson, et al. 2002. Effects of alleviation of ecological stresses on an alpine tundra community over an eight year period. *Oikos* **97**, 3–17.

Norby, R.J., J. Ledford, C.D. Reilly, et al. 2004. Fine-root production dominates response of a deciduous forest to atmospheric CO_2 enrichment. *Proceedings of the National Academy of Sciences USA* **101**, 9689–9693.

Nordin, A., P. Högberg and T. Nasholm. 2001. Soil nitrogen form and plant nitrogen uptake along a boreal forest productivity gradient. *Oecologia* **129**, 125–132.

Nordin, A., I.K. Schmidt and G.R. Shaver. 2004. Nitrogen uptake by arctic soil microbes and plants in relation to soil nitrogen supply. *Ecology* **85**, 955–962.

Northup, R.R., Z.S. Yu, R.A. Dahlgren, et al. 1995. Polyphenol control of nitrogen release from pine litter. *Nature* **377**, 227–229.

Nuñez, M.A., T.R. Horton and D. Simberloff. 2009. Lack of belowground mutualisms hinders Pinaceae invasions. *Ecology* **90**, 2352–2359.

Nykanen, H. and J. Koricheva. 2004. Damage-induced changes in woody plants and their effects on insect herbivore performance: a meta-analysis. *Oikos* **104**, 247–268.

O'Dowd, D.J., P.T. Green and P.S. Lake. 2003. Invasional 'meltdown' on an oceanic island. *Ecology Letters* **6**, 812–817.

Odum, E.P. 1969. The strategy of ecosystem development. *Science* **164**, 262–270.

Oechel, W.C., S.J. Hastings, G.L. Vourlitis, et al. 1993. Recent changes in arctic tundra ecosystems from a carbon sink to source. *Nature* **361**, 520–523.

Oechel, W.C., G.L. Vourlitis, S.J. Hastings, et al. 1995. Change in arctic CO_2 flux over two decades: effects of climate change at Barrow, Alaska. *Ecological Applications* **5**, 846–855.

Ohashi, M., L. Finér, T. Domisch, et al. 2007. Seasonal and diural CO_2 efflux from red wood ant (*Formica aquilonia*) mounds in boreal coniferous forests. *Soil Biology and Biochemistry* **39**, 1504–1511.

Ohtonen, R., H. Fritze, T. Pennanen, et al. 1999. Ecosystem properties and microbial community changes in primary succession on a glacier forefront. *Oecologia* **119**, 239–246.

Oksanen, L., S.D. Fretwell, J. Arruda, et al. 1981. Exploitation ecosystems in gradients of primary productivity. *American Naturalist* **118**, 240–261.

Olff, H. and M.E. Ritchie. 1998. Effects of herbivores on grassland plant diversity. *Trends in Ecology and Evolution* **13**, 261–265.

Olff, H., J. Huisman and B.F. van Tooren. 1993. Species dynamics and nutrient accumulation during early succession in coastal sand dunes. *Journal of Ecology* **81**, 693–706.

Olff, H., M.E. Ritchie and H.H.T. Prins. 2002. Global environmental controls of diversity in large herbivores. *Nature* **415**, 901–904.

Olofsson, J., S. Stark and L. Oksanen. 2004. Reindeer influence on ecosystem processes in the tundra. *Oikos* **105**, 386–396.

Olofsson, J., L. Oksanen, T. Callaghan, et al. 2009. Hebivores inhibit climate-driven shrub expansion on the tundra. *Global Change Biology* **15**, 2681–2693.

Orwin, K.H. and D.A. Wardle. 2005. Plant species composition affects the resistance and resilience of the soil microflora to a drying disturbance. *Plant and Soil* **278**, 205–211.

Ostfeld, R.S. and F. Keesing. 2004. Oh the locusts sang, then they dropped dead. *Science* **306**, 1488–1489.

Overpeck, J., K. Hughen, D. Hardy, et al. 1997. Arctic environmental change of the last four centuries. *Science* **278**, 1251–1256.

Pace, M.L., J.J. Cole and S.R. Carpenter. 1999. Trophic cascades revealed in diverse ecosystems. *Trends in Ecology and Evolution* **14**, 483–488.

Packer, A. and K. Clay. 2000. Soil pathogens and spatial patterns of seedling mortality in a temperate tree. *Nature* **404**, 278–281.

Packer, A. and K. Clay. 2003. Soil pathogens and *Prunus serotina* seedlings and sapling growth near conspecific trees. *Ecology* **84**, 108–119.

Pacovsky, R.S., G. Fuller, A.E. Stafford, et al. 1986. Nutrient and growth interactions in soybeans colonized with *Glomus versiforme* and *Rhizobium japonicum*. *Plant and Soil* **92**, 37–45.

Paetzold, A., M. Lee and D.M. Post. 2008. Marine resource flows to terrestrial arthropod predators of a temperate island: the role of subsidies between systems of similar productivity. *Oecologia* **157**, 653–659.

Paine, R.T. 1969. A note on trophic complexity and community stability. *American Naturalist* **103**, 91–93.

Parfitt, R.L., Ross, D.J., Coomes, D.A., et al. 2005. N and P in New Zealand soil chronosequences and relationships with foliar N and P. *Biogeochemistry* **75**, 305–328.

Parker, I.M., D. Simberloff, W.M. Lonsdale, et al. 1999. Impact: toward a framework for understanding the ecological effects of invaders. *Biological Invasions* **1**, 3–19.

Parmelee, R.W., M.H. Beare and J.M. Blair. 1989. Decomposition and nitrogen dynamics of surface weed residues in no-tillage agroecosystems under drought conditions: influence of resource quality on the decomposer community. *Soil Biology and Biochemistry* **21**, 97–103.

Parmenter, R.R. and J.A. MacMahon. 2009. Carrion decomposition and nutrient cycling in a semiarid shrub-steppe ecosystem. *Ecological Monographs* **79**, 637–661.

Parmesan, C. and G. Yohe. 2003. A globally coherent fingerprint of climate change impacts across natural systems. *Nature* **421**, 37–42.

Pärtel, M., L., Laanisto and S.D. Wilson. 2008. Soil nitrogen and carbon heterogeneity in woodlands and grasslands: contrasts between temperate and boreal regions. *Global Ecology and Biogeography* **17**, 18–24.

Parton, W., W. Silver, I. Burke, et al. 2007. Global scale similarities in nitrogen release patterns during long term decomposition. *Science* **315**, 361–364.

Pastor, J., R.J. Naiman, B. Dewey, et al. 1988. Moose, microbes and the boreal forest. *BioScience* **38**, 770–777.

Pastor, J., R.J. Dewey, R.J. Naiman, et al. 1993. Moose browsing and soil fertility in the boreal forests of Isle Royale National Park. *Ecology* **74**, 467–480.

Pauchard, A., R.A. Garcia, E. Pena, et al. 2008. Positive feedbacks between plant invasions and fire regimes: *Teline monspessulana* (L.) K. Koch (Fabaceae) in central Chile. *Biological Invasions* **10**, 547–553.

Paustian, K. 1994. Modelling soil biology and biochemical processes for sustainable agricultural research. In: *Soil Biota. Management in Sustainable Farming Systems* (C.E. Pankhurst, B.M. Doube, V.V.S.R. Gupta, et al., eds), pp. 182–193. CSIRO, Melbourne.

Pearson, S.M., M.G. Turner, L.L. Wallace, et al. 1995. Winter habitat use by large ungulates following fire in northern Yellowstone National Park. *Ecological Applications* **5**, 744–755.

Pei, S., H. Fu and C. Wann. 2008. Changes in soil properties and vegetation following exclosure and grazing in degraded Alxa desert steppe of Inner Mongolia, China. *Agriculture, Ecosystems and Environment* **124**, 33–39.

Peltzer, D., P.J. Bellingham, H. Kurokawa, et al. 2009. Punching above their weight: low-biomass non-native plant species alter soil ecosystem properties during primary succession. *Oikos* **118**, 1001–1014.

Peltzer, D.A., R.B. Allen, R.B., G.M. Lovett, et al. 2010. Effects of biological invasions on forest carbon sequestration. *Global Change Biology* **16**, 732–746.

Perez Harguindeguy, N., C.M. Blundo, D.E. Gurvich, et al. 2008. More than a sum of its parts? Assessing litter heterogeneity effects on the decomposition of litter mixtures through leaf chemistry. *Plant and Soil* **303**, 151–159.

Perry, D.A., A.P. Amaranthus, J.G. Boerschers, et al. 1989. Bootstrapping in ecosystems: internal interactions largely determine productivity and stability in biological systems with strong feedback. *BioScience* **39**, 230–237.

Perry, L.G., G.C. Thelan, W.M. Ridenour, et al. 2007. Concentrations of the allelochemical (+ /-) – catechin in *Centaurea maculosa* soils. *Journal of Chemical Ecology* **12**, 2337–2344.

Personeni, E. and Loiseau, P. 2004. How does the nature of living and dead roots affect the residence time of carbon in the root litter continuum? *Plant Soil* **267**, 129–141.

Petchey, O.L., P.T. McPhearson, T.M. Casey, et al. 1999. Environmental warming alters food-web structure and ecosystem function. *Nature* **402**, 69–72.

Petermann, J.S., A.J.F. Fergus, L.A. Turnbull, et al. 2008. Janzen-Connell effects are wide-spread and strong enough to maintain diversity in grasslands. *Ecology* **89**, 2399–2406.

Peters, D. and G. Weste. 1997. The impact of *Phytophthora cinnamomi* on six rare native tree and shrub species in the Brisbane Ranges, Victoria. *Australian Journal of Botany* **45**, 975–995.

Peters, D.C., B.T. Bestelmeyer, J.E. Herrick, et al. 2006. Disentangling complex landscapes: new insights into arid and semi-arid system dynamics. *Bioscience* **56**, 491–501.

Petersen, H. and M. Luxton. 1982. A comparative analysis of soil fauna populations and their role in decomposition processes. *Oikos* **39**, 287–388.

Phoenix, G.K., R.E. Booth, J.R. Leake, et al. 2004. Effects of enhanced nitrogen deposition and phosphorus limitation on nitrogen budgets of semi-natural grasslands. *Global Change Biology* **9**, 1309–1321.

Phoenix, G.K., W.K. Hicks, S. Cinderby, et al. 2006. Atmospheric nitrogen deposition in world biodiversity hotspots: the need for a greater global perspective in assessing N deposition impacts. *Global Change Biology* **12**, 470–476.

Pickett, S.T.A., S.L. Collins and J.J. Armesto. 1987. Models, mechanisms and pathways of succession. *Botanical Reviews* **53**, 336–371.

Piearce, T.G., N. Roggero and R. Tipping, R. 1994. Earthworms and seeds. *Journal of Biological Education* **28**, 195–202.

Pimm, S.L., G.J. Russel, J.L. Gittleman, et al. 1995. The future of biodiversity. *Science* **269**, 347–350.

Piñeiro, G., J.M. Paruelo, E.G. Jobbágy, et al. 2009. Grazing effects on belowground C and N stocks along a network of cattle exclosures in temperate and subtropical grasslands of South America. *Global Biogeochemical Cycles* **23**, GB2003.

Pinto-Tomás, A.A., M.A. Anderson, G. Suen, et al. 2009. Symbiotic Nitrogen fixation in the fungus gardens of leaf-cutter ants. *Science* **326**, 1120–1123.

Polis, G.A. 1994. Food webs, trophic cascades and community structure. *Australian Journal of Ecology* **19**, 121–136.

Polis, G.A. and S.D. Hurd. 1996. Linking marine and terrestrial food webs: allochthonous input from the ocean supports high secondary productivity on small islands and coastal communities. *American Naturalist* **147**, 396–423.

Polis, G.A., W.B. Anderson and R.D. Holt. 1997. Toward an integration of landscape and food web ecology: the dynamics of spatially subsidized food webs. *Annual Review of Ecology and Systematics* **28**, 289–316.

Polley, H.W., A.B. Frank, J. Sanabria, et al. 2008. Interannual variability in carbon dioxide fluxes and flux-climate relationships on grazed and ungrazed northern mixed-prairie. *Global Change Biology* **14**, 1620–1632.

Pollierer, M.M., R. Langel, C. Körner C, et al. 2007. The underestimated importance of belowground carbon input for forest soil animal food webs. *Ecology Letters* **10**, 729–736.

Poorter, H. and C. Remkes. 1990. Leaf area ratio and net assimilation rate of 24 wild species differing in relative growth rate. *Oecologia* **83**, 553–559.

Porazinska, D.L., R.D. Bardgett, M.B. Blaauw, et al. 2003. Relationships at the aboveground-belowground interface: Plants, soil biota, and soil processes. *Ecological Monographs* **73**, 377–395.

Post, E. and C. Pedersen. 2008. Opposing plant community responses to warming with and without herbivores. *Proceedings of the National Academy of Sciences USA* **105**, 12353–12358.

Post, E. and M.C. Forchhammer. 2008. Climate change reduces reproductive success of an Arctic herbivore through trophic mismatch. *Philosophical Transactions of the Royal Society Series B Biological Science* **363**, 2369–2375.

Post, E., R.O. Peterson, N.C. Stenseth, et al. 1999. Ecosystem consequences of wolf behavioral response to climate. *Nature* **401**, 905–907.

Power, M., D. Tilman, J.A. Estes, et al. 1996. Challenges in the quest for keystones. *BioScience* **46**, 609–620.

Powers, J.S., R.A. Montgomery, E.C. Adair, et al. 2009. Decomposition in tropical forests; a pan-tropical study of the effects of litter type, litter placement and mesofaunal exclusion across a precipitation gradient. *Journal of Ecology* **97**, 801–811.

Prentice, I.C., W. Cramer, S.P. Harisson, et al. 1992. A global biome model based on plant physiology and dominance, soil properties and climate. *Journal of Biogeography* **19**, 117–134.

Press, M.C., J.A., Potter, M.J.W. Burke, et al. 1998. Responses of a subarctic dwarf shrub heath community to simulated environmental change. *Journal of Ecology* **86**, 315–327.

Pringle, A. and E.C. Vellinga. 2006. Last chance to know? Using literature to explore the biogeography and invasion biology of the death cap mushroom *Amanita phalloides* (Vaill. Ex. fr. Fr.) link. *Biological Invasions* **8**, 1131–1144.

Pritchard, S.G., A.E. Strand, M.L. McCormack, et al. 2008. Fine root dynamics in a loblolly pine forest are influenced by Free-Air-CO_2-enrichment (FACE): a six year minirhizotron study. *Global Change Biology* **14**, 588–602.

Pueschel, D., J. Rydlova, M. Vosatka, et al. 2008. Does the sequence of plant dominants affect mycorrhiza development in simulated succession on spoil banks? *Plant and Soil* **302**, 273–282.

Quested, H.M., J.H.C. Cornelissen, M.C. Press, et al. 2003. Decomposition of sub-arctic plants with differing nitrogen economies: A functional role for hemiparasites. *Ecology* **84**, 3209–3221.

Quested, H.M., T.V. Callaghan, J.H.C. Cornelissen, et al. 2005. The impact of hemiparasitic plant litter on plant decomposition: direct, seasonal and litter mix effects. *Journal of Ecology* **93**, 97–98.

Quested, H., O. Eriksson, C. Fortunel, et al. 2007. Plant traits relate to whole-community litter quality and decomposition following land use change. *Functional Ecology* **21**, 1016–1026.

Quinn, T.P., S.M. Carlson, S.M. Gende, et al. 2009. Transportation of Pacific salmon carcasses from streams to riparian forests by bears. *Canadian Journal of Zoology* **87**, 195–203.

Raab, T.K., D.A. Lipson and R.K. Monson. 1999. Soil amino acid utilization among species of the Cyperacea: Plant and soil processes. *Ecology* **80**, 2408–2419.

Raich, J.W. and C.S. Potter. 1995. Global patterns of carbon-dioxide emissions from soils. *Global Biogeochemical Cycles* **9**, 23–36.

Ramsay, A. 1983. Bacterial biomass in ornithogenic soils in Antarctica. *Polar Biology* **1**, 221–225.

Ramsey, P.W., M.C. Rillig, K.P. Feris, et al. 2006. Choice of methods for soil microbial community analysis: PLFA maximizes power compared to CLPP and PCR-based approaches. *Pedobiologia* **50**, 275–280.

Read, D.J. 1991. Mycorrhizas in ecosystems. *Experimentia* **47**, 376–391.

Read, D.J. 1992. The mycorrhizal mycelium. In: *Mycorrhizal Functioning* (M.F. Allen, ed.), pp. 102–133. Chapman and Hall, London.

Read, D.J. 1994. Plant-microbe mutualisms and community structure. In: *Biodiversity and Ecosystem Function* (E.-D. Schulze and H.A. Mooney, eds), pp. 181–209. Springer-Verlag, Berlin.

Read, D.J. and J. Perez-Moreno. 2003. Mycorrhizas and nutrient cycling in ecosystems—a journey towards relevance? *New Phytologist* **157**, 475–492.

Read, D.J., J.R. Leake and J. Perez-Moreno. 2004. Mycorrhizal fungi as drivers of ecosystem processes in heathland and boreal forest biomes. *Canadian Journal of Botany* **82**, 1243–1263.

Reay, D.S., F. Dentener, P. Smith, et al. 2008. Global nitrogen deposition and carbon sinks. *Nature Geoscience* **1**, 430–437.

Redfield, A.C. 1958. The biological control of chemical factors in the environment. *American Scientist* **46**, 205–221.

Reeder, J.D. and Schuman, G.E. 2002. Influence of livestock grazing on C sequestration in semi-arid mixedgrass and short-grass rangelands. *Environmental Pollution* **116**, 457–463.

Reich, P.B., J. Knops, D. Tilman, et al. 2001. Plant diversity enhances ecosystem responses to elevated CO_2 and nitrogen deposition. *Nature* **410**, 809–812.

Reinhart, K.O. and R.M. Callaway. 2006. Soil biota and invasive plants. *New Phytologist* **170**, 445–457.

Reinhart, K.O., A. Packer, W.H. Van der Putten, et al. 2003. Plant-soil biota interactions and spatial distribution of black cherry in its native and invasive ranges. *Ecology Letters* **6**, 1046–1050.

Reinhart, K.O., A.A. Royo, W.H. Van der Putten, et al. 2005. Soil feedback and pathogen activity in *Prunus serotina* throughout its native range. *Journal of Ecology* **93**, 890–898.

Rejmanek, M., D.M. Richardson, S.I. Higgins, et al. 2005. Ecology of invasive species: state of the art. In: *Invasive Species: A New Synthesis* (H.A. Mooney, R.N. Mack, J.A. McNeely, et al., eds), pp. 104–161. Island Press, Washington DC.

Reuman, D.C., J.E. Cohen and C. Mulder. 2009. Human and environmental factors influence soil faunal abundance-mass allometry and structure. *Advances in Ecological Research* **41**, 45–85.

Rey, A. and P. Jarvis. 2006. Modelling the effect of temperature on carbon mineralization rates across a network of European forest sites (FORCAST). *Global Change Biology* **12**, 1894–1908.

Reynolds, H.L., A. Packer, J.D. Bever, et al. 2003. Grassroots ecology: Plant-microbe-soil interactions as drivers of plant community structure and dynamics. *Ecology* **84**, 2281–2291.

Reynolds, H.L., A.E. Hartley, K.M. Vogelsang, et al. 2005. Arbuscular mycorrhizal fungi do not enhance nitrogen acquisition and growth of old-field perennials under low nitrogen supply in glasshouse culture. *New Phytologist* **167**, 869–880.

Rhoades, D.F. 1985. Offensive-defensive interactions between herbivores and plants: their relevance in herbivore population dynamics and ecological theory. *American Naturalist* **125**, 205–238.

Richardson, D.M. 2006. *Pinus*: a model group for unlocking the secrets of alien plant invasions? *Preslia* **78**, 375–388.

Richardson, D.M., P.A. Williams and R.J. Hobbs. 1994. Pine invasions in the Southern Hemisphere: determinants of spread and invisibility. *Journal of Biogeography* **21**, 511–527.

Richardson, D.M., N. Allsop, C.M. D'Antonio, et al. 2000. Plant invasions – the role of mutualists. *Biological Reviews* **75**, 65–93.

Richardson, S.J., D.A. Peltzer, R.B. Allen, et al. 2004. Rapid development of phosphorus limitation in temperate rainforest along the Franz Josef soil chronosequence. *Oecologia* **139**, 267–276.

Richardson, S.J., D.A. Peltzer, R.B. Allen, et al. 2005. Resorption proficiency along a chronosequence: Responses among communities and within species. *Ecology* **86**, 20–25.

Richardson, S.J., R.B. Allen and J.E. Doherty. 2008. Shifts in leaf N:P ratio during resorption reflect soil P in temperate rainforest. *Functional Ecology* **22**, 738–745.

Ridder, B. 2008. Questioning the ecosystem services argument for biodiversity conservation. *Biodiversity and Conservation* **17**, 781–790.

Ridenour, W.M. and R.M. Callaway. 2001. The relative importance of allelopathy in interference: the effects of an invasive weed on a native bunchgrass. *Oecologia* **126**, 444–450.

Rietkerk, M. and J. van de Koppel. 1997. Alternative stable states and threshold effects in semi-arid grazing systems. *Oikos* **79**, 69–76.

Rietkerk, M., F. van den Bosch and J. van de Koppel. 1997. Site-specific properties and irreversible vegetation changes in semi-arid grazing systems. *Oikos* **80**, 241–252.

Rietkerk, M., S.C. Dekker, P.C. de Ruiter and J. van de Koppel. 2004. Self-organized patchiness and catastrophic shifts in ecosystems. *Science* **305**, 1926–1929.

Rillig, M.C. and D.L. Mummey. 2006. Mycorrhizas and soil structure. *New Phytologist* **171**, 41–53.

Rillig, M.C., G.W. Hernandez and P.C.D. Newton. 2000. Arbuscular mycorrhizae respond to elevated atmospheric CO_2 after long-term exposure: evidence from a CO_2 spring in New Zealand supports the resource balance model. *Ecology Letters* **3**, 475–478.

Rinnan, R., S. Stark and A. Tolanen. 2009. Response of vegetation and soil microbial communities to warming and simulated herbivory in a subarctic heath. *Journal of Ecology* **97**, 788–800.

Ripple, W.J. and R.L. Beschta. 2004. Wolves, elk, willows, and trophic cascades in the upper Gallatin Range of Southwestern Montana, USA. *Forest Ecology and Management* **200**, 161–181.

Ripple, W.J. and R.L. Beschta. 2006. Linking a cougar decline, trophic cascade, and catastrophic regime shift in Zion National Park. *Biological Conservation* **133**, 397–408.

Ripple, W.J. and R.L. Beschta. 2008. Trophic cascades involving cougar, mule deer, and black oaks in Yosemite National Park. *Biological Conservation* **141**, 1249–1256.

Risch, A.C. and D.A. Frank. 2006. Carbon dioxide fluxes in a spatially and temporally heterogeneous temperate grassland. *Oecologia* **147**, 291–302.

Risch, A.C., M.F. Jurgensen, M. Schütz, et al. 2005. The contribution of red wood ants to soil C and N pools and to CO_2 emissions in subalpine forests. *Ecology* **86**, 419–430.

Ritchie, M.E., D. Tilman and J.M.H. Knops. 1998. Herbivore effects on plant and nitrogen dynamics in oak savanna. *Ecology* **79**, 165–177.

Robinson, C.H., P. Ineson, T.G. Piearce, et al. 1992. Nitrogen mobilization by earthworms in limed peat soils under *Picea sitchensis*. *Journal of Applied Ecology* **29**, 226–237.

Robinson, C.H., J. Dighton, J.C. Frankland, et al. 1993. Nutrient and carbon dioxide release by interacting species of straw-decomposing fungi. *Plant and Soil* **151**, 139–142.

Robinson, C.H., P. Ineson, T.G. Piearce, et al. 1996. Effects of earthworms on cation and phosphate mobilisation in limed peat soils under *Picea sitchensis*. *Forest Ecology and Management* **86**, 253–258.

Rochon, J.J., C.J. Doyle, J.M. Greef, et al. 2004. Grazing legumes in Europe: a review of their status, management, benefits, research needs and future prospects. *Grass and Forage Science* **59**, 197–214.

Roser, D.J., R.D. Seppelt and N. Ashbolt. 1993. Microbiology of ornithogenic soils from the Windmill Islands, Budd Coast, continental Antarctica: microbial biomass distribution. *Soil Biology and Biochemistry* **25**, 165–175.

Ross, D.J., P.C.D. Newton and K.R. Tate. 2004. Elevated CO_2 effects on herbage and soil carbon and nitrogen pools and mineralization in a species–rich, grazed pasture on a seasonally dry sand. *Plant and Soil* **260**, 183–196.

Roulet, N.T. and T.R. Moore. 1995. The effect of forestry drainage practices on the emissions of methane from northern peatlands. *Canadian Journal of Forest Research* **25**, 491–499.

Rovira, A.D., E.I. Newman, H.J. Bowen, et al. 1974. Quantitative assessment of the rhizoplane microflora by direct microscopy. *Soil Biology and Biochemistry* **6**, 211–216.

Ruess, R.W., R.L. Hendrick and J.P. Bryant. 1998. Regulation of fine root dynamics by mammalian browsers in early successional Alaskan taiga forests. *Ecology* **79**, 2706–2720.

Sabo, J.L. and M.E. Power. 2002. River-watershed exchange: effects of riverine subsidies on riparian lizards and their terrestrial prey. *Ecology* **83**, 1860–1869.

Saetre, P. and E. Bååth. 2000. Spatial variation and patterns of the soil microbial community structure in a mixed spruce-birch stand. *Soil Biology and Biochemistry* **32**, 909–917.

Sala, O.E., F.S. Chapin, J.J. Armesto, et al. 2000. Global biodiversity scenarios for the year 2100. *Science* **287**, 1770–1774.

Salt, D.T., P. Fenwick and J.B. Whittaker. 1996. Interspecific herbivore interactions in a high CO_2 environment: root and shoot aphids feeding on *Cardamine*. *Oikos* **77**, 326–330.

Sanchez-Piñero, F. and G.A. Polis. 2000. Bottom-up dynamics of allochthonous input: direct and indirect effects of seabirds on islands. *Ecology* **81**, 3117–3232.

Sankaran, M. and D.J. Augustine. 2004. Large herbivores suppress decomposer abundance in a semiarid grazing ecosystem. *Ecology* **85**, 1052–1061.

Sankaran, M. and S.J. McNaughton. 1999. Determinants of biodiversity regulate compositional stability of communities. *Nature* **401**, 691–693.

Sankaran, M., N.P. Hanan, R.J. Scholes, et al., 2005. Determinants of woody cover in African savannas. *Nature* **438**, 846–849.

Santiago, L.S. 2007. Extending the leaf economics spectrum to decomposition: Evidence from a tropical forest. *Ecology* **88**, 1126–1131.

Santos, P.F., J. Phillips and W.G. Whitford. 1981. The role of mites and nematodes in early stages of buried litter decomposition in a desert. *Ecology* **63**, 664–669.

Saravesi, K., A. Markkola, P. Rautio, et al. 2008. Defoliation causes parallel temporal responses in a host tree and its fungal symbionts. *Oecologia* **156**, 117–123.

Sariyildiz, T. 2008. Effects of gap-size classes on long-term litter decomposition rates of beech, oak and chestnut species at high elevations in northeast Turkey. *Ecosystems* **11**, 841–853.

Sax, D.F. and S.D. Gaines. 2003. Species diversity: from global decreases to local increases. *Trends in Ecology and Evolution* **18**, 561–566.

Sax, D.F., B.P. Kinlan and K.F. Smith. 2005. A conceptual framework for comparing species assemblages in native and exotic habitats. *Oikos* **108**, 457–464.

Schädler, M. and R. Brandl. 2005. Do invertebrate decomposers affect the disappearance rate of litter mixtures? *Soil Biology and Biochemistry* **37**, 329–337.

Schädler, M., G. Jung, R. Brandl, et al. 2004. Secondary succession is influenced by below-ground insect herbivory on a productive site. *Oecologia* **138**, 242–252.

Schadt, C.W.A.P. Martin, D.A. Lipson, et al. 2003. Seasonal dynamics of previously unknown fungal lineages in tundra soils. *Science* **301**, 1359–1361.

Schellekens, J., F.N. Scatena, L.A. Bruijnzeel, et al. 2004. Stormflow generation in a small rainforest catchment in the Luquillo Experimental Forest, Puerto Rico. *Hydrological Processes* **18**, 505–530.

Scheu, S. 1987. The role of substrate-feeding earthworms (Lumbricidae) for bioturbation in a beechwood soil. *Oecologia* **72**, 192–196.

Scheu, S. 2001. Plants and generalist predators as links between the below-ground and above-ground system. *Basic and Applied Ecology* **2**, 3–13.

Scheu. S. 2003. Effects of earthworms on plant growth: patterns and perspectives. *Pedobiologia* **47**, 846–856.

Scheu, S. and D. Parkinson. 1994. Effects of invasion of an aspen forest (Canada) by *Dendrobaena octaedra* (Lumbricidae) on plant growth. *Ecology* **75**, 2348–2361.

Scheu, S. and M. Schaefer. 1998. Bottom-up control of the soil macrofauna community in a beechwood on limestone: manipulation of food resources. *Ecology* **79**, 1573–1585.

Scheu, S. and H. Setälä. 2002. Multitrophic interactions in decomposer food webs. In: *Multitrophic Level Interactions* (T. Tscharntke and B. Hawkins, eds), pp. 223–264. Cambridge University Press, Cambridge.

Scheu, S., A. Theenhaus and T.H. Jones. 1999. Links between the detritivore and the herbivore system: effects of earthworms and Collembola on plant growth and aphid development. *Oecologia* **119**, 541–551.

Scheublin, T.R., K.P. Ridgway, J.P.W. Young, et al. 2004. Nonlegumes, legumes, and root nodules harbor different arbuscular mycorrhizal fungal communities. *Applied Environmental Microbiology* **70**, 6240–6246.

Schimel, J.P. and J. Bennett. 2004. Nitrogen mineralization: Challenges of a changing paradigm. *Ecology* **85**, 591–602.

Schimel, J.P. and F.S. Chapin. 1996. Tundra plant uptake of amino acid and NH_4^+ nitrogen in situ: plants compete well for amino acid N. *Ecology* **77**, 2142–214.

Schimel, J.P. and C. Mikan. 2005. Changing microbial substrate use in Arctic tundra soils through a freeze-thaw cycle. *Soil Biology and Biochemistry* **37**, 1411–1418.

Schimel, D.S., B.H. Braswell, E.A. Holland, et al. 1994. Climatic, edaphic, and biotic controls over storage and turnover of carbon in soils, *Global Biogeochemical Cycles* **8**, 279–293.

Schimel, J.P., R.G. Cates and R.W. Ruess. 1998. The role of balsam poplar secondary chemicals in controlling soil nutrient dynamics through succession in the Alaskan taiga. *Biogeochemistry* **42**, 221–234.

Schimel, J.P., J. Bennett and N. Frierer. 2005. Microbial community composition and soil nitrogen cycling: is there really a connection? In: *Biological Diversity and Function in Soil* (R.D. Bardgett, M.B. Usher and D.W. Hopkins, eds), pp. 172–188. Cambridge University Press, Cambridge.

Schimel, J., T.C. Balser and M. Wallenstein. 2007. Microbial stress-response physiology and its implications for ecosystem function. *Ecology* **88**, 1386–1394.

Schlesinger, W.H. 1997. *Biogeochemistry: An Analysis of Global Change*, 2nd edn. Academic Press, San Diego.

Schlesinger, W.H. 2009. On the fate of anthropogenic nitrogen. *Proceedings of the National Academy of Sciences USA* **106**, 203–208.

Schlesinger, W.H. and A.M. Pilmanis. 1998. Plant-soil interactions in deserts. *Biogeochemistry* **42**, 169–187.

Schlesinger, W.H., J.F. Reynolds, G.L. Cunningham, et al. 1990. Biological feedbacks in global desertification. *Science* **247**, 1043–1048.

Schmidt, S., W.C. Dennison, G.J. Moss, et al. 2004. Nitrogen ecophysiology of Heron Island, a subtropical coral cay of the Great Barrier Reef, Australia. *Functional Plant Biology* **31**, 517–528.

Schmidt, S.K., Reed, S.C., Nemergut, D.R., et al. 2008. The earliest stages of ecosystem succession in high-elevation (5000 metres above sea level), recently deglaciated soils. *Proceedings of the Royal Society of London Series B Biological Sciences* **275**, 2793–2802.

Schmitz, O.J. 2008a. Herbivory from individuals to ecosystems. *Annual Review of Ecology, Evolution, and Systematics* **39**, 133–152.

Schmitz, O.J. 2008b. Effects of predator hunting mode on grassland ecosystem function. *Science* **319**, 952–954.

Schmitz, O.J., O. Ovadia and V. Krivan. 2004. Trophic cascades: the primary trait-mediated indirect interactions. *Ecology Letters* **7**, 153–163.

Schoenecker, K.A., F.J. Singer, R.S.C. Menezes, et al. 2002. Sustainability of vegetation communities grazed by elk in Rocky Mountain National Park. In: *Ecological Evaluation of the Abundance and Effects of Elk Herbivory in Rocky Mountain National Park 1994–1999* (F.J. Singer and L.C. Zeigenfuss, eds), pp. 187–204. US Geological Survey, Fort Collins, CO.

Schultz, J.C. and I.T. Baldwin. 1982. Oak leaf quality declines in response to defoliation by gypsy moth larvae. *Science* **217**, 149–151.

Schulze, E.-D. and H.A. Mooney (eds). 1993. *Biodiversity and Ecosystem Function*. Springer, Berlin.

Schuman, G.E., J.D. Reeder, J.T. Manley, et al. 1999. Impact of grazing management on the carbon and nitrogen balance of a mixed-grass rangeland. *Ecological Applications* **9**, 65–71.

Schütz, M., A.C. Risch, G. Achermann, et al. 2006. Phosphorus translocation by red deer on a subalpine grassland in the central European Alps. *Ecosystems* **9**, 624–633.

Schuur, E.A.G., J.G. Vogel, K.G. Crummer. 2009. The effect of permafrost thaw on old carbon release and net carbon exchange from tundra. *Nature* **459**, 556–559.

Schweitzer, J.A., J.K. Bailey, B.J. Rehill, et al. 2004. Genetically based trait in a dominant tree affects ecosystem processes. *Ecology Letters* **7**, 127–134.

Schweitzer, J.A., J.K. Bailey, D.G. Fischer, et al. 2008. Plant-soil-microorganism interactions: Heritable relationship between plant genotype and associated soil microorganisms. *Ecology* **89**, 773–781.

Seastedt, T.R. 1984. The role of microarthropods in decomposition and mineralization processes. *Annual Review of Entomology* **29**, 25–46.

Seastedt, T.R. and G.A. Adams. 2001. Effects of mobile tree islands on alpine tundra soil. *Ecology* **82**, 8–17.

Seitzinger, S.P., J. Harrison, J. Bohlke, et al. 2006. Denitrification across landscapes and waterscapes: a synthesis. *Ecological Applications.* 16, 2064–2090.

Selosse, M.A., F. Richard, X.H. He, et al. 2006. Mycorrhizal networks: des liaisons dangereuses? *Trends in Ecology and Evolution* **21**, 621–628.

Serreze, M.C., J.E. Walsh, F.S. Chapin, et al. 2000. Observational evidence of recent change in the northern high-latitude environment. *Climatic Change* **46**, 159–207.

Setälä, H. 1995. Growth of birch and pine seedlings in relation to grazing by soil fauna on ectomycorrhizal fungi. *Ecology* **76**, 1844–1851.

Setälä, H. and V. Huhta. 1991. Soil fauna increase *Betula pendula* growth: laboratory experiments with coniferous forest floor. *Ecology* **72**, 665–671.

Setälä, H. and M.A. McLean. 2004. Decomposition rate of organic substrates in relation to the species diversity of soil saprophytic fungi. *Oecologia* **139**, 98–107.

Setälä, H., V.G. Marshall and J.A. Trofymow. 1996. Influence of body size of soil fauna on litter decomposition and ^{15}N uptake by poplar in a pot trial. *Soil Biology and Biochemistry* **28**, 1661–1675.

Sharma, S., Z. Szele, R. Schilling, et al. 2006. Influence of freeze-thaw on the structure and function of microbial communities in soil. *Applied and Environmental Microbiology* **72**, 48–54.

Sharma, S., S. Courtier and S.D. Côte. 2009. Impacts of climate change on the seasonal distribution of migratory caribou. *Global Change Biology* **15**, 2549–2562.

Sharpley, A.N. and J.K. Syers. 1976. Potential role of earthworm cats for the phosphorus enrichment of run-off waters. *Soil Biology and Biochemistry* **8**, 341–346.

Sharpley, A.N., J.K. Syers and J.A. Springett. 1979. Effect of surface casting earthworms on the transport of phosphorus and nitrogen in surface runoff from pasture. *Soil Biology and Biochemistry* **11**, 459–462.

Shaver, G.R., L.C. Johnson, D.H. Cades, et al. 1998. Biomass and CO_2 flux in wet sedge tundras: Responses to nutrients, temperature, and light. *Ecological Monographs* **68**, 75–97.

Shaver, G.R., S.M. Bret-Harte, M.H. Jones, et al. 2001. Species composition interacts with fertilizer to control long-term change in tundra productivity. *Ecology* **82**, 3163–3181.

Shaw, M.R., E.S. Zavaleta, N.R. Chiariello, et al. 2002. Grassland responses to global environmental changes suppressed by elevated CO_2. *Science* **298**, 1987–1990.

Shearer, B.L., C.E. Craine, R.G. Fairman, et al. 1998. Susceptibility of plant species in coastal dune vegetation of southwestern Australia to killing by *Armillaria luteobubalina*. *Australian Journal of Botany* **46**, 321–334.

Shrestha, G. and P.D. Stahl. 2008. Carbon accumulation and storage in semi-arid sagebrush steppe: Effects of long-term grazing exclusion. *Agriculture, Ecosystems and Environment* **125**, 173–181.

Siemann, E., W.P. Carson, W.E. Rogers, et al. 2003. Reducing herbivory using insecticides. In: *Insects and Ecosystem Function* (W.W. Weisser and E. Siemann, eds), pp. 303–327. Springer-Verlag, Berlin.

Silfver, T., J. Mikola, H. Roininen, et al. 2007. Leaf litter decomposition differs among genotypes in a local *Betula pendula* population. *Oecologia* **152**, 707–714.

Silver, W.L., D.J. Herman and M.K. Firestone. 2001. Dissimilatory nitrate reduction to ammonium in tropical forest soils. *Ecology* **82**, 2410–2416.

Silver, W.L., A.W. Thompson, M.K. Firestone, et al. 2005. Nitrogen retention and loss in tropical plantations and old growth forests. *Ecological Applications* **15**, 1604–1614.

Simard, S.W., M.D. Jones and D.M. Durall. 2002. Carbon and nutrient fluxes within and between mycorrhizal plants. In: *Mycorrhizal Ecology* (M.G.A. Van der Heijden and I.R. Sanders, eds), pp. 33–74. Ecological Studies 157. Springer Verlag, Heidelberg.

Sinsabaugh, R.L., C.L. Lauber, M.N. Weintraub, et al. 2008. Stoichiometry of soil enzyme activity at a global scale. *Ecology Letters* **11**, 1252–1264.

Sitch, S., B. Smith, I.C. Prentice, et al. 2003. Evaluation of ecosystem dynamics, plant geography and terrestrial carbon cycling in the LPJ dynamic global vegetation model. *Global Change Biology* **9**, 161–185.

Six, J., S.D. Frey, R.K. Thiet, et al. 2006. Bacterial and fungal contributions to carbon sequestration in agroecosystems. *Soil Science Society of America Journal* **70**, 555–569.

Sjögersten, S., R. Van der Wal and S.J. Woodin. 2008. Habitat type determines herbivory controls over CO_2 fluxes in a warmer arctic. *Ecology* **89**, 2103–2116.

Slade, E.M., D.J. Mann, J.F. Villanueva, et al. 2007. Experimental evidence of the effects of dung beetle functional group richness and composition on ecosystem function in a tropical forest. *Journal of Animal Ecology* **76**, 1094–1104.

Slaytor, M. 2000. Energy metabolism in the termite and its gut microbiota. In: *Termites: Evolution, Sociality, Symbioses, Ecology* (T. Abe, D.E. Bignell and M. Higashi, eds), pp. 307–332. Kluwer Academic Press, Dordrecht.

Smith, P., D. Martino, Z. Cai, et al. 2008. Greenhouse gas mitigation in agriculture. *Philosophical Transactions of the Royal Society Series B Biological Sciences* **363**, 789–813.

Smith, R.S., R.S. Shiel, R.D. Bardgett, et al. 2008. Long-term change in vegetation and soil microbial communities during the phased restoration of traditional meadow grassland. *Journal of Applied Ecology* **45**, 670–679.

Smith, S.E. and D.J. Read. 1997. *Mycorrhizal Symbiosis*, 2nd edn. Academic Press, London.

Smith, V.R. and M. Steenkamp. 1990. Climatic change and its ecological implications at a subantarctic island. *Oecologia* **85**, 14–24.

Smith, V.R., N.L. Avenant and S.L. Chown. 2002. The diet and impact of house mice on a sub-Antarctic island. *Polar Biology* **25**, 703–715.

Snyder, W.E. and E.W. Evans. 2006. Ecological effects of invasive arthropod generalist predators. *Annual Review of Ecology and Systematics* **37**, 95–122.

Solan, M., B.J. Cardinale, A.L. Downing, et al. 2004. Extinction and ecosystem function in the marine benthos. *Science* **306**, 1177–1187.

Sørensen, L.I., M.M. Kytöviita, J. Olfsson, et al. 2008. Soil feedback on plant growth in a subarctic grassland as a result of repeated defoliation. *Soil Biology and Biochemistry* **40**, 2891–2897.

Sørensen, L.I., J. Mikola, M.M. Kytöviita, et al. 2009. Trampling and spatial heterogeneity explain decomposer abundances in a sub-arctic grassland subjected to simulated reindeer grazing. *Ecosystems* **12**, 830–842.

Soudzilovskaia, N.A., V.G. Onipchenko, J.H.C. Cornelissen, et al. 2007. Effects of fertilisation and irrigation on 'foliar afterlife' in alpine tundra. *Journal of Vegetation Science* **18**, 755–766.

Sousa-Souto, L., J.H. Schoereder and C.E.G.R. Schaeffer. 2007. Leaf-cutting ants, seasonal burning and nutrient distribution in Cerrado vegetation. *Austral Ecology* **32**, 758–765.

Sousa-Souto, L., J.H. Schoereder, C.E.G.R. Schaeffer, et al. 2008. Ant nests and soil nutrient availability: the negative impact of fire. *Journal of Tropical Ecology* **24**, 639–646.

Spain, A.V. and J.G. McIvor. 1988. The nature of herbaceous vegetation associated with termitaria in north-eastern Australia. *Journal of Ecology* **76**, 181–191.

Spear, D. and S.L. Chown. 2009. Non-indigenous ungulates as a threat to biodiversity. *Journal of Zoology* **279**, 1–17.

Speed, J.D.M., S.J. Woodin, H. Tømmervik, et al. 2009. Predicting habitat utilization and extent of ecosystem disturbance by an increasing herbivore population. *Ecosystems* **12**, 349–359.

Spehn, E.M., M. Scherer-Lorenzen, B. Schmid, et al. 2002. The role of legumes as a component of biodiversity in a cross-European study of grassland biomass nitrogen. *Oikos* **98**, 205–218.

Sprent, J.I. and R. Parsons. 2000. Nitrogen fixation in legume and non-legume trees. *Field Crops Research* **65**, 183–196.

Srivastava, D.S. and R.L. Jefferies. 1986. A positive feedback: herbivory, plant growth, salinity, and the desertification of an Arctic salt-marsh. *Journal of Ecology* **84**, 31–42.

St John M.G., D.H. Wall and H.W. Hunt. 2006. Are soil mite assemblages structured by the identity of native and invasive alien grasses? *Ecology* **87**, 1314–1324.

Staddon, P.L., N. Ostle and A.H. Fitter. 2003. Earthworm extraction by electroshocking does not affect canopy CO_2 exchange, root respiration, mycorrhizal fungal abundance or mycorrhizal fungal vitality. *Soil Biology and Biochemistry* **35**, 421–426.

Staddon, P.L., I. Jakonsen and H. Blum. 2004. Nitrogen input mediates the effects of free-air CO_2 enrichment on mycorrhizal fungal abundance. *Global Change Biology* **10**, 1687–1688.

Stadler, B. and A.F.G. Dixon. 2005. Ecology and evolution of aphid-ant interactions. *Annual Reviews of Ecology, Evolution and Systematics* **36**, 345–372.

Stadler, B., S.T. Solinger and B. Michalzik. 2001. Insect herbivores and the nutrient flow from the canopy to the soil in coniferous and deciduous forests. *Oecologia* **126**, 104–113.

Stamp, N.E. and M.D. Bowers. 1996. Consequences for plantain chemistry and growth when herbivores are attacked by predators. *Ecology* **77**, 535–549.

Stampe, E.D. and C.C. Daehler. 2003. Mycorrhizal species identity affects plant community structure and invasion: a microcosm study. *Oikos* **100**, 362–372.

Standish, R.J., P.A. Williams, A.W. Robertson, et al. 2004. Invasion by a perennial herb increases decomposition rate and alters nutrient availability in warm temperate lowland forest remnants. *Biological Invasions* **6**, 71–81.

Stark, S., D.A. Wardle, R. Ohtonen, et al. 2000. The effect of reindeer grazing on decomposition, mineralisation and soil biota in a dry oligotrophic Scots pine forest. *Oikos* **90**, 301–310.

Stark, S., J. Tuomi, R. Strömmer, et al. 2003. Non-parallel changes in soil microbial carbon and nitrogen dynamics due to reindeer grazing in northern boreal forests. *Ecography* **26**, 51–59.

Steinbeiss, S., H. Bessler, C. Engels, et al. 2008. Plant diversity positively affects short-term soil carbon storage in experimental grasslands. *Global Change Biology* **14**, 2937–2949.

Steinmann, K.R., T.W. Siegwolf, M. Saurer, et al. 2004. Carbon fluxes to the soil in a mature temperate forest assessed by ^{13}C isotope tracing. *Oecologia* **141**, 489–501.

Stephan, A., A.H. Meyer and B. Schmid. 2000. Plant diversity affects culturable soil bacteria in experimental grassland communities. *Journal of Ecology* **88**, 988–998.

Stevens, C.J., N.D. Dise, J.O. Mountford, et al. 2004. Impact of nitrogen deposition on the species richness of grasslands. *Science* **303**, 1876–1879.

Stevenson, B.G. and D.L. Dindal. 1987. Insect effects on decomposition of cow dung in microcosms. *Pedobiologia* **30**, 81–92.

Stinson, K.A., S.A. Campbell, J.R. Powell, et al. 2006. Invasive plant suppresses the growth of native tree seedlings by disrupting belowground mutualisms. *PLOS Biology* **4**, 727–731.

Stowe, L.G. 1979. Allelopathy and its influence on the distribution of plants in an Illinois old field. *Journal of Ecology* **67**, 1065–1085.

Strickland, M.S., E. Osburn, C. Lauber, et al. 2009. Litter quality is in the eye of the beholder: initial decomposition rates as a function of inoculum characteristics. *Functional Ecology* **23**, 627–636.

Strong, D.R. 1992. Are trophic cascades all wet? Differentiation and donor control in speciose ecosystems. *Ecology* **73**, 747–754.

Sturm, M., C. Racine and K. Tape. 2001. Climate change: increasing shrub abundance in the Arctic. *Nature* **411**, 546–547.

Sturm, M., T. Douglas, C. Racine and G.E. Liston. 2005. Changing snow and shrub conditions affect albedo with global implications. *Journal of Geophysical Research-Biogeosciences* **110**, G01004.

Styrsky, J.D. and M.D. Eubanks. 2007. Ecological consequences of interactions between ants and honeydew-producing insects. *Proceedings of the Royal Society London B.*, **274**, 151–164.

Subler, S., C.M. Baranski and C.A. Edwards. 1997. Earthworm additions increased short-term nitrogen availability and leaching in two grain-crop agroecosystems. *Soil Biology and Biochemistry* **29**, 413–421.

Suding, K.N., I.W. Ashton, H. Bechtold, et al. 2008. Plant and microbe contribution to community resilience in a directionally changing environment. *Ecological Monographs* **78**, 313–329.

Susiluoto, S., T. Rasilo, J. Pumpanen, et al. 2008. Effects of grazing on the vegetation structure and carbon dioxide exchange of a Fennoscandian fell ecosystem. *Arctic, Antarctic and Alpine Research* **40**, 422–431.

Swift, M.J., O.W. Heal and J.M. Anderson. 1979. *Decomposition in Terrestrial Ecosystems.* Blackwell, Oxford.

Syers, J.K., P.E.H. Gregg and A.G. Gillingham. 1980. Phosphorus uptake and return in grazed sheep pastures. II. Above-ground components of the phosphorus cycle. *New Zealand Journal of Agricultural Research* **23**, 323–330.

Takatsuki, S. 2009. Effects of sika deer on vegetation in Japan: a review. *Biological Conservation* **142**, 1922–1929.

Tansley, A.G. 1935. The use and abuse of vegetational terms and concepts. *Ecology* **16**, 284–307.

Tao, J., X. Chen, M. Liu, et al. 2009. Earthworms change the abundance and community structure of nematodes and protozoa in a maize residue amended rice–wheat rotation agroecosystem. *Soil Biology and Biochemistry*, **41**, 898–904.

Tape, K., M. Sturm, C. Racine. 2006. The evidence for shrub expansion in Northern Alaska and the Pan-Arctic. *Global Change Biology* **12**, 686–702.

Taylor, B.R., W.F.J. Pardons and D. Parkinson. 1989. Nitrogen and lignin content as predictors of decomposition rate: a microcosm test. *Ecology* **70**, 97–104.

Taylor, S.W., A.L. Carroll, R. Alfaro, et al. 2006. Forest, climate and mountain pine beetle outbreak dynamics in western Canada. In: *The Mountain Pine Beetle: A Synthesis of*

Biology, Management and Impacts in Lodgepole Pine (L. Safranyik and B. Wilson, eds), pp. 67–94. Natural Resources Canada, Canadian Forest Service, Victoria.

Templer, P.H., W.L. Silver, J. Pett-Ridge, et al. 2008. Plant and microbial controls on nitrogen retention and loss in a humid tropical forest. *Ecology* **89**, 3030–3040.

Terborgh, J.K. and J.A. Estes (eds). 2010. *Trophic Cascades.* Island Press, Washington DC, in press.

Terborgh, J., L. Lopez, P.N. Nunez, et al. 2001. Ecological meltdown in predator-free forest fragments. *Science* **294**, 1923–1926.

Terborgh, J., K. Feeley, M. Silman, et al. 2006. Vegetation dynamics of predator-free land-bridge islands. *Journal of Ecology* **94**, 253–263.

Teste, F.P. and S.W. Simard. 2008. Mycorrhizal networks and distance from mature trees alter patterns of competition and facilitation in dry Douglas-fir forests. *Oecologia* **159**, 193–203.

Teuben, A. and H.A. Verhoef. 1992. Direct contribution by soil arthropods to nutrient availability through body and faecal nutrient content. *Biology and Fertility of Soils* **14**, 71–75.

Thébault, E. and M. Loreau. 2003. Food web constraints on biodiversity-ecosystem functioning relationships. *Proceedings of the National Academy of Sciences USA* **100**, 14949–14954.

Thing, H. 1984. Feeding ecology of the West Greenland caribou (*Rangifer tarandus groenlandicus*) in the Sisimiut-Kangerlussuaq region. *Danish Review of Game Biology* **12**, 1–51.

Thomas, C.D., A. Cameron, R.E. Green, et al. 2004. Extinction risk from climate change. *Nature* **427**, 145–148.

Thompson, C.H. 1981. Podzol chronosequences on coastal sand dunes of eastern Australia. *Nature* **291**, 59–61.

Thompson, K.H., A. Green and A.M. Jewels. 1994. Seeds in soil and worm casts from a neutral grassland. *Functional Ecology* **8**, 29.35.

Thompson, K., Hodgson, J.G. and T.C.G. Rich. 1995. Native and alien invasive plants: more of the same? *Ecography* **18**, 390–402.

Thorpe, A.S., G.T. Thelan, A. Diaconu, et al. 2009. Root exudate is allelopathic in invaded community but not in native community: field evidence for the novel weapons hypothesis. *Journal of Ecology* **97**, 641–645.

Throop, H.L. and M.T. Lerdau. 2004. Effects of nitrogen deposition on insect herbivory: Implications for community and ecosystem processes. *Ecosystems* **7**, 109–133.

Throop, H.L. and S.R. Archer. 2008. Shrub (*Prosopis velutina*) encroachment in a semidesert grassland: spatial–temporal changes in soil organic carbon and nitrogen pools. *Global Change Biology* **14**, 2420–2431.

Thuiller, W., S. Lavorel, M.B. Araújo, et al. 2005. Climate change threats to plant diversity in Europe. *Proceedings of the National Academy of Sciences USA* **102**, 8245–8250.

Tierney, T. and J.H. Cushman. 2006. Temporal changes in native and exotic vegetation and soil characteristics following disturbances by feral pigs in a Californian grassland. *Biological Invasions* **8**, 1073–1089.

Tilman, D. 1999. The ecological consequences of changes in biodiversity: the search for general principles. *Ecology* **80**, 1455–1474.

Tilman, D., D. Wedin and J. Knops. 1996. Productivity and sustainability influenced by biodiversity in grassland ecosystems. *Nature* **379**, 718–720.

Tilman, D., J. Knops, D. Wedin, et al. 1997. The influence of functional diversity and composition on ecosystem processes. *Science* **277**, 1300–1302.

Tilman, D., P.B. Reich, J.M.H. Knops. 2006. Biodiversity and ecosystem stability in a decade-long grassland experiment. *Nature* **441**, 629–632.

Tiunov, A.V. and S. Scheu. 1999. Microbial respiration, biomass, biovolume and nutrient status in burrow walls of *Lumbricus terrestris* L. (Lumbricidae). S*oil Biology and Biochemistry* **31**, 2039–2048.

Tiunov, A.V. and S. Scheu. 2005. Facilitative interactions ratherthan resource partitioning drive diversity-functioning relationships in laboratory fungal communities. *Ecology Letters* **8**, 618–625.

Tjoelker, M.G., J.M. Craine, D. Wedin, et al. 2005. Linking leaf and root trait syndromes among 39 grassland and savannah species. *New Phytologist* **167**, 493–508.

Torsvik, V., L. Ovreas and T.F. Thingstad. 2002. Prokaryotic diversity—Magnitude, dynamics, and controlling factors. *Science* **296**, 1064–1066.

Towne, E.G. 2000. Prairie vegetation and soil nutrient responses to ungulate carcasses. *Oecologia* **122**, 232–239.

Towns, D.R., D.A. Wardle, C.P.H. Mulder, et al. 2009. Predation of seabirds by invasive rats: multiple indirect consequences for invertebrate communities. *Oikos* **118**, 420–430.

Townsend, A.R., C.C. Cleveland, G.P. Asner, et al. 2007. Controls over foliar N : P ratios in tropical rain forests. *Ecology* **88**, 107–118.

Tracy, B.F. and D.A. Frank. 1998. Herbivore influence on soil microbial biomass and N mineralization in a northern grassland ecosystem: Yellowstone National Park. *Oecologia* **114**, 556–562.

Trenbath, B.R. 1974. Biomass productivity of mixtures. *Advances in Agronomy* **26**, 177–210.

Treseder, K.K. 2008. Nitrogen additions and microbial biomass: a meta-analysis of ecosystem studies. *Ecology Letters* **11**, 1111–1120.

Treseder, K.K. and P.M. Vitousek. 2001. Potential ecosystem-level effects of genetic variation among populations of *Metrosideros polymorpha* from a soil fertility gradient in Hawaii. *Oecologia* **126**, 266–275.

Trumbore, S. 2006. Carbon respired by terrestrial ecosystems–recent progress and challenges. *Global Change Biology* **12**, 141–153.

Tscharntke, T., A.M. Klein, A. Kruess, et al. 2005. Landscape perspectives on agricultural intensification and biodiversity—ecosystem service management. *Ecology Letters* **8**, 857–874.

Turetsky, M.R. 2003. The role of bryophytes in carbon and nitrogen cycling. *Bryologist* **196**, 395–409.

Turnbull, M.H., D.T. Tissue, K.L. Griffin, et al. 2005. Respiration characteristics in temperate rainforest tree species differ along a long-term soil-development chronosequence. *Oecologia* **143**, 271–279.

Turner, B.L. 2008. Resource partitioning for soil phosphorus: a hypothesis. *Journal of Ecology* **96**, 698–702.

Tylianakis, J.M., R.K. Didham, J. Bascompte, et al. 2008. Global change and species interactions in terrestrial ecosystems. *Ecology Letters* **11**, 1351–1363.

Uroz, S., M. Buée, C. Murat, et al. 2010. Pyrosequencing reveals a contrasted bacterial diversity between oak rhizosphere and surrounding soil. *Environmental Microbiology Reports*, in press.

Van Auken, O.W. 2000. Shrub invasions of North American semiarid grasslands. *Annual Reviews of Ecology and Systematics* **31**, 197–215.

Van Calster, H., R. Vandenberghe, M. Ruysen, et al. 2008. Unexpectedly high 20th century floristic losses in a rural landscape in northern France. *Journal of Ecology* **96**, 927–936.

Van der Heijden, M.G.A. and T.R. Horton. 2009. Socialism in soil? The importance of mycorrhizal fungal networks for facilitation in natural ecosystems. *Journal of Ecology* **97**, 1139–1150.

Van der Heijden, M.G.A., T. Boller, A. Wiemken, et al. 1998a. Different arbuscular mycorrhizal fungal species are potential determinants of plant community structure. *Ecology* **79**, 2082–2091.

Van der Heijden, M.G.A., J.N. Klironomos, M. Ursic, et al. 1998b. Mycorrhizal fungal diversity determines plant biodiversity, ecosystem variability and productivity. *Nature* **396**, 72–75.

Van der Heijden, M.G.A., R. Bakker, J. Verwaal, et al. 2006a. Symbiotic bacteria as a determinant of plant community structure and plant productivity in dune grassland. *FEMS Microbiology Ecology* **56**, 178–187.

Van der Heijden, M.G.A., R. Streitwolf-Engel, R. Riedl, et al. 2006b. The mycorrhizal contribution to plant productivity, plant nutrition and soil structure in experimental grassland. *New Phytologist* **172**, 739–752.

Van der Heijden, M.G.A., R.D. Bardgett and N.M. van Straalen. 2008. The unseen majority: soil microbes as drivers of plant diversity and productivity in terrestrial ecosystems. *Ecology Letters* **11**, 296–310.

Vanderhoeven, S., N. Dassonville and P. Meerts. 2005. Increased topsoil nutrient concentrations under invasive exotic plants in Belgium. *Plant and Soil* **275**, 169–179.

van de Koppel, J., M. Rietkerk and F.J. Weissing. 1997. Catastrophic vegetation shifts and soil degradation in terrestrial grazing ecosystems. *Trends in Ecology and Evolution* **12**, 352–356.

van de Koppel, J., R.D. Bardgett, J. Bengtsson, et al. 2005. Trophic interactions in a changing world: The role of spatial scale. *Ecosystems* **8**, 801–807.

Van der Krift, T.A.J., P.J. Kuikman and F. Berendse. 2002. The effect of living plants on root decomposition of four grass species. *Oikos* **96**, 36–45.

Vandermeer, J. 1990. *The Ecology of Intercropping*. Cambridge University Press, Cambridge.

Van der Putten, W.H. 2003. Plant defense belowground and spatiotemporal processes in natural vegetation. *Ecology* **84**, 2269–2280.

Van der Putten, W.H. 2005. Plant-soil bedback and plant diversity affect the composition of plant communities. In *Biological Diversity and Function in Soils* (R.D. Bardgett, M.B. Usher and D.W. Hopkins, eds), pp. 250–272. Cambridge University Press, Cambridge.

Van der Putten, W.H. 2009. A multitrophic perspective on the functioning and evolution of facilitation in plant communities. *Journal of Ecology* **97**, 1131–1138.

Van der Putten, W.H., L.E.M. Vet and B.A.M. Peters. 1993. Plant-specific soil-borne diseases contribute to succession in foredune vegetation. *Nature* **362**, 53–56.

Van der Putten, W.H., L. Vet, J.A. Harvey, et al. 2001. Linking above- and belowground multitrophic interactions of plants, herbivores and their antagonists. *Trends in Ecology and Evolution* **16**, 547–554.

Van der Putten, W.H., J.N. Klironomos and D.A. Wardle. 2007. Microbial ecology of biological invasions. *The ISME Journal* **1**, 28–37.

Van der Putten, W.H., R.D. Bardgett, P.C. de Ruiter, et al. 2009. Empirical and theoretical challenges in aboveground-belowground ecology. *Oecologia* **161**, 1–14.

Van der Wal, A., J.A. van Veen, W. Smant, et al. 2006. Fungal biomass development in a chronosequence of land abandonment. *Soil Biology and Biochemistry*, **38**, 51–60.

Van der Wal, R. 2006. Do herbivores cause habitat degradation or vegetation state transition? Evidence from the tundra. *Oikos* **114**, 117–186.

Van der Wal, R. and R.W. Brooker. 2004. Mosses mediate grazer impacts on grass abundance in arctic ecosystems. *Functional Ecology* **18**, 77–86.

Van der Wal, R., R. Brooker, E. Cooper, et al. 2001. Differential effects of reindeer on high Arctic lichens. *Journal of Vegetation Science* **12**, 705–710.

Van der Wal, R., I.S.K. Pearce, R. Brooker, et al. 2003. Interplay between nitrogen deposition and grazing causes habitat degradation. *Ecology Letters* **6**, 141–146.

Van der Wal, R., R.D. Bardgett, K.A. Harrison, et al. 2004. Vertebrate herbivores and ecosystem control: cascading effects of faeces on tundra ecosystems. *Ecography* **27**, 242–252.

Van der Wal, R., S. Sjögersten, S.J. Woodin, et al. 2007. Spring feeding by pink-footed geese reduces carbon stocks and sink strength in tundra ecosystems. *Global Change Biology* **13**, 539–545.

Van Grunsven, R.H.A., W.H. Van der Putten, T.M. Bezemer, et al. 2007. Reduced plant-soil feedback of plant species expanding their range as compared to natives. *Journal of Ecology* **95**, 1050–1057.

Van Grunsven, R.H.A., W.H. Van der Putten, T.M. Bezemer, et al. 2010. Plant–soil interactions in the expansion and native range of a poleward shifting plant species. *Global Change Biology* **16**, 380–385.

Van Mantgem, P.J., N.L. Stephensen, J.C. Byrne, et al. 2009. Widespeard increase of tree mortality rates in western United States. *Science* **323**, 521–524.

Van Straalen, N.M. and D. Roelofs. 2006. *An Introduction to Ecological Genomics*. Oxford University Press, Oxford.

Van Wijnen, H.J. and R. Van der Wal. 1999. The impact of herbivores on nitrogen mineralisation rate: Consequences for salt-marsh succession. *Oecologia* **118**, 225–231.

Vasconcelis, H.L. and J.M. Cherrett. 1995. Changes in leaf-cutting ant populations (Formicidae: Attini) after the clearing of mature forest in Brazilian Amazonia. *Studies on Neotropical Fauna and Environment* **30**, 107–113.

Vasconcelis, H.L., E.H.M. Vierea-Neto, F.M. Mundim, et al. 2006. Roads alter the colonization dynamics of a keystone herbivore in neotropical savannas. *Biotropica* **38**, 661–665.

Vázquez, D. 2002. Multiple effects of introduced mammalian herbivores in a temperate forest. *Biological Invasions* **4**, 175–191.

Veen, G.F., Olff, H., Duyts, et al. 2010. Vertebrate herbivores influence soil nematodes by modifying plant communities. *Ecology*, in press.

Venette, R.C. and S.D. Cohen. 2006. Potential climate suitability for establishment of *Phytophthora ramorum* within the contiguous United States. *Forest Ecology and Management* **231**, 18–26.

Verchot, L.V., P.M. Groffman and D.A. Frank. 2002. Landscape versus ungulate control of gross mineralisation and gross nitrification in semiarid grassland of Yellowstone National Park. *Soil Biology and Biochemistry* **34**, 1691–1699.

Verchot, L.V., P.R. Moutinho and E.A. Davidson. 2003. Leaf-cutting ant (*Atta sexdens*) and nutrient cycling: deep soil inorganic nitrogen stocks, mineralization, and nitrification in Eastern Amazonia. *Soil Biology and Biochemistry* **35**, 1219–1222.

Verstraete, M.M., R.J. Scholes and M.S. Smith. 2009. Climate and desertification: looking at an old problem through new lenses. *Frontiers in Ecology and the Environment* **7**, 421–428.

Viketoft, M. 2008. Effects of six grassland plant species on soil nematodes: A glasshouse experiment. *Soil Biology and Biochemistry* **40**, 906–915.

Vilcheck, G. 1997. Arctic ecosystem stability and disturbance: a West-Siberian case history. In: *Disturbance and Recovery in Arctic Lands, an Ecological Perspective* (R.M.M. Crawford, ed.), pp. 179–189. Kluwer Academic Press, Dordrecht.

Vile, D., B. Shipley and E. Garnier. 2006. Ecosystem productivity can be predicted from potential relative growth rate and species abundance. *Ecology Letters* **9**, 1061–1067.

Visser, M.E. and C. Both. 2005. Shifts in phenology due to global climate change: the need for a yardstick. *Proceedings of the Royal Society B-Biological Sciences* **272**, 2561–2569.

Vitousek, P.M. 2004. *Nutrient Cycling and Limitation: Hawai'i as a Model System*. Princeton University Press, Princeton, NJ.

Vitousek, P.M. and R.W. Howarth. 1991. Nitrogen limitation on land and in the sea – how can it occur. *Biogeochemistry* **13**, 87–115.

Vitousek, P.M. and P.A. Matson. 1988. Nitrogen transformations in a range of tropical forest soils. *Soil Biology and Biochemistry* **20**, 361–367.

Vitousek, P.M. and L.R. Walker. 1989. Biological invasion by *Myrica faya* in Hawaiii: plant demography, nitrogen fixation, ecosystem effects. *Ecological Monographs* **59**, 247–265.

Vitousek, P.M., L.R. Walker and L.D. Whiteaker. 1987. Biological. Invasion by Myerica faya alters ecosystem development in Hawaii. *Science* **238**, 802–804.

Vitousek, P.M., J.D. Aber, R.W. Howarth, et al. 1997a. Human alteration of the global nitrogen cycle: sources and consequences. *Ecological Applications* **7**, 737–750.

Vitousek, P.M., C.M. D'Antonio, L.L. Loope, et al. 1997b. Introduced species: A significant component of human-caused global change. *New Zealand Journal of Ecology* **21**, 1–16.

Vitousek, P.M., H.A. Mooney, J. Lubchenco, et al. 1997c. Human domination of the Earth's ecosystems. *Science* **277**, 494–499.

Vivanco, L. and A.T. Austin. 2008. Tree species identity alters forest litter decomposition through long term plant and soil interactions in Patagonia, Argentina. *Journal of Ecology* **96**, 727–736.

Vogelsang, K.M., H.L. Reynolds and J.D. Bever. 2006. Mycorrhizal fungal identity and richness determine the diversity and productivity of a tallgrass prairie system. *New Phytologist* **172**, 554–562.

Vossbrink, C.R., D.C. Coleman and T.A. Wooley. 1979. Abiotic and biotic factors in litter decomposition in a semiarid grassland. *Ecology* **60**, 265–271.

Vreeken-Buijs, M.J., M. Geurs, P.C. de Ruiter, et al. 1997. The effects of bacterivorous mites and amoebae on mineralization in a detrital based below-ground food web: microcosm experiment and simulation of interactions. *Pedobiologia* **41**, 481–493.

Vtorov, I.P. 1993. Feral pig removal: effects on soil microarthropods in a Hawaiian rain forest. *Journal of Wildlife Management* **57**, 875–880.

Wachinger, G., S. Fiedler, K. Zepp, et al. 2000. Variability of soil methane production on the micro-scale: spatial association with hotspots of organic material and Archaeal populations. *Soil Biology and Biochemistry* **32**, 1121–1130.

Wait, D.A., D.P. Aubrey and W.B. Anderson. 2005. Seabird guano influences on desert islands: soil chemistry and herbaceous species richness and productivity. *Journal of Arid Environments* **60**, 681–695.

Waldrop, M.P., D.R. Zak and R.L. Sinsabaugh. 2004. Microbial community responses to nitrogen deposition in northern forested ecosystems. *Soil Biology and Biochemistry* **36**, 1443–1451.

Walker, J., C.H. Thompson, P. Reddell, et al. 2001. The importance of landscape age in influencing landscape health. *Ecosystem Health* **7**, 7–14.

Walker, L.R. 1989. Soil nitrogen changes during primary succession on a floodplain in Alaska, U.S.A. *Arctic and Alpine Research* **21**, 341–349.

Walker, L.R. and R. del Moral. 2003. *Primary Succession and Ecosystem Rehabilitation.* Cambridge University Press, Cambridge.

Walker, M.D., C.H. Wahren, R.D. Hollister RD, et al. 2006. Plant community responses to experimental warming across the tundra biome. *Proceedings of the National Academy of Sciences USA* **103**, 1342–1346.

Walker, T.W. and J.K. Syers. 1976. The fate of phosphorus during pedogenesis. *Geoderma* **15**, 1–19.

Wall, D.H. 2007. Global change tipping points: Above- and below-ground biotic interactions in a low diversity ecosystem. *Philosophical Transactions of the Royal Society of London Series B Biological Sciences* **362**, 2291–2306.

Wall, D.H., M.A. Bradford, M.G. St John, et al. 2008. Global decomposition experiment shows soil animal impacts on decomposition are climate-dependent. *Global Change Biology* **14**, 2661–2677.

Wallenstein, M.D., S. McMahon and J.P. Schimel. 2007. Bacterial and fungal community structure in Arctic tundra tussock and shrub soils. *Fems Microbiology Ecology* **59**, 428–435.

Walsingham, J.M. 1976. Effect of sheep grazing on the invertebrate population of agricultural grassland. *Proceedings of the Royal Society of Dublin* **11**, 297–304.

Walther, G.R., E. Post, P. Convey, et al. 2002. Ecological responses to recent climate change. *Nature* **416**, 389–395.

Walther, G.R., S. Berger and M.T. Sykes. 2005. An ecological 'footprint' of climate change. *Proceedings of the Royal Society of London Series B Biological Sciences* **272**, 1427–1432.

Ward, C.M. 1988. Marine terraces of the Waitutu district and their relation to the late Cenozoic tectonics of the southern Fiordland region, New Zealand. *Journal of the Royal Society of New Zealand* **18**, 1–28.

Ward, S.E., R.D. Bardgett, N.P. McNamara, et al. 2007. Long-term consequences of grazing and buring on northern peatland carbon dynamics. *Ecosystems* **10**, 1069–1083.

Ward, S.E., R.D. Bardgett, N.P. McNamara, et al. 2009. Plant functional group identity influences short-term peatland ecosystem carbon flux: evidence from a plant removal experiment. *Functional Ecology* **23**, 454–462.

Wardle, D.A. 1992. A comparative assessment of factors which influence microbial biomass carbon and nitrogen levels in soils. *Biological Reviews* **67**, 321–358.

Wardle, D.A. 1999. Is "sampling effect" a problem for experiments investigating biodiversity – ecosystem function relationships? *Oikos* **87**, 403–407.

Wardle, D.A. 2002. *Communities and Ecosystems: Linking the Aboveground and Belowground Components.* Princeton University Press, Princeton, NJ.

Wardle, D.A. 2006. The influence of biotic interactions on soil biodiversity. *Ecology Letters* **9**, 870–886.

Wardle, D.A. 2010. Trophic cascades, aboveground—belowground linkages, and ecosystem functioning. In: *Trophic Cascades* (J. Terborgh and J. Estes, eds). Island Press, Washington DC, in press.

Wardle, D.A. and G.W. Yeates. 1993. The dual importance of competition and predation as regulatory forces in terrestrial ecosystems: evidence from decomposer food-webs. *Oecologia* **93**, 303–306.

Wardle, D.A. and K.S. Nicholson. 1996. Synergistic effects of grassland plant species on soil microbial biomass and activity: implications for ecosystem-level effects of enriched plant diversity. *Functional Ecology* **10**, 410–416.

Wardle, D.A. and P. Lavelle. 1997. Linkages between soil biota, plant litter quality and decomposition. In: *Driven by Nature—Plant Litter Quality and Decomposition* (K.E. Giller and G. Cadisch, eds), pp. 107–124. CAB International, Wallingford.

Wardle, D.A. and W. Van der Putten. 2002. Biodiversity, ecosystem functioning and above-ground-belowground linkages. In: *Biodiversity and Ecosystem Functioning* (M. Loreau, S. Naeem and P. Inchausti, eds), pp. 155–168. Oxford University Press, Oxford.

Wardle, D.A. and R.D. Bardgett. 2004. Human-induced changes in densities of large herbivorous mammals: consequences for the decomposer subsystem. *Frontiers in Ecology and the Environment* **2**, 145–153.

Wardle, D.A. and O. Zackrisson. 2005. Effects of species and functional group loss on island ecosystem properties. *Nature* **435**, 806–810.

Wardle, D.A. and M. Jonsson. 2010. Biodiversity loss in real ecosystems: a response to Duffy. *Frontiers in Ecology and the Environment* **8**, 10–11.

Wardle, D.A., K.S. Nicholson and A. Rahman. 1995. Ecological effects of the invasive weed species *Senecio jacobaea* L. (ragwort) in a New Zealand pasture. *Agriculture, Ecosystems and Environment* **56**, 19–28.

Wardle, D.A., K.I. Bonner and K.S. Nicholson. 1997a. Biodiversity and plant litter: experimental evidence which does not support the view that enhanced species richness improves ecosystem function. *Oikos* **79**, 247–258.

Wardle, D.A., O. Zackrisson, G. Hörnberg, et al. 1997b. Influence of island area on ecosystem properties. *Science* **277**, 1296–1299.

Wardle, D.A., G.M. Barker, K.I. Bonner, et al. 1998a. Can comparative approaches based on plant ecophysiological traits predict the nature of biotic interactions and individual plant species effects in ecosystems? *Journal of Ecology* **86**, 405–420.

Wardle, D.A., M.-C. Nilsson, C. Gallet, et al. 1998b. An ecosystem level perspective of allelopathy. *Biological Reviews* **73**, 305–319.

Wardle, D.A., H.A. Verhoef and M. Clarholm. 1998c. Trophic relationships in the soil microfood-web: predicting the responses in a changing global environment. *Global Change Biology* **4**, 713–727.

Wardle, D.A., K.I. Bonner, G.M. Barker, et al. 1999. Plant removals in perennial grassland: vegetation dynamics, decomposers, soil biodiversity and ecosystem properties. *Ecological Monographs* **69**, 535–568.

Wardle, D.A., K.I. Bonner and G.M. Barker. 2000. Stability of ecosystem properties in response to above-ground functional group richness and composition. *Oikos* **89**, 11–23.

Wardle, D.A., G.M. Barker, G.W. Yeates, et al. 2001. Impacts of introduced browsing mammals in New Zealand forests on decomposer communities, soil biodiversity and ecosystem properties. *Ecological Monographs* **71**, 587–614.

Wardle, D.A., K.I. Bonner and G.M. Barker. 2002. Linkages between plant litter decomposition, litter quality, and vegetation responses to herbivores. *Functional Ecology* **16**, 585–595.

Wardle, D.A., G. Hörnberg, O. Zackrisson, et al. 2003a. Long term effects of wildfire on ecosystem properties across an island area gradient. *Science* **300**, 972–975.

Wardle, D.A., M.-C. Nilsson, O. Zackrisson, et al. 2003b. Determinants of litter mixing effects in a Swedish boreal forest. *Soil Biology and Biochemistry* **35**, 827–835.

Wardle, D.A., G.W. Yeates, G.M. Barker, et al. 2003c. Island biology and ecosystem functioning in epiphytic soil communities. *Science* **301**, 1717–1720.

Wardle, D.A., G.W. Yeates, W. Williamson, et al. 2003d. The response of a three trophic level soil food web to the identity and diversity of plant species and functional groups. *Oikos* **102**, 45–56.

Wardle, D.A., R.D. Bardgett, J.N. Klironomos, et al. 2004a. Ecological linkages between aboveground and belowground biota. *Science* **304**, 1629–1633.

Wardle, D.A., L.R. Walker and R.D. Bardgett. 2004b. Ecosystem properties and forest decline in contrasting long-term chronosequences. *Science* **305**, 509–513.

Wardle, D.A., G.W. Yeates, W.M. Williamson, et al. 2004c. Linking aboveground and belowground communities: the indirect influence of aphid species identity and diversity on a three trophic level soil food web. *Oikos* **107**, 283–294.

Wardle, D.A., W.M. Williamson, G.W. Yeates, et al. 2005. Trickle-down effects of aboveground trophic cascades on the soil food web. *Oikos* **111**, 348–358.

Wardle, D.A., G.W. Yeates, G.M. Barker, et al. 2006. The influence of plant litter diversity on decomposer abundance and diversity. *Soil Biology and Biochemistry* **38**, 1052–1062.

Wardle, D.A., P.J. Bellingham, C.P.H. Mulder, et al. 2007. Promotion of ecosystem carbon sequestration by invasive predators. *Biology Letters* **3**, 479–482.

Wardle, D.A., A. Lagerström and M.-C. Nilsson. 2008a. Context dependent effects of plant species and functional group loss on vegetation invasibility across an island area gradient. *Journal of Ecology* **96**, 1174–1186.

Wardle, D.A., M.-C. Nilsson and O. Zackrisson. 2008b. Fire-derived charcoal causes loss of forest humus. *Science* **320**, 629.

Wardle, D.A., S.K. Wiser, R.B. Allen, et al. 2008c. Aboveground and belowground effects of single tree removals after forty years in a New Zealand temperate rainforest. *Ecology* **89**, 1232–1245.

Wardle, D.A., R.D. Bardgett, L.R. Walker, et al. 2009a. Among- and within-species variation in plant litter decomposition in contrasting long term chronosequences. *Functional Ecology* **23**, 442–453.

Wardle, D.A., P.J. Bellingham, K.I. Bonner, et al. 2009b. Indirect effects of invasive predators on plant litter quality, decomposition and nutrient resorption on seabird-dominated islands. *Ecology* **90**, 452–464.

Warren, M.S., J.K. Hill, J.A. Thomas, et al. 2001. Rapid responses of British butterflies to opposing forces of climate and habitat change. *Nature* **414**, 65–69.

Warrington, S. and J.B. Whittaker. 1985. An experimental field study of different levels of insect herbivory induced by *Formica rufa* predation on sycamore (*Acer pseudoplatanus*). I. *Lepidoptera larvae. Journal of Applied Ecology* **22**, 775–785.

Wasilewska, L. 1994. The effect of age of meadows on succession and diversity in soil nematode communities. *Pedobiologia* **38**, 1–11.

Webb, D.P. 1977. Regulation of deciduous forest litter decomposition by soil arthropod feces. In: *The role of Arthropods in Forest Ecosystems* (W.J. Mattson, ed.), pp. 57–69. Springer-Verlag, Berlin.

Wedin, D. and D. Tilman. 1993. Competition among grasses along a nitrogen gradient: initial conditions and mechanisms of competition. *Ecological Monographs* **63**, 199–229.

Weedon, J.T., W.C. Cornwell, J.H.C. Cornelissen, et al. 2009. Global meta-analysis of wood decomposition rates: a role for trait variation among tree species? *Ecology Letters* **12**, 45–56.

Weigelt, A., R. Bol and R.D. Bardgett. 2005. Preferential uptake of soil nitrogen forms by grassland plant species. *Oecologia* **142**, 627–635.

Weintraub, M.N. and J.P. Schimel. 2005. Nitrogen cycling and the spread of shrubs control changes in the carbon balance of arctic tundra ecosystems. *Bioscience* **55**, 408–415.

Welker, J.M., W.D. Bowman and T.R. Seastedt, T.R. 2001. Environmental change and future directions in alpine research. In: *Structure and Function of an Alpine Ecosystem* (W.D. Bowman and T.R. Seastedt, eds), pp. 304–322. Oxford University Press, Oxford.

Welker, J.M., J.T. Fahnestock, K.L. Povirk, et al. 2004. Alpine grassland CO_2 exchange and nitrogen cycling: Grazing history effects, Medicine Bow Range, Wyoming, USA. *Arctic, Antarctic and Alpine Research* **36**, 11–20.

Wertz, S., V. Degrange, J.I. Prosser, et al. 2007. Decline of soil microbial diversity does not influence the resistance and resilience of key soil microbial functional groups following a model disturbance. *Environmental Microbiology* **9**, 2211–2219.

Westoby, M. 1998. A leaf-height-seed (LHS) plant ecology strategy scheme. *Plant and Soil* **199**, 213–227.

White, S.L., R.E. Sheffield, S.P. Washburn, et al. 2001. Spatial and time distribution of dairy cattle excreta in an intensive pasture system. *Journal of Environmental Quality* **30**, 2180–2187.

White, T.R.C. 1978. The importance of relative shortage of food in animal ecology. *Oecologia* **33**, 71–86.

Whitehead, D., N.T. Boelman, M.H. Turnbull, et al. 2005. Photosynthesis and reflectance indices for rainforest species in ecosystems undergoing progression and retrogression along a soil fertility chronosequence in New Zealand. *Oecologia* **144**, 233–244.

Whitham, T.G., W.P. Young, G.D. Martinsen, et al. 2003. Community and ecosystem genetics: A consequence of the extended phenotype. *Ecology* **84**, 559–573.

Whitham, T.G., S.P. Di Fazio, J.A. Schweitzer, et al. 2008. Extending genomics to natural communities and ecosystems. *Science* **320**, 492–495.

Whitman, W.B., D.C. Coleman and W.J. Wiebe. 1998. Prokaryotes: the unseen majority. *Proceedings of the National Academy of Sciences USA* **95**, 6578–6583.

Whittaker, R.H. 1956. Vegetation of the great smoky mountains. *Ecological Monographs* **26**, 1–69.

Widden, P. and D. Hsu. 1987. Competition between *Trichoderma* species: effects of temperature and litter type. *Soil Biology and Biochemistry* **19**, 89–93.

Wilkinson, C.E., M.D. Hocking and T.E. Reimchen. 2005. Uptake of salmon-derived nitrogen by mosses and liverworts in coastal British Colombia. *Oikos* **108**, 85–98.

Williams, D.W. and A.M. Liebhold. 2002. Climate change and the outbreak ranges of two North American bark beetles. *Agricultural and Forest Entomology* **4**, 87–99.

Williamson, W.M., D.A. Wardle and G.W. Yeates. 2005. Changes in soil microbial and nematode communities during ecosystem retrogression across a long term chronosequence. *Soil Biology and Biochemistry* **37**, 1289–1301.

Willott, S.J., A.J. Miller, L.D. Incoll, et al. 2000. The contribution of rabbits (*Oryctolagus cuniculus* L.) to soil fertility in semi-arid Spain. *Biology and Fertility of Soils* **31**, 379–384.

Wilson, G.W.T., C.W. Rice, M.C. Rillig, et al. 2009. Soil aggregation and carbon sequestration are tightly correlated with the abundance of arbuscular mycorrhizal fungi: results from long-term field experiments. *Ecology Letters* **12**, 452–461.

Wilson, S.D. and C. Nilsson. 2009. Arctic alpine vegetation change over 20 years. *Global Change Biology* **15**, 1676–1684.

Wohlfahrt, G., A. Anderson-Dunn, M. Bahn, et al. 2008. Biotic, abiotic, and management controls on the net ecosystem CO_2 exchange of European mountain grassland ecosystems. *Ecosystems* **11**, 1338–1351.

Wold, E.N. and R.J. Marquis. 1997. Induced defense in white oak: effects on herbivores and consequences for the plant. *Ecology* **78**, 1356–1369.

Wolfe, B.E. and J.N. Kliromonos. 2005. Breaking new ground: soil communities and exotic plant invasion. *BioScience* **55**, 477–487.

Wolfe, B.E., V.L. Rogers, K.A. Stinson, et al. 2008. The invasive plant *Alliaria petiole* (garlic mustard) inhibits ectomycorrhizal fungi in its introduced range. *Journal of Ecology* **96**, 777–783.

Wolters, V., W.L. Silver, D.E. Bignell, et al. 2000. Effects of global changes on above- and belowground biodiversity in terrestrial ecosystems: implications for ecosystem functioning. *BioScience* **50**, 1089–1098.

Woodward, F.I. and M.R. Lomas. 2004. Vegetation-dynamics – simulating responses to climate change. *Biological Reviews* **79**, 643–670.

Woodward, F.I., Lomas, M.R. and C.K. Kelly. 2004. Global climate and the distribution of plant biomes. *Proceedings of the Royal Society of London Series B Biological Sciences* **359**, 1465–1476.

Wookey, P.A., R. Aerts, R.D. Bardgett, et al. 2009. Ecosystem feedbacks and cascade processes: understanding their role in the responses of arctic and alpine ecosystems to environmental change. *Global Change Biology* **15**, 1153–1172.

Worthy, T.H. and R.N. Holdaway. 2002. *The Lost World of the Moa*. Canterbury University Press, Christchurch.

Wright, D.C., R. Van der Wal, S. Wanless, et al. 2010. The influence of seabird nutrient enrichment and grazing on the structure and function of island soil food webs. *Soil Biology and Biochemistry* **42**, 592–600.

Wright, I.J., P.B. Reich, M. Westoby, et al. 2004. The worldwide leaf economics spectrum. *Nature* **428**, 821–827.

Wu, T., E. Ayres, G. Li, et al. 2009. Molecular profiling of soil animal diversity in natural ecosystems: Incongruence of molecular and morphological results. *Soil Biology and Biochemistry* **41**, 849–857.

Wurst, S. and W.H. Van der Putten. 2007. Root herbivore identity matters in plant-mediated interactions between root and shoot herbivores. *Basic and Applied Ecology* **8**, 491–499.

Wurst, S.R. Langel, A. Reineking, et al. 2003. Effects of earthworms and organic litter distribution on plant performance and aphid reproduction. *Oecologia* **137**, 90–96.

Wurst, S.R. Langel and S. Scheu. 2005. Do endogeic earthworms change plant competition? A microcosm study. *Plant and Soil* **271**, 123–130.

Wyman, R.L. 1998. Experimental assessment of salamanders as predators of detrital food webs: effects on invertebrates, decomposition and the carbon cycle. *Biodiversity and Conservation* **7**, 641–650.

Yamada, A., T. Inoue, D. Wiwatwitaya, et al. 2005. Carbon mineralization by termites in tropical forests, with emphasis on fungas combs. *Ecological Research* **20**, 453–460.

Yamada, D., O. Imura, K. Shi, et al. 2007. Effect of tunneler dung beetles on cattle dung deposition, soil nutrients and herbage growth. *Grassland Science* **53**, 121–129.

Yang, L.H. 2004. Periodical cicadas as resource pulses in North American forests. *Science* **306**, 1565–1567.

Yang, L.H., J.L. Bastow, K.O. Spense, et al. 2008. What can we learn from pulsed resources? *Ecology* **89**, 621–634.

Yeates, G.W. 1979. Soil nematodes in terrestrial ecosystems. *Journal of Nematology* **11**, 213–229.

Yeates, G.W. 1981. Soil nematode populations depressed in the presence of earthworms, *Pedobiologia* **22**, 191–195.

Yeates, G.W. and P.M. Williams. 2001. Effects of three invasive weeds and invasive site factors in New Zealand. *Pedobiologia* **45**, 367–383.

Yokoyama, K. and H. Kai. 1993. Distribution and flow of nitrogen in cow dung-soil system colonized by paracoprid dung beetles. *Edaphologia* **50**, 1–10.

Yokoyama, K., H. Kai, T. Koga, et al. 1991. Nitrogen mineralization and microbial populations in cow dung, dung balls and underlying soil affected by paracoprid dung beetles. *Soil Biology and Biochemistry* **23**, 649–653.

Zaady, E., P.M. Groffman, M. Shachak, et al. 2003. Consumption and release of nitrogen by the harvester termite *Anacanthotermes ubachi* navas in northern Negev desert, Israel. *Soil Biology and Biochemistry* **35**, 1299–1303.

Zackrisson, O., M.-C. Nilsson and D.A. Wardle. 1996. Key ecological function of charcoal from wildfire in the boreal forest. *Oikos* **77**, 10–19.

Zackrisson, O., M.-C. Nilsson, A. Jäderlund, et al. 1999. Nutritional effects of seed fall during mast years in boreal forest. *Oikos* **84**, 17–26.

Zackrisson, O., T.H. DeLuca, M.C. Nilsson, et al. 2004. Nitrogen fixation increases with successional age in boreal forests. *Ecology* **85**, 3327–3334.

Zak, D.R., K.S. Pregitzer, P.S. Curtis, et al. 1993. Elevated atmospheric CO_2 and feedback between carbon and nitrogen cycles. *Plant and Soil* **151**, 105–117.

Zak, D.R., D. Tilman, R.R. Parmenter, et al. 1994. Plant production and soil microorganisms in late-successional ecosystems: a continental-scale study. *Ecology* **75**, 2333–2347.

Zak, D.R., P.M. Grofman, K.S. Pregitzer, et al. 1990. The vernal dam: plant–microbe competition for nitrogen in northern hardwood forests. *Ecology* **71**, 651–656.

Zak, D.R., W.E. Holmes, D.C. White, et al. 2003. Plant diversity, soil microbial communities, and ecosystem function: Are there any links? *Ecology* **84**, 2042–2050.

Zak, D.R., C.B. Blackwood and M.P. Waldrop. 2006. A molecular dawn for biogeochemistry. *Trends in Ecology and Evolution* **21**, 288–295.

Zaller, G. and J.A. Arnone. 1999. Interactions between plant species and earthworm casts in a calcareous grassland under elevated CO_2. *Ecology* **80**, 837–881.

Zavaleta, E.S. and K.B. Hulvey. 2004. Realistic species losses disproportionately reduce grassland resistance to biological invaders. *Science* **306**, 1175–1177.

Zeller, V., R.D. Bardgett and U. Tappeiner. 2001. Site and management effects on soil microbial properties of subalpine meadows: A study of land abandonment along a north-south gradient in the European Alps. *Soil Biology and Biochemistry* **33**, 639–650.

Zhang, Q. and J.C. Zak. 1998. Potential physiological activities of fungi and bacteria in relation to plant litter decomposition along a gap size gradient in a natural subtropical forest. *Microbial Ecology* **35**, 172–179.

Zhou, G.Y., L.L. Guan, X.H. Wei, et al. 2008. Factors influencing leaf litter decomposition: an intersite decomposition experiment across China. *Plant and Soil* **311**, 61–72.

Zibilske, L.M. and J.M. Bradford. 2007. Oxygen effects on carbon, polyphenols, and nitrogen mineralization potential in soil. *Soil Science Society of America Journal* **71**, 133–139.

Zimmer, M. and W. Topp. 1998. Microorganisms and cellulose digestion in the gut of *Porcellio scaber* (Isopoda: Oniscidea). *Journal of Chemical Ecology* **24**, 1397–1408.

Zimmer, M. and W. Topp. 2002. The role of coprophagy in nutrient release from feces of phytophagous insects. *Soil Biology and Biochemistry* **34**, 1093–1099.

Zimov, S.A., V.I. Chuprynin, A.P. Oreshko, et al. 1995. Steppe-tundra transition: a herbivore-driven biome shift at the end of the Pleistocene. *American Naturalist* **146**, 765–794.

Zogg, D.G., D.R. Zak, K.S. Pregitzer, et al. 2000. Microbial immobilization and the retention of anthropogenic nitrate in a northern hardwood forest. *Ecology* **81**, 1858–1866.

Zou, J.W., W.E. Rogers, S.J. DeWalt, et al. 2006. The effect of Chinese tallow tree (*Sapium sebiferum*) ecotype on soil-plant system carbon and nitrogen processes. *Oecologia* **150**, 272–281.

Index

Printed in the United States
By Bookmasters